FLUCTUATIONS AND FRACTAL STRUCTURE

Fluctuations and Fractal Structure

Ringberg Workshop on Multiparticle Production

Ringberg Castle, Germany

June 25 - 28, 1991

Editors

R. C. Hwa
University of Oregon, Eugene, USA

W. Ochs and N. Schmitz
Max – Planck – Institut für Physik, Munich, Germany

World Scientific
Singapore • New Jersey • London • Hong Kong

Published by

World Scientific Publishing Co Pte Ltd
5 Toh Tuck Link, Singapore 596224
USA office: 27 Warren Street, Suite 401-402, Hackensack, NJ 07601
UK office: 57 Shelton Street, Covent Garden, London WC2H 9HE

Library of Congress Cataloging-in-Publication data is available.

FLUCTUATIONS AND FRACTAL STRUCTURE
Ringberg Workshop on Multipartlcle Production

ISBN-13 9789810208103
ISBN-10 9810208103

ORGANIZING COMMITTEE

Rudolph C. Hwa	University of Oregon Eugene, Oregon, USA
Wolfgang Ochs	Max-Planck-Institut für Physik Munich, Germany
Norbert Schmitz	Max-Planck-Institut für Physik Munich, Germany

CONFERENCE SECRETARY

Barbara Kollmer	Max-Planck-Institut für Physik Munich, Germany

PREFACE

The Ringberg Workshop on Multiparticle Production is in a series which started in Aspen (USA, 1986) and was continued in Jinan (China, 1987), Perugia (Italy, 1988) and Santa Fe (USA, 1990). These workshops are devoted to the discussion of theoretical and experimental results in the field of strong interaction hadron physics which, despite the overall success of the standard model of particle physics, is not yet well understood and requires detailed models which cannot be based entirely on the fundamental theory. It was expected that a gathering of experimental and theoretical physicists actively working in the field in a small workshop with about 30 participants and plenty of time for discussion would be best suited to communicate the experimental results from the various collision processes and the sometimes controversial opinions about the best theoretical approach.

In the last two years the main interest was in the structural properties of multiplicity fluctuations, in particular, whether the data give evidence for an underlying fractal structure, also called "intermittency". This interest suggested the subtitle of this year's workshop "Fluctuations and Fractal Structure".

Since the last workshop in Santa Fe, which was already devoted to the same subject, the perhaps most noteworthy progress came from the experimental side. Here new methods have been applied to unravel the fractal structures in the complex multidimensional phase space of high energy hadronic final states, and detailed comparisons of high statistics data with hadronisation models, in particular for e^+e^- annihilation, have been performed. The experimental results will be contained in the first part of these proceedings followed by the theoretical papers which discuss the structure of correlations, the multifractal structure of final states and present predictions from specific phenomenological hadronisation models and field theoretical models.

The workshop was held in June 25 - 28, 1991 at the Ringberg Castle of the Max-Planck Society in the scenic area near Lake Tegernsee in the Bavarian Alps, about 60 km south of Munich, Germany. The comfortable interior of the castle, once used by Duke Luitpold in Bavaria, has provided the relaxed atmosphere to continue discussions after the talks, whereas in the evenings most inspiration has been found in the "Room of the Witches". A glance at Bavarian history was possible during an excursion to the castle of King Ludwig II at Herreninsel of Lake Chiemsee and contemporary Bavarian folk music could be heard one evening in the castle at a Bavarian conference dinner.

We would like to thank very cordially Mr. A. Hörmann, our host at the castle, and his crew for providing us this pleasant environment for our workshop, for their hospitality and in particular for the excellent dishes. We are grateful to Mrs. Barbara Kollmer, our conference secretary, for her intelligent and efficient help in organizing the workshop and during the meeting, and to Mrs. Edeltraud Haag for her valuable assistance in editing these proceedings. Finally our thanks are due to the Dr. Ernst-Rudolf-Schloeßmann-Stiftung for financial support.

R. C. Hwa
W. Ochs
N. Schmitz

CONTENTS

RECENT DELPHI RESULTS ON MULTIPLICITY FLUCTUATIONS IN HADRONIC FINAL STATES FROM THE DECAY OF THE Z^0 BOSON[1]

ALESSANDRO DE ANGELIS
Istituto di Fisica dell'Universita' di Udine and INFN Trieste
Via Fagagna 208, 33100 Udine, Italy

NATALE DEMARIA
Istituto di Fisica dell'Universita' di Torino and INFN Torino
Via Giuria 1, 10125 Torino, Italy

ABSTRACT

An analysis of the fluctuations in the phase space distribution of hadrons produced in the decay of the Z^0 boson has been carried out, using the method of factorial moments. The high statistics collected by the DELPHI experiment at LEP during 1990 (around 80,000 events after the cuts) allowed studies not only for the global event sample, but also in intervals of p_t and multiplicity, and for different jet topologies.

1. Introduction

This paper presents results of the study of fluctuations in the phase space distribution of hadrons produced in e^+e^- collisions around the Z^0 energy, using the DELPHI detector at LEP. It follows, complements and extends to a larger statistics previous studies from the same collaboration [1].

In order to provide a quantitative test of anomalous multiplicity fluctuations (spikes) in variable intervals of rapidity, Bialas and Peschanski[2] proposed in 1986 to analyze the distributions of multiplicity in terms of normalized factorial moments. Given an experimental distribution of particles in the rapidity interval from $-Y/2$ to $Y/2$, the interval Y is divided into M equal subintervals, each of size $\delta y = Y/M$. By defining N to be the number of particles in the whole rapidity interval, n_m to be the number of particles in the m-th bin ($m = 1...M$), the factorial moment of (integer) rank j of the distribution with respect to the partition is defined as

$$F_j(\delta y) = \frac{M^{j-1}}{< N >^j} < \sum_{m=1}^{M} n_m(n_m - 1)...(n_m - j + 1) > \qquad (1)$$

where the average is taken over many events. The factorial moment of rank j for a rapidity interval δy acts as a filter for selecting events with j particles or more

---

[1]**Presented by A. De Angelis.**

in at least one bin and is therefore highly sensitive to events with large density fluctuations.

Simple models representing the hadronization process as a random cascade with selfsimilar structure predict a powerlike increase of the factorial moments in the limit in which the bin size δy goes to zero, i.e.

$$F_j \propto (1/\delta y)^{f_j} \quad (2)$$

and the validity of the above relation was taken by the authors of Ref. [2] as definition of *intermittency*, a term mutuated from hydrodinamics, as most of the mathematical techniques used in this field [3]. We will take in the following (2) as definition of intermittency; it should be underlined, however, that a universally accepted definition does not exist in the literature[2].

The first direct measurement of factorial moments in e^+e^- annihilations was published by the TASSO collaboration [4], at a centre of mass energy of around 35 GeV, and claimed an intermittency effect that could not be explained by the JETSET Parton Shower Monte Carlo [5] (JETSET PS in the following), nor by the Marchesini-Webber [6] and the Hoyer model [7]. The TASSO work was confirming the results of an indirect analysis of the HRS data [8] at $\sqrt{s}$ =29 GeV (more recently, the HRS collaboration has provided a direct analysis [9], whose conclusions are in agreement with the previous one, but no explicit comparison with Monte Carlos is done). The predictions of various models for e^+e^- interactions differ considerably at high energies [10]. This situation motivated an investigation of possible intermittency effects in e^+e^- annihilation at the Z^0 peak. This analysis [1] was carried out by making use of data collected with the DELPHI detector [11] at the e^+e^- storage ring LEP during its first runs at the Z^0 resonance. The results of the DELPHI analysis [1] were essentially that the presence of intermittency, defined as in (2), was questionable when searched in the rapidity distribution of final-state hadrons, evident when searched in a 2-dimensional projection of the phase space, but in any case the results were compatible with the predictions of parton shower models. After that, CELLO [12] reported an agreement with Parton Shower models at the same energy as TASSO, and OPAL [13] and ALEPH [14] at the same energy as DELPHI.

Intermittency has motivated during the last years many theoretical hypotheses and experimental investigations, for which exhaustive reviews have been recently published [15]. Often, theoretical works hypothesize new physics for explaining the phenomenon; many authors pointed out however that self-similar cascading mechanisms [2, 16, 17] or models in which short-range correlations derived from the data are introduced *ad hoc* [18] can reproduce the effect.

[2]In a vague definition of intermittent are put signals for which a traditional statistical description through probability density function, mean value, root mean square, etc., fails.

2. Event selection

The sample of events used in the analysis was collected by the DELPHI detector at the LEP e^+e^- collider during 1990. A description of the DELPHI detector can be found elsewhere [1]. Features of the apparatus relevant for the analysis of multihadronic final states (with emphasis on the detection of charged particles) are outlined in Ref. [19], as well as the cuts used for the selection of hadronic events.

Only charged tracks reconstructed by means of the central detectors (with polar angle between 25° and 155°) were used in the present analysis. A total of 80,000 events accumulated during 1990 satisfied the cuts. Events due to beam-gas scattering, to $\gamma\gamma$ interactions and to decays of the Z^0 into $\tau^+\tau^-$ have been estimated to be less than 0.3% of the selected sample.

The resolving power of the detector was tested by means of Monte Carlo, and by studying the two-track density as a function of the rapidity y and of the area in the (y,ϕ) space[3] in the experimental data. We obtain indications that the limits of our experimental investigations (see below) are well above the experimental resolution for two-track particle density. We did not find any suitable indicator of the goodness of n-track resolution, and thus we assumed this quantity to scale as $\sqrt{n}$.

The Monte Carlo simulation DELSIM [20] was used to correct the data for the geometrical acceptance, kinematical cuts, resolution, particle interaction with the detector material and other detector imperfections. A sample of Z^0 decays, approximately equal in size to the sample of real events for reasons stated below, was generated with JETSET 6.3 PS [5], and followed through this detailed simulation of the detector. From the samples of accepted and generated events, correction factors

$$C(\delta y) = \frac{F(\delta y)_{generated}}{F(\delta y)_{accepted}}$$

were computed. These factors were then used to correct the quantities calculated from the real data. The generated event sample contained all final state particles with a lifetime above 10^{-9} s before any tracking was done through the detector. The accepted event sample contained all final state particles observed after tracking the fully simulated events through the DELPHI detector. Simulated raw data were then processed through the same reconstruction and analysis chain as the real data.

3. Analysis and results

In the following, factorial moments of projections of the phase space of hadrons originated from the decay of the Z^0 are compared with the predictions of QCD-based Monte Carlo programs.

The comparison is mainly done with JETSET PS, that has proven, after one year of activity of LEP, to well reproduce the hadronic final states from the decay of

[3]The rapidity y and the azimuth ϕ are defined, where not explicitly stated, with respect to the sphericity axis.

the Z^0, both from the point of view of the description of shape variables [19, 21, 22] and, more important for our study, of multiplicity (global and in restricted intervals of rapidity [23]). Parameters of JETSET PS have been eventually retuned according to Ref. [21]. In some cases, comparisons were made also to other Monte Carlos, based on parton cascades or on the exact second order QCD matrix element followed by string fragmentation. In particular:

1. the ARIADNE [24] Monte Carlo, with parameters optimized as in Ref. [21];
2. the JETSET 7.2 Monte Carlo with a matrix element calculation up to $O(\alpha_s^2)$, and the optimization of parameters described in [25] (JETSET ME retuned).

The essential features of these Monte Carlos are summarized in Ref. [10].

3.1 Projections of Phase Space Onto 1 Dimension

First, we extended the previous DELPHI work to the present statistics, by plotting the factorial moments of the rapidity distribution between -2 and +2 and comparing with the prediction by JETSET PS default, JETSET PS retuned, JETSET ME retuned, ARIADNE. The small deviations with respect to JETSET PS default, for which an indication was existing in the previous paper (i.e., a larger value of the factorial moments with respect to the values predicted by the Monte Carlo), become now statistically significant. The discrepancy does not however go in the direction of a spectacular power-law rise of factorial moments, not reproduced in the Monte Carlo. Data in Fig. 1 are corrected via Monte Carlo for the resolution of the detector; the correction factors differ from 1.0 by less than 5% on average, as one can see in the upper part of the figures. Errors keep into account the effects of an uncertainty of $\pm 2\%$ in the probability of γ conversions in the detector.

The retuning of the fragmentation parameters causes a drastic change both in the magnitude of the factorial moments and in the slope of their distribution for small values of δy. However, even after retuning, the ME model fails to reproduce quantitatively the data. For all ranks, the logarithms of the factorial moments appear to grow as the number of subdivisions increases. A slope for this growth cannot be uniquely defined.

Expression (1) is in principle a biased estimator of the "true" value of the factorial moments. For this reason, the amount of simulated data in Fig. 1 was chosen to be of approximately the same size as the amount of real data. However, the effect of bias has been checked by comparing the factorial moments of a Monte Carlo sample (JETSET PS at generator level) of the size of $\simeq 10,000$ events with a ten times larger sample, and a sample of real data from 10,000 events with a three times larger sample. Results are in good agreement, making us confident that at the Z^0 energy a statistics of $\simeq 10,000$ events is adequate to make the bias negligible up to the factorial moment of rank 5 (at least for projections of the phase space onto one dimension, and maximum number of subdivisions used in this note).

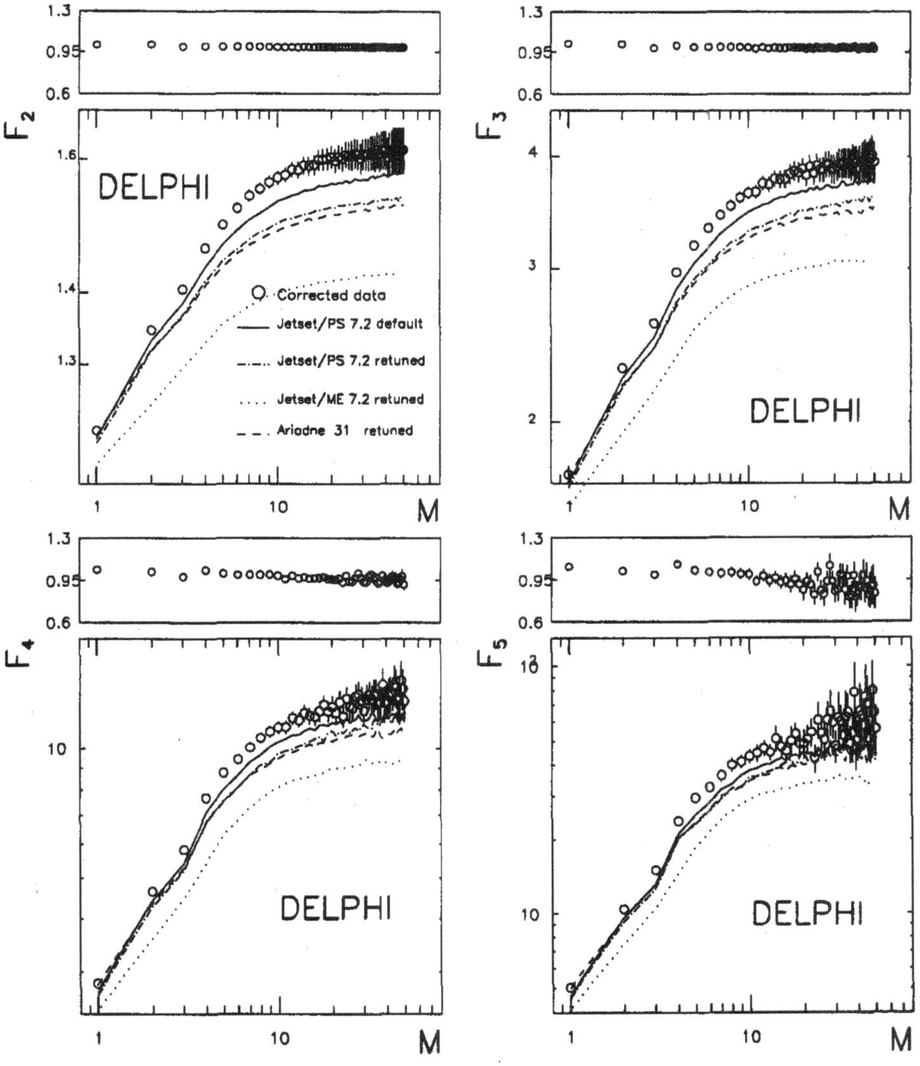

Figure 1: *Dependence of the factorial moments of rank 2 (a), 3 (b), 4 (c) and 5 (d) on the number M of subdivisions of the rapidity interval, compared to several Monte Carlos.*

The central region of the rapidity interval was chosen because, in this region, the density of particles is almost uniform. One thus avoids by this choice the problem of non-uniform population of the phase space, that can fake the signal from anomalous clusters we are looking for. The price to pay is that one excludes from the study the particles closer to the core of the jet. To overcome this problem, it was recently suggested by Ochs [26] and by Bialas and Gazdzicki [27] to study a distribution $\tilde{y}$ that is the y distribution rebinned in such a way that the population is, in average, uniform. Using self-similar models for the hadronization, the authors find in their work that factorial moments defined with respect to this variable follow more closely the power law (2).

The study of factorial moments in the $\tilde{y}$ distribution corresponding to the y region between -5 and +5 did not show results qualitatively different with respect to the study of factorial moments of the y distribution.

In the following, unless otherwise stated, we always apply the flattening procedure analyzing the p_t, the multiplicity and the jet topology dependence of factorial moments, working in the rapidity region between -5 and +5: this guarantees a good way to compare the results, being independent of how the choice of the cut can modify the average y distribution.

3.1.1 Dependence on p_t

The NA22 collaboration has recently reported on a striking disagreements between data and hadronic Monte Carlos for the track at low p_t [29]. Although p_t has of course a different physics meaning in hadronic and e^+e^- collisions, we tried the same exercise, mainly:

- To evidence if the description of the p_t sector in JETSET PS is satisfactory.
- Motivated by the fact that the low-p_t region is almost free from effects related to hard gluon radiation, that has been demonstrated [14] to be the cause of the most relevant part of the behavior of factorial moments.

We divided the full p_t range into 3 regions, chosen in such a way that the average number of particles for each region is the same, to remove a possible contamination of the result from mathematical properties of factorial moments. The 3 regions were respectively $p_t < 0.255$, $0.255 < p_t < 0.532$, $0.532 < p_t < 2.$ GeV/c.

When dividing the final state particles into p_t slices, it becomes of primary importance, due to the correlation between p_t and y, to apply to the variable y a transformation that flattens the distribution. In each p_t interval the y distribution is in fact far from being flat.

The results obtained are plotted in Fig. 2, and they evidence discrepancies from JETSET PS more pronounced than in the case of no slicing in p_t. This suggests that the treatment of correlations in the p_t sector is not adequate in JETSET PS. The effects of an uncertainty of $\pm 2\%$ in the probability of γ conversions in the detector are larger in the low-p_t spectrum, as expected from the average small momentum of electrons produced. We find that, at low p_t, the factorial moments follow a power law behaviour in the $\delta\tilde{y} < 0.2$ range. The high p_t region presents larger fluctuations, but they saturate at low $\delta\tilde{y}$, being compatible with the fact that the first rise is due to the spikes caused by the presence of the hard gluon jets, and fluctuations are then milded when digging inside the 3rd jet.

3.1.2 Dependence on Multiplicity

In the case of the dependence on the multiplicity of the event, it is more difficult to draw physics conclusions, due to the fact that the mathematical effect due to the use of a different number of tracks for computing factorial moments makes the result fuzzy.

The results on factorial moments of different intervals in the observed multiplicity N ($N \leq 15$, $15 < N \leq 20$, $N > 20$) are displayed in Fig. 3 for uncorrected data. The discrepancies are concentrated mainly in the region of low multiplicities, where the contamination from non-hadronic events is larger. Discrepancies are milded when the y distribution, instead than the $\tilde{y}$ distribution, is studied.

3.1.3 Dependence on Jet Topology

Jets were defined according to the JADE/E0 algorithm [30], with a y_{cut} value of 0.04. Factorial moments with respect to $\tilde{y}$ are plotted in Fig. 4, for 2-jet and 3-jet events. The fact that factorial moments are higher in 3-jet events is compatible with being due to the spike in rapidity caused by hard gluon jets. We do not observe, in any case, striking disagreements with respect to JETSET PS.

It can be of interest to investigate more deeply in 2-jet events, to see if the structure of fluctuations inside a jet is correctly reproduced by JETSET PS. Results from 2-jet events selected by means of a y_{cut} value of 0.01 do not differ qualitatively from the previous ones.

In 3-jet events, selected with $y_{cut} = 0.04$, we calculated also factorial moments for the tracks belonging to jet 1, jet 2 and jet 3, ordered by energy. We calculate the jet energy:

1. by defining for each jet the jet direction as the direction of the sum of the momenta of the tracks clustered in the jet

2. by balancing the projections into the event plane, that is the one identified by the first and the second eigenvectors of the sphericity tensor, and assuming that the total energy is the center of mass energy.

The results, for data and Monte Carlo, are plotted in Fig. 5. It is well known that the less energetic jet has a greater probability to be the gluon jet than the other two: it is our purpose to determine if there are differences between gluon and

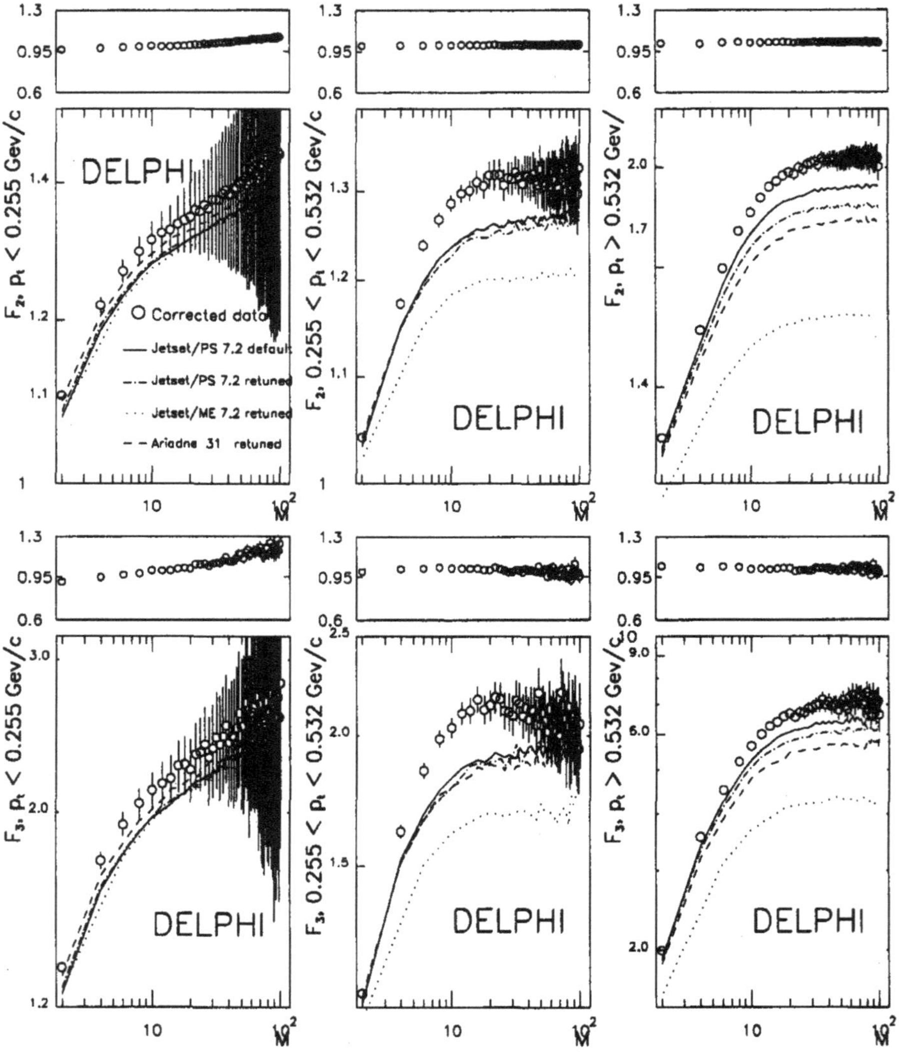

Figure 2: *Factorial moments of rank 2 (a) and 3 (b) of the $\tilde{y}$ distribution for the three intervals of p_t (see text). DELPHI corrected data (white circles) compared with JETSET PS default (solid line).*

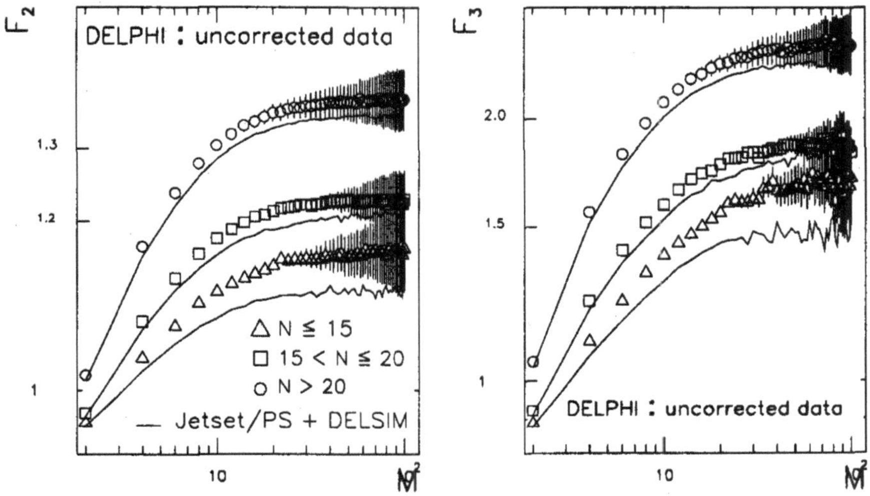

Figure 3: *Factorial moments of rank 2 (a) and 3 (b) of the $\tilde{y}$ distribution for three ranges of multiplicity (see text). DELPHI uncorrected data (white circles) compared with JETSET PS 6.3 default + DELSIM (solid line).*

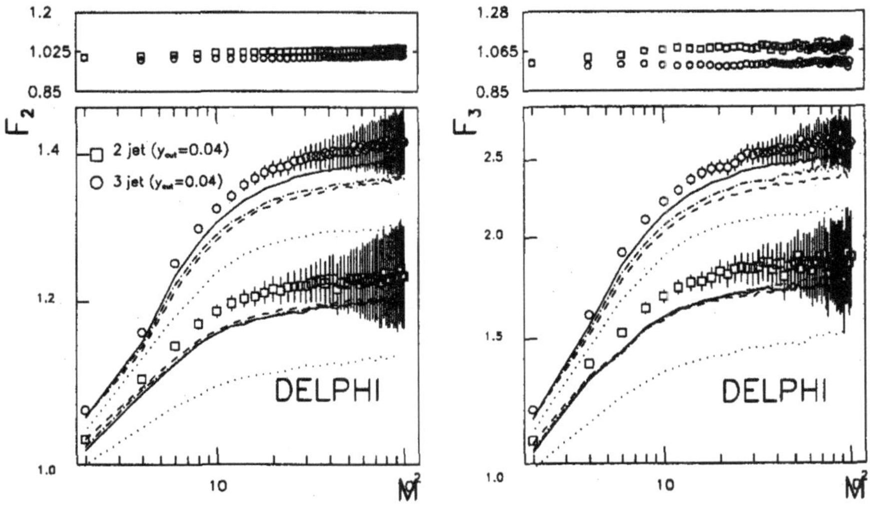

Figure 4: *Factorial moments of rank 2 (a) and 3 (b) of the $\tilde{y}$ distribution for 2-jet and 3-jet events (y_{cut} of 0.04). DELPHI corrected data (white circles) compared to models as in Fig. 2.*

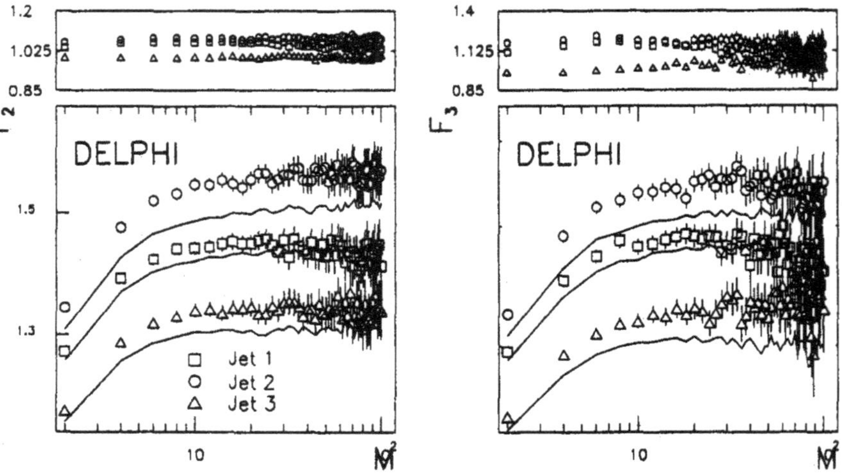

Figure 5: *Factorial moments of rank 2 (a) and 3 (b) of the first, second and third jet, ordered by energy. DELPHI corrected data (open symbols) compared with JETSET PS default (solid line).*

quark jets. We find that jet 3 presents weaker fluctuations. A disagreement with JETSET PS is visible in few standard deviations for jet 2 and jet 3.

3.1.4 Effects of Dalitz Pairs and of Bose-Einstein Correlations

We verified by a Monte Carlo study based on 50,000 events that switching on and off the Dalitz decay of the π^0 does not change the values of factorial moments out of the errors.

The study of the Bose-Einstein effect is more subtle, due to the lack of a satisfactory description in the Monte Carlos. On one hand, a test of comparison of factorial moments for same sign particles with random sign particles did not show differences [1]; on the other hand, this test cannot be conclusive since it does not keep into account the fact that non Bose-Einstein correlations are different for the two sets of particles.

To reproduce inside JETSET PS our results on Bose-Einstein correlations [31], we need to adjust the "strength parameter" λ to an unphysical value around 2.5. In addition, this adjustment affects the single particle inclusive distributions (rapidity in particular), causing them not to reproduce our data. We thus cannot thrust a Monte Carlo comparison for studying the effect of Bose-Einstein correlations.

In conclusion, we did not find any satisfactory probe to disentangle the effect of Bose-Einstein correlations on factorial moments.

3.2 Projections of Phase Space Onto Higher Dimensions

Recent theoretical works [28] suggest that saturation of factorial moments at $\delta y \simeq 0.1$ is an effect of smearing of fluctuations due to the projection of phase space into a one-dimensional subspace. Stronger intermittency effects should then appear in two-dimensional and three-dimensional spectra.

When intermittency is investigated with respect to the 2-dimensional (y, ϕ) distribution, the existence of a positive slope according to Eq. 2 becomes in fact no more questionable. Also in this case, however, the predictions of JETSET PS are in good agreement with the data [1].

When increasing the dimension of the phase-space in which one is studying fluctuations, the problem of the non-uniformity of the distribution in the phase space becomes of primary importance. One could naively think that, due to the fact that the ϕ distribution is not correlated to the y distribution and, in average, flat, the distribution of $(\tilde{y}, \phi)$ is flat. Unfortunately, on an event by event basis, the ϕ distribution is nonuniform, being peaked at the azimuth ϕ_{s2} of the second eigenvector of the momentum tensor[4].

To solve this problem, we considered in all the events the values of the azimuth starting from ϕ_{s2}, and we applied a transformation 'a la Bialas-Gazdzicki'. After this transformation, the population of the $\widetilde{(y, \phi)}$ space is flat. In the same way, for the 3-dimensional distribution, we calculated a flat bin-per-bin $\widetilde{(y, \phi, p_t)}$ distribution over the full phase space.

The results on factorial moments of the 2- and 3- dimensional projections of phase space are plotted in Fig. 6, and tabulated in Table 1. Errors keep into account the effects of an uncertainty of $\pm 2\%$ in the probability of γ conversions in the detector.

Factorial moments in three dimensions display intermittent behavior in the full phase space available. This means that, in the full phase space, the fluctuations related to the presence of the hard gluon jets as seen "from far" have the same behavior with respect to factorial moments as the fluctuations of the hadrons inside a jet.

3.3 Factorial Moments and 2-particle Correlations

Carruthers and Sarcevitz [18] suggested that the behavior of factorial moments can be understood on the basis of conventional short-range correlations for the 2-particle correlations, plus the Linked Pair 'ansatz' (LPA, requiring one free parameter for each order of the factorial moment) to go from the 2-particle correlations to higher-order correlations.

In the literature, 2-particle correlations are usually fitted in the central rapidity

---

[4]This was made evident by plotting factorial moments of the ϕ distribution and comparing to factorial moments of a $\tilde{\phi}$ distribution. Factorial moments are higher in the first than in the second case.

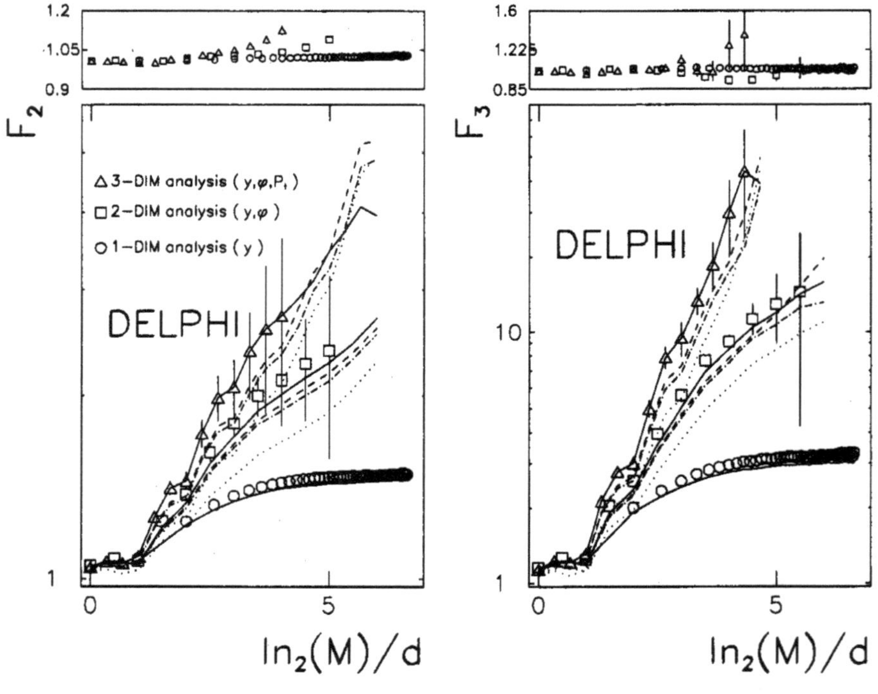

Figure 6: *Factorial moments of rank 2 (a) and 3 (b) of the* $\tilde{y}$, $\widetilde{(y,\phi)}$ *and* $\widetilde{(y,\phi,p_t)}$ *distributions. DELPHI corrected data (open symbols) compared to to models as in Fig. 2.*

$\ln_2 \frac{M}{d}$	Rank 2		
	1-d	2-d	3-d
0		$1.051 \pm 0.001 \pm 0.009$	$1.043 \pm 0.001 \pm 0.007$
1	$1.081 \pm 0.002 \pm 0.002$	$1.071 \pm 0.001 \pm 0.011$	$1.090 \pm 0.002 \pm 0.004$
2	$1.245 \pm 0.003 \pm 0.003$	$1.376 \pm 0.004 \pm 0.036$	$1.449 \pm 0.006 \pm 0.049$
3	$1.369 \pm 0.004 \pm 0.004$	$1.803 \pm 0.008 \pm 0.090$	$2.06 \pm 0.02 \pm 0.21$
4	$1.441 \pm 0.004 \pm 0.008$	$2.12 \pm 0.01 \pm 0.24$	$2.71 \pm 0.05 \pm 0.92$
5	$1.471 \pm 0.005 \pm 0.015$	$2.36 \pm 0.03 \pm 0.79$	
6	$1.479 \pm 0.006 \pm 0.025$		
$\ln_2 \frac{M}{d}$	Rank 3		
	1-d	2-d	3-d
0		$1.159 \pm 0.003 \pm 0.027$	$1.136 \pm 0.003 \pm 0.021$
1	$1.274 \pm 0.004 \pm 0.006$	$1.249 \pm 0.005 \pm 0.037$	$1.323 \pm 0.007 \pm 0.012$
2	$2.008 \pm 0.014 \pm 0.001$	$2.56 \pm 0.02 \pm 0.15$	$3.01 \pm 0.05 \pm 0.18$
3	$2.56 \pm 0.02 \pm 0.02$	$5.62 \pm 0.09 \pm 0.28$	$9.4 \pm 0.5 \pm 1.4$
4	$2.96 \pm 0.03 \pm 0.03$	$9.16 \pm 0.28 \pm 0.49$	$30.1 \pm 6.4 \pm 7.7$
5	$3.14 \pm 0.04 \pm 0.07$	$13.0 \pm 0.9 \pm 4.0$	
6	$3.23 \pm 0.05 \pm 0.13$		

Table 1: *Factorial moments of rank 2 and 3 of the $\tilde{y}$, $(\widetilde{y,\phi})$ and $(y,\widetilde{\phi,p_t})$ distributions. Statistical errors and systematics from γ conversions separated.*

region to the formula

$$r_2(\Delta y) = 1 + \gamma_2 e^{-\Delta y/\xi} \tag{3}$$

where r_2 is the ratio of the two-particle density to the product of the single-particle densities for a couple of particles, Δy is the absolute value of the difference of the rapidities, and γ_2 and ξ are the parameters of the fit. Our two-particles correlations data in the central rapidity region (with the cuts $|y_{1,2}| < 2$ and $|y_1 + y_2| < 2$) display disagreement to expression (3) in the low-δy region.

The best fit to expression (3) gives $\gamma_2 = 0.968 \pm 0.004$, $1/\xi = 0.748 \pm 0.003$. When applying the formulae of Ref. [18] using the values of γ_2 and ξ fitted to the two-particle correlations, we obtain a not satisfactory agreement to the data , also for F_2. We remind that that F_2 should in principle, if approximation (3) and the assumption of translational invariance in the central rapidity region were true, contain the same information as the two fit parameters.

If we proceed in the opposite way, and we use the expressions of Ref. [18] to *parametrize* factorial moments, the fitted values γ_2 and ξ fitted from F_2 disagree with the ones previously determined ($\hat{\gamma}_2 = 0.640 \pm 0.001$, $1/\hat{\xi} = 0.724 \pm 0.001$). However, we find that also this representation of correlations is not accurate enough to describe factorial moments of ranks higher than three.

This is an evidence of non-trivial $n > 2$-particle correlations in e^+e^- annihilations.

4. Conclusions

Confirming our previous results [1], a more detailed analysis, based on a statistics ten times larger, evidenced that present Monte Carlos based on Parton Showers provide a reasonably good description of fluctuations in the phase space of hadrons produced in the decay of the Z^0. Discrepancies observed are of the same order as the effect of retuning parameters.

The presence of intermittency, questionable in 1-dimensional projections of phase space, becomes evident in the full phase space, showing a similarity of the fluctuations that give rise to jets with the fluctuations of hadrons inside jets.

We have observed the evidence of nontrivial higher($n > 2$)-order particle correlations in e^+e^- annihilations.

5. Acknowledgements

The material presented in this talk comes from the effort of many members of the DELPHI collaboration. We want to thank all them, and in particular Hermann Fürstenau (Karsluhe), Xarris Lambropoulos (Athens), Franz Mandl (Wien) and George Theodosiou (Athens).

We are greatly indebted to our technical staffs and collaborators and funding agencies for their support in building the DELPHI detector and to the many members of the LEP division for the excellent performance of the LEP machine.

Discussions with A. Bialas, N. Geddes, A. Giovannini, R. Hwa, W. Kittel, N. Lieske, W. Ochs, R. Peschanski and J. Seixas helped a lot in clarifying the problems related to the interpretation of results.

Finally, one of us (A.D.A.) wants to thank the secretary and the organizers of this workshop for the nice and fruitful days spent in Ringberg.

References

1. P. Abreu et al. (DELPHI Coll.), *Phys. Lett.* **B247** (1990) 137.
   A. De Angelis, *Mod. Phys. Lett.* **A5** (1990) 2395.
2. A. Bialas and R. Peschanski, *Nucl. Phys.* **B273** (1986) 703.
3. G. Paladin and A. Vulpiani, *Phys. Rep.* **56** (1987) 147.
4. W. Braunschweig et al. (TASSO Coll.), *Phys. Lett.* **B231** (1989) 548.
5. T. Sjöstrand, *Comp. Phys. Comm.* **27** (1982) 243, ibid. **28** (1983) 229.
   T. Sjöstrand and M. Bengtsson, *Comp. Phys. Comm.* **43** (1987) 367.
6. G. Marchesini and B.R. Webber, *Nucl. Phys.* **B238** (1984) 1.
7. P. Hoyer et al., *Nucl. Phys.* **B161** (1979) 349.
8. B. Buschbeck et al., *Phys. Lett.* **B215** (1988) 788. Based on data of the HRS collaboration, assuming that multiplicity follows the negative binomial distribution.

9. S. Abachi et al. (HRS Coll.), "Study of intermittency in e^+e^- annihilations at 29 GeV", contributed to the XXV conference on High Energy Physics, Singapore, August 1990.

10. QCD generators for LEP, Proc. of the 1989 Workshop on Z^0 Physics at LEP 1, B. Bambah et al., CERN 89-08, Vol. 3, p.143, Eds. G. Altarelli, R. Kleiss and C. Verzegnassi.

11. P. Aarnio et al. (DELPHI Coll.), *Nucl. Instr. and Meth.* **A303** (1991) 233.

12. H.J. Behrend et al. (CELLO Coll.), *Phys. Lett.* **B256** (1991) 97.
O. Podobrin, DESY 91-028, April 1991.
O. Podobrin, this workshop.

13. M.Z.Akrawy et al. (OPAL Coll.), *Phys. Lett.* **B262** (1991) 351.
N. Geddes, this workshop.

14. D. Decamp et al. (ALEPH Coll.), CERN-PPE/91-121, July 1991.
N. Lieske and V. Raab, this workshop.

15. L. Van Hove, *Mod. Phys. Lett.* **A4** (1989) 1867.
A. Bialas, CERN-TH.5791/90, July 1990.
R. Peschanski, CERN-TH.5891/90, October 1990.

16. A. Bialas and R. Peschanski, *Nucl. Phys.* **B308** (1988) 857.

17. W. Ochs and J. Wosiek, *Phys. Lett.* **B214** (1988) 617.
W. Ochs and J. Wosiek, *Phys. Lett.* **B232** (1989) 271.

18. P. Carruthers and I. Sarcevic, *Phys. Rev. Lett.* **63** (1989) 1562.

19. P. Aarnio et al. (DELPHI Coll.), *Phys. Lett.* **B240** (1990) 271.

20. DELSIM User Manual, DELPHI 87-96 PROG-99, Geneva, July 1989.
DELSIM Reference Manual, DELPHI 87-98 PROG-100, Geneva, July 1989.

21. M.Z. Akrawy et al. (OPAL Coll.), *Z. Phys.* **C47** (1990) 505.

22. D. Decamp et al. (ALEPH Coll.), *Phys. Lett.* **B234** (1990) 209.

23. P. Abreu et al. (DELPHI Coll.), *Z. Phys.* **C50** (1991) 185.
P. Abreu et al. (DELPHI Coll.), CERN-PPE/91-78.

24. L. Lonnblad, LU TP 90-10, Lund 1990.

25. W. de Boer, H. Fürstenau and J.H. Köhne, **C49** (1991) 141.

26. W. Ochs, *Z. Phys.* **C50** (1991) 339.

27. A. Bialas and M. Gazdzicki, *Phys. Lett.* **B252** (1990) 483.

28. W. Ochs, MPI-PAE/Pth 35/90, June 1990.
A. Bialas and J. Seixas, *Phys. Lett.* **B250** (1990) 161.

29. N.M. Agababyan et al. (EHS/NA22 Coll.), *Phys. Lett.* **B261** (1991) 165.

30. W. Bartel et al. (JADE Coll.), *Z. Phys.* **C33** (1986) 23.
S.Bethke et al., *Phys. Lett.* **B213** (1988) 235.

31. DELPHI Coll., "Bose-Einstein Correlations at LEP", in preparation.

INTERMITTENCY IN ELECTRON – POSITRON ANNIHILATION AT 91GEV

Neil Geddes
Rutherford Appleton Laboratory
Chilton
Oxon.
England

Abstract

Results of recent studies of intermittency in the multihadronic decays of the Z^0 are presented. The analysis is based on data recorded by the OPAL collaboration at LEP during 1990. The data are well reproduced by a range of QCD based fragmentation models and detailed study of these models indicates that the observed "intermittency" is a result of the jet structure of the events.

1 Introduction

Over recent years the analysis of fluctuations in multiplicity distributions in multiparticle production has been a very active field [1]. In particular, following the suggestions of Bialas and Peschanski [2], there has been much interest in non-statistical fluctuations in small phase space regions. They pointed out that studying fluctuations/correlations in the particle rapidity distribution as a function of resolution or (phase space) scale, eg. via factorial moment analysis (see below), offered a valuable tool which could reveal information about the sources of observed fluctuations. Of particular interest is the possible existence of "self-similar" fluctuations, that is, fluctuations which recur in a similar fashion at all resolution scales and which could be related to possible fractal structure in hadronic final states. For such fluctuations particular scaling behaviour is predicted for the factorial moments. This has been called "intermittency". The scaling behaviour is observed, over limited ranges of phase space resolution, in hadron-hadron [4] and nucleon-nucleon [3] collisions where current models of hadron production appear unable to account for observations [7].

Early, indirect, studies in electron-positron annihilation [5] found stronger fluctuations in rapidity distributions than for hadron-hadron or nucleon-nucleon interactions and again the data were not reproduced by model (JETSET version 6.3 [6]) calculations. Subsequent direct measurements [13], however, revealed good qualitative agreement between the same models and the data, although there remained qualitative differences. In the most recent studies in electron-positron annihilation excellent agreement has been seen between models and data [8][9][10]. This apparent change can be traced to two sources; firstly improvements in the model treatment of known correlations such as resonance decays and Bose-Einstein effect [10] and secondly the different relative importance of competing processes for the effects observed at different energies. Differences remain between data and model predictions in hadron-hadron and nucleon-nucleon collisions. However, much progress has recently been made on understanding the effects of contributions from photon conversions [12], Bose-Einstein correlations, double counting etc. [12] in these complicated environments. This work may point the way to improved understanding of the observed effects.

2 Factorial Moment Analysis

The analysis of factorial moments, originally proposed by Bialas and Peschanski [2] is usually applied to the particle rapidity distributions and may be summarised as follows. The central rapidity region Y is divided into M equal width (δy) bins. The normalised factorial moment of order j is then defined as

$$F_j(M) = \frac{1}{\langle \bar{n}_m \rangle^j} \langle \frac{1}{M} \sum_{m=1}^{M} n_m(n_m - 1)...(n_m - j + 1) \rangle \tag{1}$$

with:

$$\bar{n}_m = \frac{1}{M} \sum_{m=1}^{M} n_m$$

where the sum is over all bins, n_m is the number of particles in bin m for a given event and the angle brackets imply an average over all events. (This definition of the moments is not unique and several alternative ,non equivalent, definitions are possible [15].) The factorial moments as defined in Eq. 1 have the following properties:

1. Contributions to the numerator for moments of order j come only from bins containing j or more particles, hence, the moments act as a filter for highly populated bins.

2. The moments $F_j(M)$ are independent of the number of bins, M, into which the rapidity interval is divided if the number of particles in any bin is governed simply by the mean bin population and random (Poisson) fluctuations.

3. Non statistical fluctuations or correlations in the distribution of particles will produce moments which increase with increasing number of bins while the bin size δy is larger than the characteristic correlation length. Beyond this the moments will saturate and remain constant with increasing M. Non statistical fluctuations on many different scales will lead to a continued increase of the moments with M. Intermittent self-similar fluctuations at all scales will lead to a power law dependence of F_j upon M:

$$F_j \; \alpha \; M^{f_j} \qquad\qquad \text{with } f_j > 0 \qquad (2)$$

In practice this behaviour will be smeared out when the bin size becomes comparable with the detector resolution, at which point the moments will saturate.

The development of the factorial moments with resolution scale or, equivalently, number of bins M, thus provides information on the nature, strength and range of any correlations. The factorial moment analysis need not be restricted to the rapidity distribution. Any phase space distribution may be used provided the inclusive distribution (averaged over many events) varies only slowly over the range studied. Several procedures have been suggested for transforming quantities in order to fulfil this requirement [16][17], however, these procedures often introduce further complications [10] and are not considered here.

3 Intermittency in OPAL

3.1 *Data Analysis*

The OPAL collaboration has performed a factorial moment analysis using a large sample ($\sim$ 145000) of multihadronic decays of the Z^0 recorded during the 1990 LEP run. The analysis is described in detail in [8] and is outlined only briefly here. The analysis is restricted to charged particles as the OPAL detector provides superior momentum resolution and two particle separation for charged particles. All particles are assumed to be pions. Multihadronic events are selected for analysis if:

1. They contain at least 5 charged tracks.
2. The visible energy in charged particles is at least 20% of the total CM energy.
3. The magnitude of any momentum imbalance in the event is less than 40% of the total CM energy.
4. The event axis, taken to be the event sphericity axis determined using all charged particles, has $|\cos(\theta)| < 0.7$, where θ is the polar angle with respect to the electron-positron beam axis.

The first three criteria suppress backgrounds from non-multihadronic Z^0 decays and the fourth requires that the event be well contained in the active volume of the detector. The factorial moment analysis has been performed on the 1-dimensional rapidity distribution and the 2-dimensional rapidity (y) versus azimuthal angle (ϕ) distribution. Here ϕ is the azimuthal angle about the event axis. In the 2-d case the rapidity-azimuth distribution is divided into $\sqrt{M}y \times \sqrt{M}\phi = M$ bins. The factorial moments are then calculated as in Eq. 1. Particle rapidity y is calculated as

$$y = \ln \frac{(E + p_l)}{(E - p_l)} \tag{3}$$

Where E is particle energy and p_l is the component of particle momentum along the chosen event axis. Following previous analyses the sphericity axis is taken as the event axis (except where explicitly stated otherwise) and the analysis is restricted to the region $|y| < 2$ to ensure a smooth inclusive distribution. In fact the inclusive rapidity distribution has a dip ($\sim 20\%$) about $y = 0$. The effect of this dip on the factorial moments was tested by Monte Carlo and found to be negligible. The whole ϕ range, $0° < \phi < 360°$ is used.

The effects of the biases introduced by the detector or event selection criteria have been evaluated using a detailed simulation of the OPAL detector [18]. The momentum resolution of the OPAL central detector system allows study of the rapidity distribution down to distances of 0.01 units of rapidity, although detailed understanding of detector effects becomes important for rapidity bin sizes of less than 0.1 units.

3.2 *Intermittency in 1-d*

The 2nd to 5th factorial moments, not corrected for effects due to the detector acceptance or analysis procedure, are shown in figure 1. The moments initially increase with M but then appear to saturate for $M > 20$. This is well within the resolution range of the OPAL detector. Also shown in figure 1 are the predictions

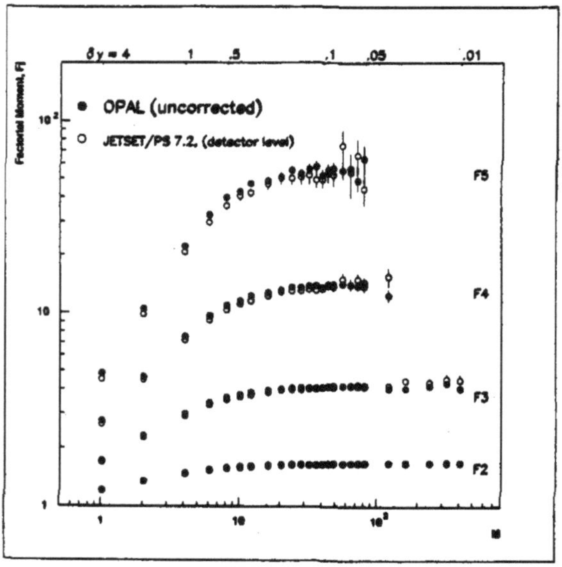

Figure 1: The uncorrected 2nd to 5th factorial moments (solid points), compared with the predictions of the JETSET Monte Carlo program including detailed detector simulation (open points).

from the JETSET Monte Carlo program (version 7.2) [6], after detailed simulation of the OPAL detector and analysis procedure. The main parameters of the Monte Carlo have been tuned to give a good description of the OPAL data [20], however, the factorial moments were not included in this "tuning". The Jetset Monte Carlo gives a very good description of the data, qualitatively and quantitatively. A similar check against data at the detector level has been performed using the HERWIG Monte Carlo program [19] with full detector simulation. Similar agreement between data and Monte Carlo was observed.

The excellent agreement between the JETSET/detector simulation prediction and the data allows us to use the Monte Carlo to correct the data for the effects of detector and analysis biases. Distributions corrected for these effects are considered below. The correction factors applied to the data are only ever smoothly varying functions of M and never exceed 50%. The corrected 2nd to 5th factorial moments are shown in figure 2 together with the predictions of the following Monte Carlo programs:

- JETSET version 7.2 [6], using the parton shower (PS) followed by string formation and fragmentation.

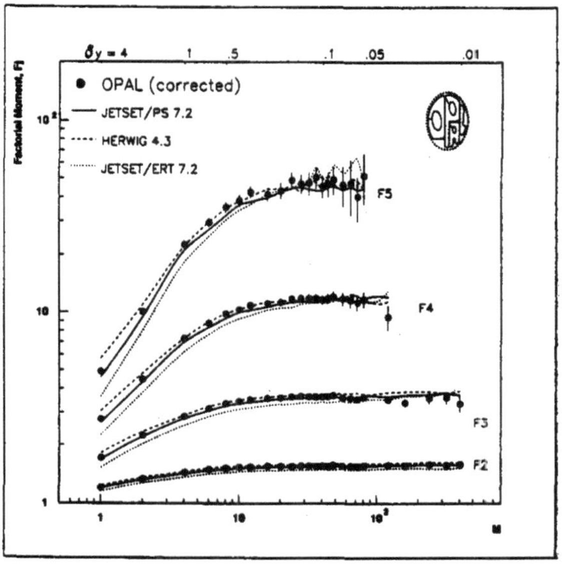

Figure 2: The corrected 2nd to 5th factorial moments (points) compared with the predictions of the JETSET/PS (solid line), JETSET/ERT (dotted line) and HERWIG (dashed line) Monte Carlo programs.

- HERWIG version 4.3 [19], using a parton shower followed by cluster fragmentation.
- JETSET version 7.2 [6], using the second order matrix element formalism of Ellis, Ross and Terrano (ERT) [21] followed by string formation and fragmentation.

All three models have been tuned to reproduce the global event shape distributions in the same way [20].

All three models give a good qualitative description of the data in figure 2. The divergence of the model predictions for small M is a reflection of their slightly different predictions for the total event multiplicity distributions. HERWIG and JETSET/ERT (as used here) predict slightly more or less high multiplicity events, respectively, than JETSET/PS and this is reflected in the factorial moments for large rapidity bins. Differences between the models and the data for large M are not large compared with the sensitivity of the predictions to the various model parameters.

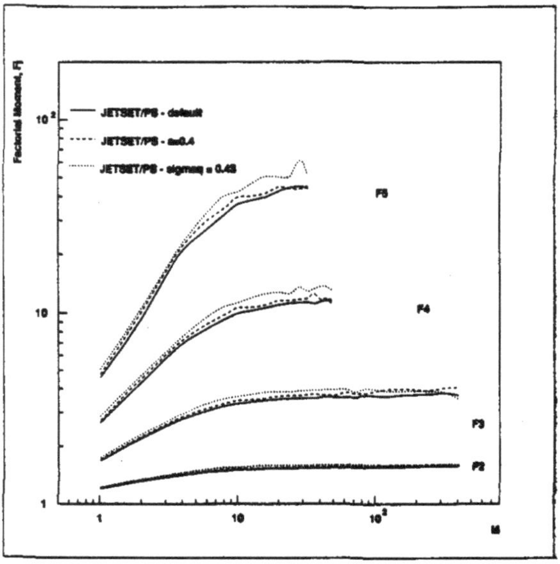

Figure 3: The (OPAL tuned) JETSET/PS predictions for the 2nd to 5th factorial moments (solid lines) compared with the predictions of the model when the tuned parameters are allowed to vary by 2σ.

This is illustrated in figure 3 which shows the tuned JETSET/PS predictions for the 2nd to 5th factorial moments together with the predictions after changing two fragmentation parameters by 2σ [20] from their optimum values. The parameters changed were; a (PARJ(41) in JETSET) one of the parameters governing the form of the longitudinal fragmentation function, and σ_q (PARJ(21)) the width of the Gaussian p_t spectrum for primary particles.

One significant source of uncertainty in the measurement of the factorial moments arises from errors in the determination of the event/sphericity axis. These problems can be avoided by choosing a fixed reference axis, such as the electron–positron beam axis. Figure 4 shows the corrected 2nd to 5th factorial moments of the rapidity distribution calculated with respect to the electron beam direction. The same event sample as described above was used, which unfortunately results in an inclusive rapidity distribution which is not slowly varying but shows significant structure. This structure affects both the magnitude of the measured moments and the details of their development with M. Part of the difference between figures 2 and 4 can be attributed to these effects illustrating the need for a good Monte Carlo simulation to account for trivial distortions introduced by the analysis. Further differences between figures 2 and 4 arise from the restriction to $| y |< 2$. All tracks passing the track/event

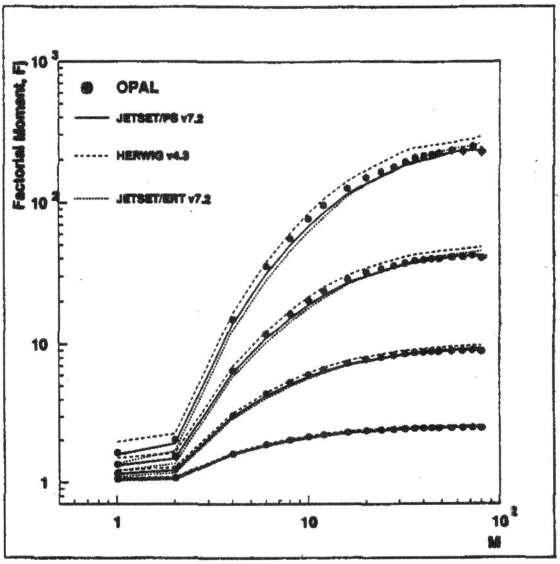

Figure 4: The corrected 2nd to 5th factorial moments (points) of the rapidity distribution calculated with respect to the electron beam axis, compared with the predictions of the JETSET/PS (solid line), JETSET/ERT (dashed line) and HERWIG (dotted line) Monte Carlo programs.

selection criteria pass this restriction for rapidity calculated with respect to the beam axis. However, with respect to the sphericity axis this cut tends to select only those tracks from multi jet events which populate the central rapidity region. All these effects are evidently well modelled by JETSET/PS, JETSET/ERT and HERWIG, the predictions of which are also included in figure 4. All three models reproduce the data well.

3.3 *Intermittency in 2-d*

The corrected factorial moments of the 2-d y versus ϕ distribution, using the event sphericity axis are the reference axis, are shown in figure 5 compared to the predictions of the three models. The data show a much stronger rise with increasing M than in the 1-d case and the moments do not appear to saturate over the range studied, which is limited by the size of the available event sample rather than by the detector resolution. There would appear to be more structure in the development of the moment with M than implied by eq. 2, although some scaling is apparent. All three

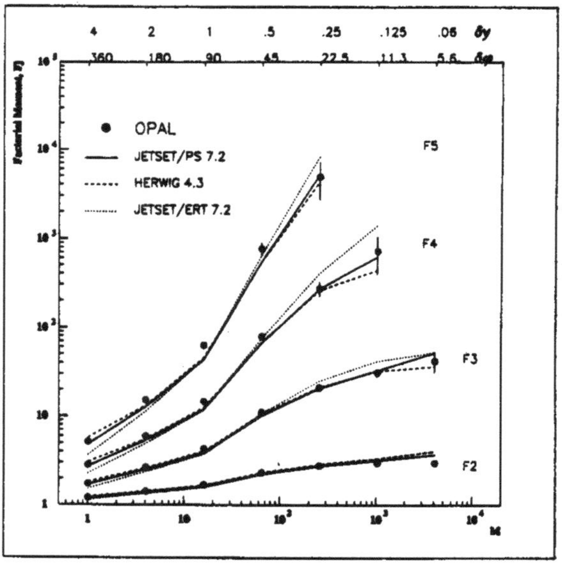

Figure 5: The corrected 2nd to 5th factorial moments (points) of the rapidity versus azimuthal angle distribution (taking the sphericity axis as the reference axis, compared with the predictions of the JETSET/PS (solid line), JETSET/ERT (dotted line) and HERWIG (dashed line) Monte Carlo programs.

QCD models again reproduce the data very well.

4 Intermittency in the Models

As all three of the models used above reproduce the data very well their predictions have been examined in detail. In all three models the hadronisation process can be broken into 3 stages:

1. A perturbative stage where an initial quark-antiquark pair may split to produce further partons. This may be achieved either by a parton shower or a matrix element approach.
2. A non perturbative stage where the partons from step 1 are converted into hadrons. Several approaches exist for this stage; for example, string fragmentation, cluster fragmentation or independent jet fragmentation.

3. Decays of the primary particles/resonances produced in step 2.

The relative importance of each stage is discussed below. The dominant source of the observed structure in the factorial moments is found to be the jet structure of the events, arising in stages 1 and 2.

The jet contributions are most clearly seen in the moments of the 2-dimensional y versus ϕ distribution, for which the increase of the moments with M is strongest. Figure 6 shows the predictions of the JETSET/PS Monte Carlo compared with the predictions of the same model but with the parton shower stage suppressed (ie. 2–parton events only). Clearly the absence of the parton showering stage removes the "intermittency" signal almost completely. As the model is essentially a 2–jet model few, if any, jets will now fall completely within the rapidity window of the analysis ($| y |< 2$). Without the perturbative stage the only remaining effect, at large M, arises from resonance decays. This contribution is known to be significant at large M from lower energy e^+e^- results [10], although the relative size of the contribution is much reduced at current energies. This also appears to be true for the contribution from Bose–Einstein correlations which do not appear to contribute significantly at 91 GeV although they do appear important at lower energies [10]. In figure 6 the parton shower suppression was achieved via simply switching off this stage during the JETSET generation, no other parameters/options were altered. Obviously this model would then not reproduce the global properties of the data very well. A more detailed analysis by Raab [11], however, has shown that even retuning the model (without the parton shower) to reproduce better the overall event shape distributions fails to introduce a significant signal.

Also shown in figure 6 are the moments from JETSET with other features suppressed:

1. **No decays** - All resonance and other particle decays are suppressed. This has only a limited effect on the observed signal, consistent with the above discussion.
2. **No p_t** - No p_t is given to quark–antiquark pairs produced during string fragmentation. This tends to increase the observed signal. This is easily understood as jets produced from any initial hard parton branching will now be more tightly collimated than the standard case, giving much stronger local fluctuations and larger factorial moments.

That the major effect is from initial hard parton branchings rather than an extended parton shower is evident from the success of the matrix element Monte Carlo in reproducing the data. In figure 7 the contributions to the observed signal from the JETSET/ME Monte Carlo from 2 parton ($q\bar{q}$), 2 and 3 parton ($q\bar{q}$ and $q\bar{q}g$) and

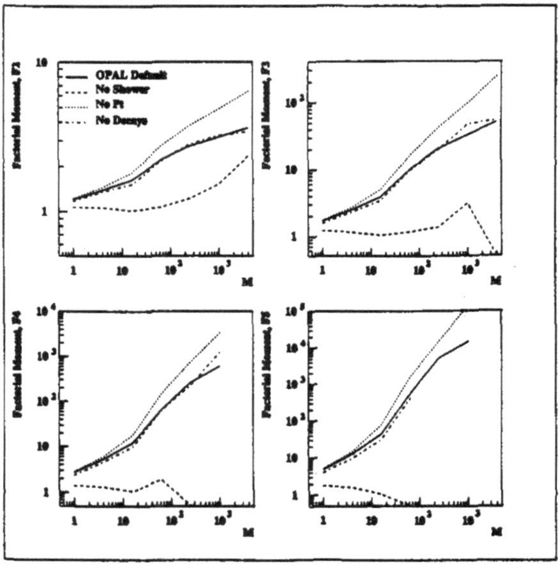

Figure 6: Various predictions from JETSET/PS for 2nd to 5th factorial moments of the y v ϕ distribution. Solid line: OPAL default, dashed line: no perturbative stage (parton shower), dotted line: no fragmentation p_t, dash-dotted line: all particle decays suppressed

2, 3 and 4 parton ($q\bar{q}$, $q\bar{q}g$, $q\bar{q}q\bar{q}$ and $q\bar{q}gg$) events are shown separately. The major effect is evidently from the 3 parton, $q\bar{q}g$ events, although the 4 parton events do contribute to the observed signal. The increase of $F2$ at large M for the 2 parton events is the contribution from resonance decays. Of course to reproduce the data it is essential that the limited number of partons present after these initial branchings hadronise further as the observed signals can not be reproduced with only 4 particles. The mechanism for the subsequent hadronisation step is not important (ie strings, independent jets extended parton shower plus cluster fragmentation) provided that the development has limited P_t with respect to these "principle" partons.

The analysis may be modified to reduce sensitivity to the presence of third or fourth jets, for example, as in the 1-d analysis using the beam axis as the reference axis presented in the previous section. In this case all jets for selected events fall within the window $\mid y \mid < 2$ and, hence, 2 jet events will also contribute to the measured moments. For given multiplicities the factorial moments, by virtue of the factorial in the numerator of eq. 1, are biased towards events with fewer, hence higher multiplicity, jets. Thus in this case the presence of more than two jets tends to reduce the factorial moments. It can then appear that the observed intermittency signal then

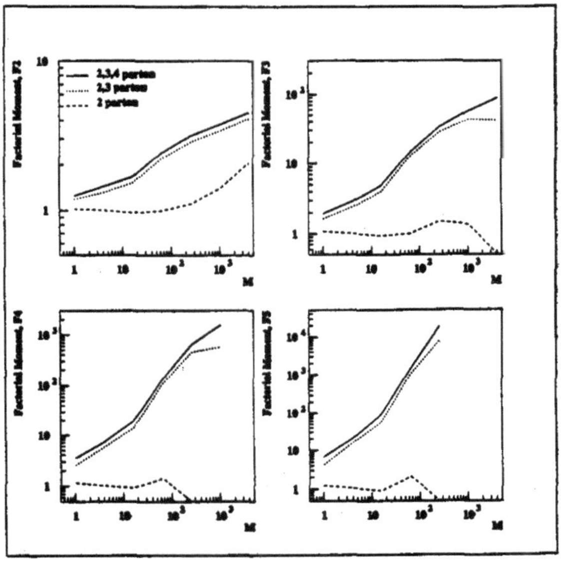

Figure 7: Various predictions from JETSET/ERT for 2nd to 5th factorial moments of the y v ϕ distribution. Dashed line: 2 parton ($q\bar{q}$) events, dotted line: 2 and 3 parton ($q\bar{q}$ and $q\bar{q}g$) events, and solid line 2, 3 and 4 parton ($q\bar{q}$, $q\bar{q}g$, $q\bar{q}q\bar{q}$ and $q\bar{q}gg$) events.

arises primarily during step 2 and that the effects of step 1 are destructive rather than the dominant feature as above. However, the major effect remains the presence of a limited number of well collimated jets of particles within the analysis window.

The major contribution to the observed behaviour of the factorial moments thus arises in steps 1 and 2 and depends only upon the production of a limited number of well collimated jets of particles. The relative importance of these two steps depends upon the details of the analysis and the model under consideration. That there is no clear separation between the contributions from 1 and 2 may be expected given that the division is to a large extent an artifact of the models.

5 Multiplicity and p_t Dependence

The factorial moments of rapidity distributions measured in hadron–hadron collisions show significant variation with total event multiplicity and particle p_t which is not well reproduced by model calculations [4]. These dependences in e^+e^- annihilation are investigated further below.

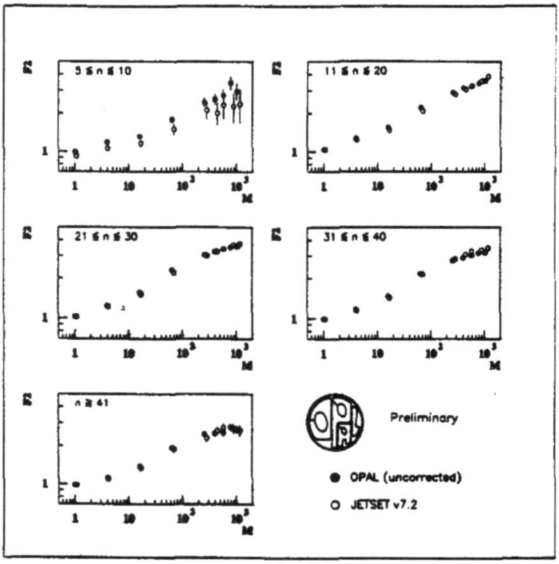

Figure 8: The uncorrected 2nd factorial moment (solid points) for 5 different event samples with different total charged multiplicity, n, compared with the predictions of the JETSET/PS Monte Carlo including detector simulation (open points).

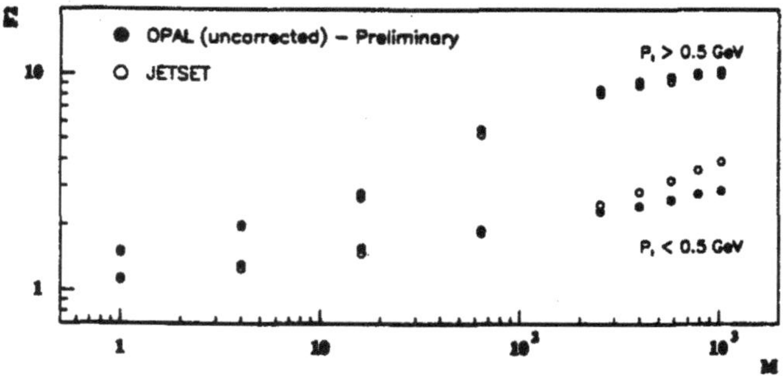

Figure 9: The uncorrected 2nd factorial moment (solid points) separately for particles with high and low p_t (momentum component transverse to event sphericity axis) compared with the predictions of the JETSET Monte Carlo including detector simulation (open points)

Figure 8 shows the 2nd factorial moment for 5 different ranges of total event multiplicity. The data are shown are detector level, with no corrections applied. Also shown are the predictions of the JETSET/PS Monte Carlo, including detailed detector simulation. No strong dependence of the moments upon charged multiplicity is observed and the Monte Carlo reproduces the data very well. A slight discrepancy may be evident for the lowest multiplicity events (the limited background events remaining in the data will be concentrated in this sample. There is no evidence for new phenomena giving rise to large local fluctuations in high multiplicity events. Similar results are obtained for the higher moments, although with more limited statistical accuracy.

Figure 9 shows the factorial moments for high p_t and low p_t particles separately; where p_t is the transverse momentum of the particle with respect to the sphericity axis. The "intermittency" is much stronger for the high p_t particles. This is consistent with the discussion of the previous section as higher p_t particles will tend to be found in multijet events. The predictions of JETSET, also shown in figure 9, are again in good agreement with the data, except that the predictions for low p_t particles for large M lie significantly above the data. Similar results are found for the higher moments, although the discrepancy for low p_t particles at large M are reduced for the higher moments. Further investigation with JETSET reveals that the dominant contribution to the rise in the Monte Carlo prediction for low p_t and large M comes from resonance decays. The differences between data and JETSET could thus be explained by problems in the detailed simulation of these decays and their reconstruction.

6 Conclusions

The properties of the factorial moments of rapidity distributions of hadronic final states produced in e^+e^- at 91 GeV are extremely well reproduced by QCD based Monte Carlo programs. The observed behaviour of the moments is dominated by the jet structure of the events. Contributions from other sources such as resonance decays, although present, are not as important as at lower energies.

The detailed behaviour of the factorial moments depends critically upon the details of any analysis and, consequently, it is often difficult to isolate the most important features. Although the moments are dominated by the jet structure of the events it is not clear that they offer a good tool for detailed investigation of this structure.

7 Acknowledgements

I would like to thank the conference organisers for providing such a pleasant location for the workshop. The Ringberg castle provided a beautiful setting for a productive workshop. I would also like to extend thanks to all my colleagues on OPAL and the CERN SL division without whom none of the results presented here would have been possible.

References

[1] See for example
"Intermittency in High Energy Collisions", Proceedings of the Santa Fe Workshop, March 1990, edited by F. Cooper, R. C. Hwa and I. Sarcevic (World Scientific, Singapore 1991) and these proceedings.

[2] A. Bialas and R. Peschanski, Nucl. Phys. B273 (1986) 703.
A. Bialas and R. Peschanski, Phys. Lett. B207 (1988) 59.
A. Bialas and R. Peschanski, Nucl. Phys. B308 (1988) 857.

[3] R. Holynski *et al.* Phys. Rev. Lett. 62 (1989) 733.
R. Albrecht *et al.* Phys. Lett. B221 (1989) 427.

[4] I. Ainenko *et al.* (NA22), Phys Lett. B222 (9189) 306.
M. Adamus *et al.* (NA22), Phys. Lett. B185 (1987) 200.
C. Albajar *et al.* (UA1), Nucl. Phys. B345 (1990) 1.

[5] B. Buschbeck, P. Lippa and R. Peschanski, Phys. Lett. B215 (1988) 788.

[6] T. Sjöstrand, Comp. Phys. Comm. 39 (1986) 347.
M. Bengtsson and T. Sjöstrand, Comp. Phys. Comm. 43 (1987) 367.
M. Bengtsson and T. Sjöstrand, Nucl. Phys. B289 (1987) 810.

[7] For recent reviews see, for example,
L. van Hove, Mod. Phys. Lett. A4 (1989) 1867.
B. Buschbeck and P. Lipa, Mod. Phys. Lett. A4 (1989) 1871.
R.Peschanski, CERN-TH5891/90, To be published in J. Mod. Phys.

[8] M. Akrawy *et al.* (OPAL), Phys Lett. B262 (1991) 351-361.
N. Geddes, Proceedings of XXVI Rencontre de Moriond, March 1991

[9] P. Abreu *et al.* (DELPHI) Phys. Lett. B214 (1990) 137.
A. De Angelis, CERN-PPE/90-129.
A. De Angelis, these proceedings

[10] H.Behrend *et al.* (CELLO) Phys. Lett. B256 (1991) 97
O. Podobrin, DESY 91-028
O. Podobrin, these proceedings

[11] N. Lieske, these proceedings.
V. Raab, these proceedings.

[12] See talks by E. Stenlund and P. Lippa in these proceedings.

[13] W. Braunschweig *et al.* (TASSO), Phys. Lett. B231 (1989) 548.

[14] W.Ochs and J.Wosiek, Phys. Lett. B214 (1988) 617.
W.Ochs and J.Wosiek, Phys. Lett. B232 (1989) 271.

[15] C. B. Chiu and R. .C. Hwa, OITS 466, May 1991.

[16] W. Ochs, MPI preprint MPI-PAE/PTh 63/90 (1990).

[17] A. Bialas and M. Gazdzicki, Phys. Lett. B252 (1990) 483.

[18] D. .R .Ward, Procedings of the MC91 Workshop, Amsterdam, April 1991.

[19] G. Marchesini and B. Webber, Nucl. Phys. B310 (1988) 461.
G. Marchesini and B. Webber, Cavendish-HEP-88/7.

[20] M. Akrawy *et al.* (OPAL), Z. Phys. C47 (1990) 505.

[21] R. K. Ellis, D. A. Ross, A. E. Terrano, Nucl. Phys. B178 (1981) 421.

AN INVESTIGATION INTO INTERMITTENCY

NATHALIE LIESKE
Department of Physics, Imperial College,
London SW7 2BZ, United Kingdom

Abstract

The results of an investigation based on ALEPH data, $e^+e^- \rightarrow$ hadrons at $\sqrt{s} = 91$ GeV, into fluctuations in rapidity space are presented. It is found that the behaviour of the factorial moments is well represented by the Lund parton shower model. An estimate is made of the scale of fluctuations needed to describe the data. Differential moments are introduced and are used to demonstrate that within the average represented by the traditional factorial moments the pattern of fluctuations is itself a strong function of rapidity. This pattern is shown to be primarily associated with the emission of hard gluons. The implied structure between the hadron clusters and partons is explored by the use of transverse mass moments. It is concluded that the principal features of the one-dimensional factorial moments, differential moments and transverse mass moments have their origin in the $O(\alpha_s)$ matrix element, kinematic boundaries and the method of selection of the event axis.

1 Introduction

The approach taken here uses the fact that the Lund parton shower (PS) Monte Carlo [1] describes the intermittency effect seen in the data. The Monte Carlo model is studied and used to understand the underlying physics creating the effect. This investigation focuses on why the one dimensional moments in rapidity space behave as they do.

2 Event Selection and Simulation

The ALEPH detector is described in detail elsewhere [2]. The analysis is based on a sample of ALEPH 1990 data on e^+e^- at energies close to the Z resonance peak. The detailed detector cuts applied are discussed elsewhere [3]. Selected events were required to have at least 5 such tracks in the appropriate rapidity range; after cuts, approximately 80K events were analysed. The thrust axis was found from the accepted tracks.

For direct comparison with data, Monte Carlo events (approximately 70K) with complete detector and experimental event selection were generated using as input JETSET 6.3 specially tuned [4] to match the data on inclusive and event shape variables. These are referred to as GALEPH [GEANT + ALEPH] events after the name of the simulation program. The investigations were based on generator level Monte Carlo events created with the same parameter values; it was confirmed that the physics conclusions were not changed in adopting this procedure, [3]. Appropriate variations in the parameters could then be introduced at the generator level and comparisons made between the hadron and the parton level without the need for correction for detector effects.

3 Global Factorial Moments

The rapidities with respect to the thrust axis of all charged particles in a hadronic event are calculated assuming the pion mass. The interval Δy, $-2 \leq y \leq 2$, is divided into M equal bins of width $\Delta y/M$. From the number, n_m, of particles in the mth bin, the quantities $(S_i)_m = n_m(n_m - 1)\ldots(n_m - i + 1)$ are found for each bin.

Two slightly different definitions of the factorial moments [5] referred to here as standard moments and normalized moments [6] $F_i^s(M)$ and $F_i^n(M)$ respectively.

$$F_i^s(M) := \frac{\langle \frac{1}{M} \sum_{m=1}^{M} (S_i)_m \rangle \; M^i}{\langle N \rangle^i} \qquad (1)$$

$$F_i^n(M) := \langle \frac{\frac{1}{M} \sum_{m=1}^{M} (S_i)_m \; M^i}{N(N-1)\ldots(N-i+1)} \rangle \qquad (2)$$

N is the total number of particles in the Δy range, typically 12. The normalisation of the standard moments includes the mean multiplicity of events, and so they are sensitive to the shape of the distribution in N, whereas this dependency is removed in the normalized moments.

Figure 1 shows that the Lund PS model agrees well with the data for the standard and normalized moments. The general conclusion is that the use of the Lund model could lead to the understanding of which effects are important in contributing to the signal detected.

3.1 A Simple Model

To give an indication of the magnitude of the effects needed to account for such moments, comparison is made with a simple model of a fluctuation process in which, within a constant $\langle dN/dy \rangle$ there are two contributions to the hadrons,

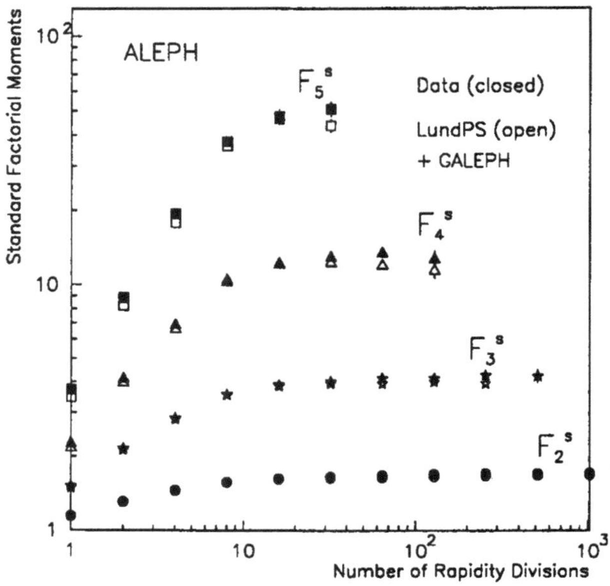

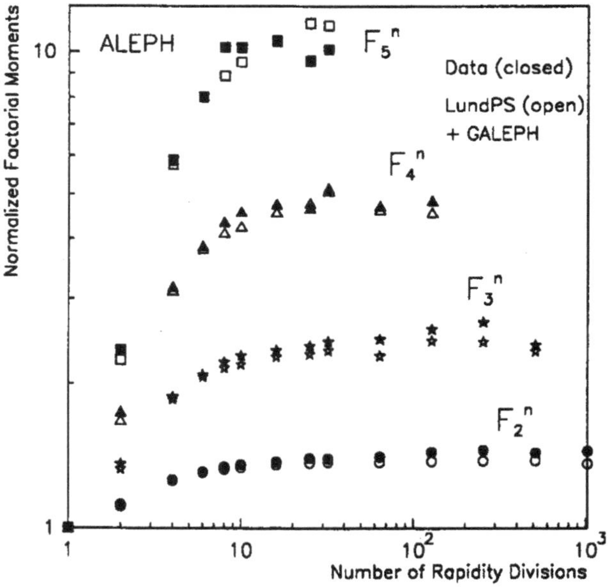

Figure 1: Comparison of Data and Monte Carlo for Standard and Normalized 1-D Factorial Moments In Rapidity

each of which is Poissonian. One process contributes uniformly over the full y range of interest with a total number of particles n_1. The second component extends over a fraction f of the full y range, within which it raises the mean level by a factor β. The position of this component is to be randomly distributed over the full y range. One could picture some physics process which superimposes a local clustering of particles and which itself could occur anywhere within the y range, on top of the main, structureless background. The model assumes that every event is of this nature.

Taking $\langle N \rangle = 12$ to match the data over the interval $|y| \leq 2$, there are effectively just two parameters of the model, f and β. The fraction, f, controls the value of M at which the moments start to saturate with increasing M, roughly at $M = 1/f$. The parameter β controls the asymptotic values of the moments at high M, though not their relative magnitudes. For a Poisson distribution both the standard and the normalized moments equal 1.0 for $M = 1$, whereas the moments are higher in the data, reflecting a non- Poissonian distribution. No attempt is made to simulate a non-Poisson distribution in total N, and therefore comparison of the simple model is made to the normalized moments. With appropriate choices of β and f, quite a reasonable description of the data is possible. This is shown in figure 2 with $\beta = 3.5$ and $f = 0.35$. The agreement, including the relative magnitudes, is surprisingly good in view of the simplicity of the assumptions.

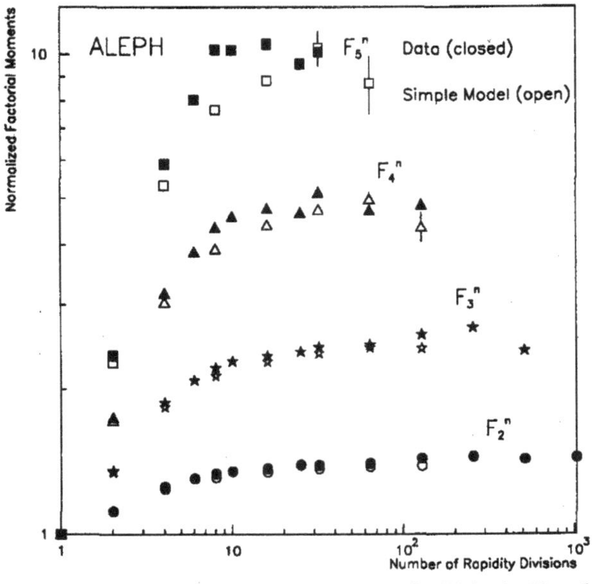

Figure 2: 1-D Normalized Moments Compared With A Simple Two-Parameter Model

The parameters of the simple model indicate the magnitude of the effects sufficient to account for the observed behaviour of the moments, namely, that there be some process or processes that can increase the density of the tracks over about 1.4 units of rapidity space by an average factor of about 3.5 – not necessarily constant of course, as the process itself could be subject to fluctuations. This magnitude would have to be even greater if only a fraction of the events were affected.

4 Differential Moments

4.1 Method

The moments which have been used so far do not contain information on the distribution in y of the fluctuations, because they are averaged over the entire range. The differential moments are introduced in order to study the y dependence of the fluctuations and are calculated as follows: the value, $(S_i)_m$ as defined earlier, is given for each of the M individual bins across the y range. The average over all events,

$$(f_i)_m = \langle (S_i)_m \rangle / \langle n_m \rangle^i \tag{3}$$

is taken. The denominator, the mean number of tracks in the bin, ensures that for a Poisson distribution $f_i = 1.0$ for all i. Since each y bin is analysed with respect to its own mean, there is no longer any need for a relatively flat y distribution, and the differential moments can be used to study the fluctuations in regions where the rapidity density is varying rapidly. Therefore, a range $-4 \leq y \leq 4$ is used, which includes the majority of particles produced. Attention is focused on the third moment, f_3; the statistical errors are small and any effect of gamma conversions is reduced due to the weighting of the third moment. The average value of a moment at a given y will be unchanged under changes in M, or bin width, if the particles within that bin are uncorrelated. This is important in following the evolution of the structure as a function of M.

4.2 Results and Interpretation

The differential moment f_3 is shown in figure 3 over an extended range of y. Beyond $|y| = 2.0$, where the density is falling very rapidly with increasing $|y|$, f_3 is almost constant with a value of about 1.5 which is close to the Poissonian value of 1. Main attention is directed to the central region of $|y| < 2.0$, where two peaks appear near $|y| = 0.6$. The horizontal line shows the corresponding global moment $F_3^s(M = 16)$. It is seen that within this average there is a rapid and striking variation in the pattern of fluctuations as measured by f_3.

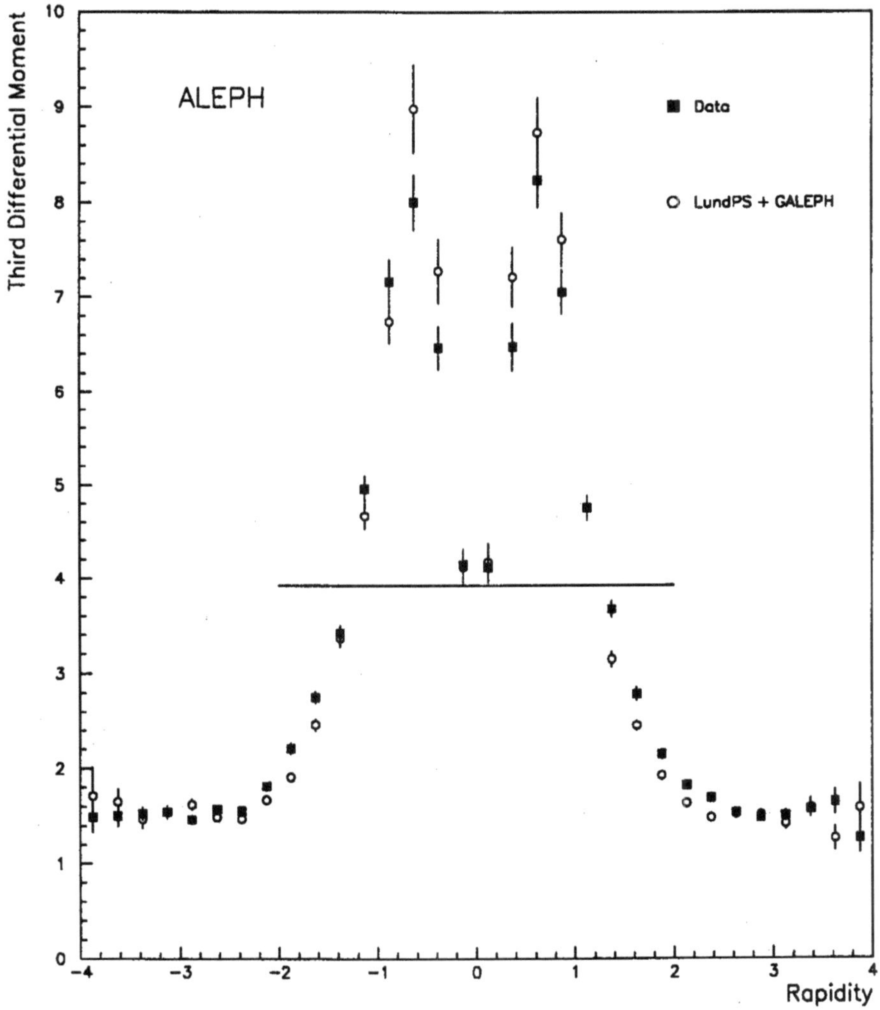

Figure 3: Comparison of Data and Monte Carlo for the Third Differential Moment. The solid line indicates the value of the corresponding global moment $F_3^s(16)$ over the range $-2 < y < 2$.

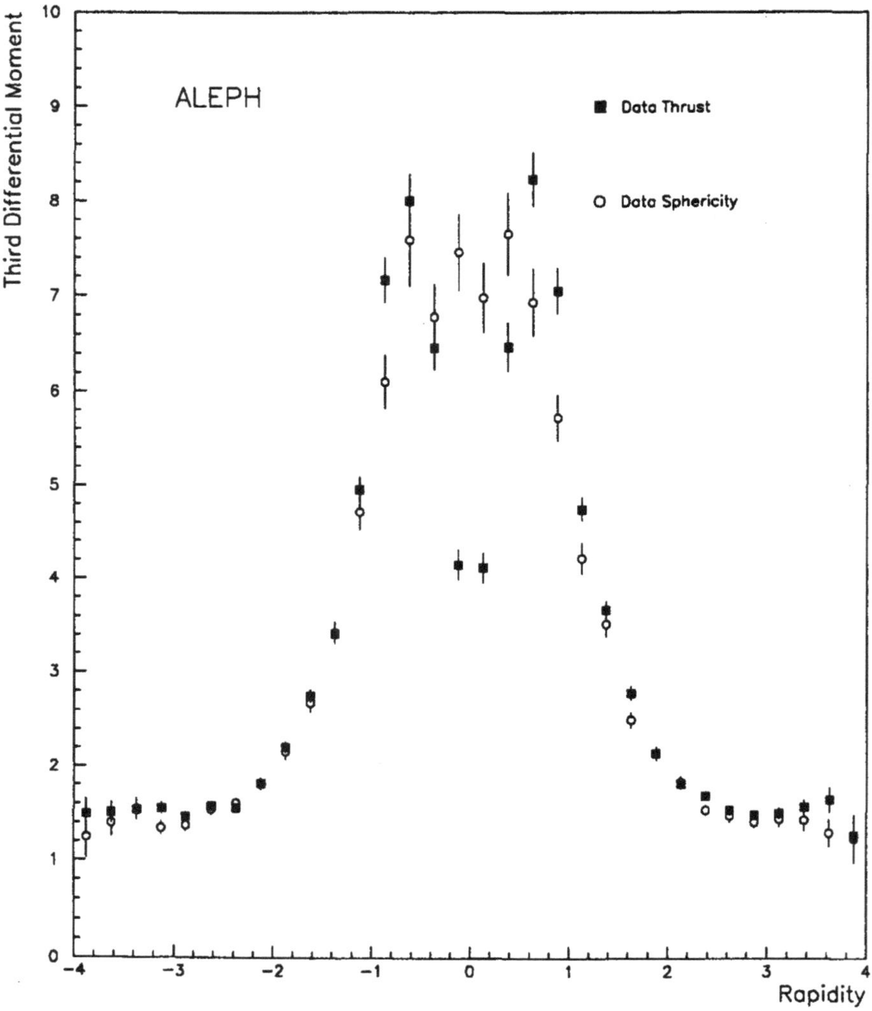

Figure 4: Dependence of the Differential Moment on the Use of Thrust or Sphericity Axis for the Calculation of Rapidity. It can be seen that in the central region, $|y| < 1$, the dip structure disappears if sphericity is used. The explanation is given in section 6.

Investigations were carried out using the Monte Carlo to determine the source of the correlations. At generator level the magnitudes of the moments are decreased, relative to those with experimental detector and cut simulation, within the region $|y| < 1$, [3], but the structure has the same general shape thus retaining the same physics. The rest of the section deals with these analyses.

The effect of variations in particle density as a function of flavour was small; Monte Carlo studies show that this is greatest around $|y| = 2$ for $\bar{b}b$ events, which are only 1/5 of the total. In the moments, light and heavy quarks investigated alone showed the same general structure. It was also verified that the assignment of the pion mass to all charged tracks had a negligible effect on the moments.

A high particle multiplicity on one side of an event in rapidity was usually associated with high gluon energy flow on the same side. This relationship was most directly tested by switching off the perturbative QCD effects. This prevented the formation of the parton shower in the JETSET model so that the string which fragments into hadrons was directly stretched between the quark and the antiquark. With all fragmentation parameters unchanged, including the break-up probability per unit proper time, the effect on the moments can be seen in figure 5a. There is a significant though much smaller structure at $|y| = 2.6$. This is principally the effect of the admixture of $\bar{b}b$ in the simulation which produces a high particle multiplicity in this region when there is a hard fragmentation function and no gluon radiation.

The total mean number of charged particles in this last test fell from about 20 to 12. The significance of clustering must be compared with that expected from Poisson statistics for a given mean. To correct for this, the Lund fragmentation parameter, a, [1] was increased from 0.5 to 4.0, giving a mean of 20. Since other string fragmentation parameters were left unchanged, effects such as particle composition, resonance decays, flavour admixture and the fragmentation process itself should be present in similar proportions in the non-perturbative and Lund PS distributions. Figure 5b gives the result. The differential moments decrease because residual fluctuations become relatively smaller with respect to the increased means. The absolute values are also closer to Poissonian. Jet-like structures produced by perturbative QCD are the most important effect.

These results should not be taken to exclude contributions to intermittency from non-perturbative processes in general. The investigation indicates that the main behaviour of the global and differential moments in rapidity has its origin at the parton level and that their study can be a simple way of examining parton-induced effects in the event sample. The exploration of the relationship between partons and hadrons as it affects intermittency is described in the next two sections.

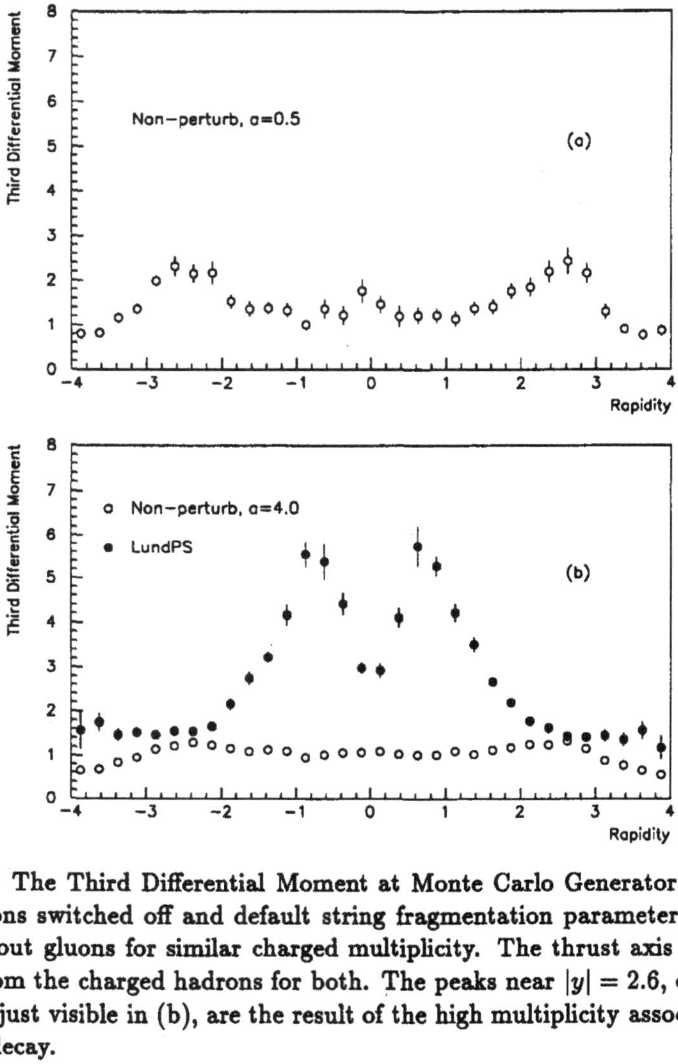

Figure 5: The Third Differential Moment at Monte Carlo Generator Level: (a) with gluons switched off and default string fragmentation parameters, (b) with and without gluons for similar charged multiplicity. The thrust axis was determined from the charged hadrons for both. The peaks near $|y| = 2.6$, clear in (a) but only just visible in (b), are the result of the high multiplicity associated with b quark decay.

5 Transverse Mass

The previous investigation has dealt solely with fluctuations in hadron number density, this being the main tool of intermittency analysis. However, hadrons with a given rapidity cover a wide spectrum of energies particularly in the important central region where rapidity becomes essentially a function of polar angle alone.

It is proposed here that the source of the fluctuations comes from the superposition of events with a gluon contribution to the y distribution, and therefore the significance of a fluctuation should lie also in the associated energies, rather than merely the number, of hadrons in a rapidity bin. In order to pursue this idea the simplest procedure would be to weight each hadron in a given rapidity bin by its energy, effectively replacing hadron number by total hadron energy. This however introduces a major disadvantage in studying y dependence because energy density is not invariant under boosts along the rapidity axis as is the number density. This problem is avoided by the following procedure. Hadrons are allocated to bins in y as before and the quantity: $M_t = \sqrt{\sum(E+P_{||})\sum(E-P_{||})}$ for the set of tracks in each bin is calculated. This quantity is referred to as the transverse mass of the set of hadrons and reduces to the total energy at $y=0$; it is invariant under boosts in y. It can be applied at both the hadron and the parton level.

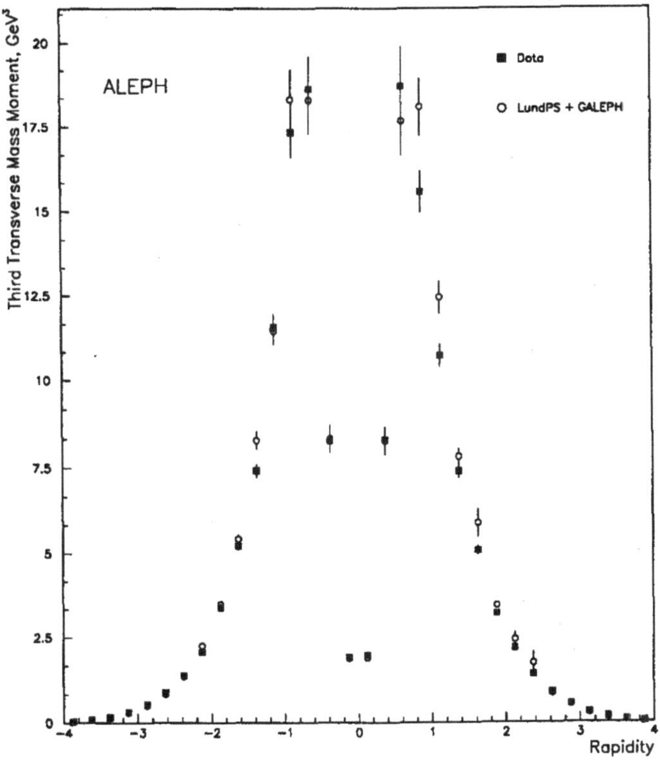

Figure 6: Third Transverse Mass Moment for Data and Monte Carlo at the Hadron Level as a function of Rapidity

Attention is again directed to the third moment, simply defined as $\langle M_t{}^3 \rangle$ and which is shown for data and Monte Carlo as a function of y in figure 6. The general structure is even clearer in the transverse mass moments where the fall at high y seen in the differential moments is accentuated. The Monte Carlo can now be used to make a direct comparison between hadrons and partons at generator level. For the most direct comparison, the same axis is used for events at both levels, ie. the thrust determined on the basis of charged hadrons. Parton energies and rapidities are determined from their momenta prior to fragmentation. The gluon mass is taken as zero and the default Lund values are used for the quarks. The moments for the hadrons are shown in figure 7a and for the partons in 7b. The similarity in the shape of the transverse mass moments is confirmation that the structure seen in the various moments at hadron level has its origin at the parton level. Monte Carlo studies have shown that the major source of the relative magnitude of difference between the hadrons and partons is mainly due to missing transverse mass resulting from the use of charged hadrons only. If, as before, the perturbative processes are switched off, the moments become very small (figure 7a,b); for the hadrons there remains a small effect from the intrinsic transverse momentum from the hadronisation, resonance decays and choice of the thrust axis. The discontinuities in the moments in figure 7b, just below $|y| = 3.0$, are produced by b quark events, which have a maximum rapidity at this value because of the large b quark mass.

6 The Origin of the Structure

It has been shown with the differential moments and confirmed with the transverse mass moments that the origin of the fluctuations in rapidity space is at the parton level. The purpose of this section is to demonstrate how the structure seen in the moments is present at $O(\alpha_s)$.

When the $O(\alpha_s)$ cross section for the process $e^+e^- \longrightarrow \bar{q}qg$ is transformed to the $m_t - y$ plane, the cross-section for the emission of a gluon becomes:

$$\frac{d\sigma}{\sigma} = \frac{4\alpha_s}{3\pi} \left({x_1}^2 + {x_3}^2\right) \frac{dm_t dy}{m_t} \tag{4}$$

where the fractional energies, x_i, are defined as $x_1 = E_{quark}/E_{beam}$ refering to the quark, x_2 to the gluon and x_3 to the antiquark, all assumed massless. The scaled transverse mass is $m_t = M_t/E_{beam} = x_2 \sin\theta_{21} = x_3 \sin\theta_{31}$, where θ_{ij} is the angle between partons i and j as defined when the axis is taken along and in the direction of the quark.

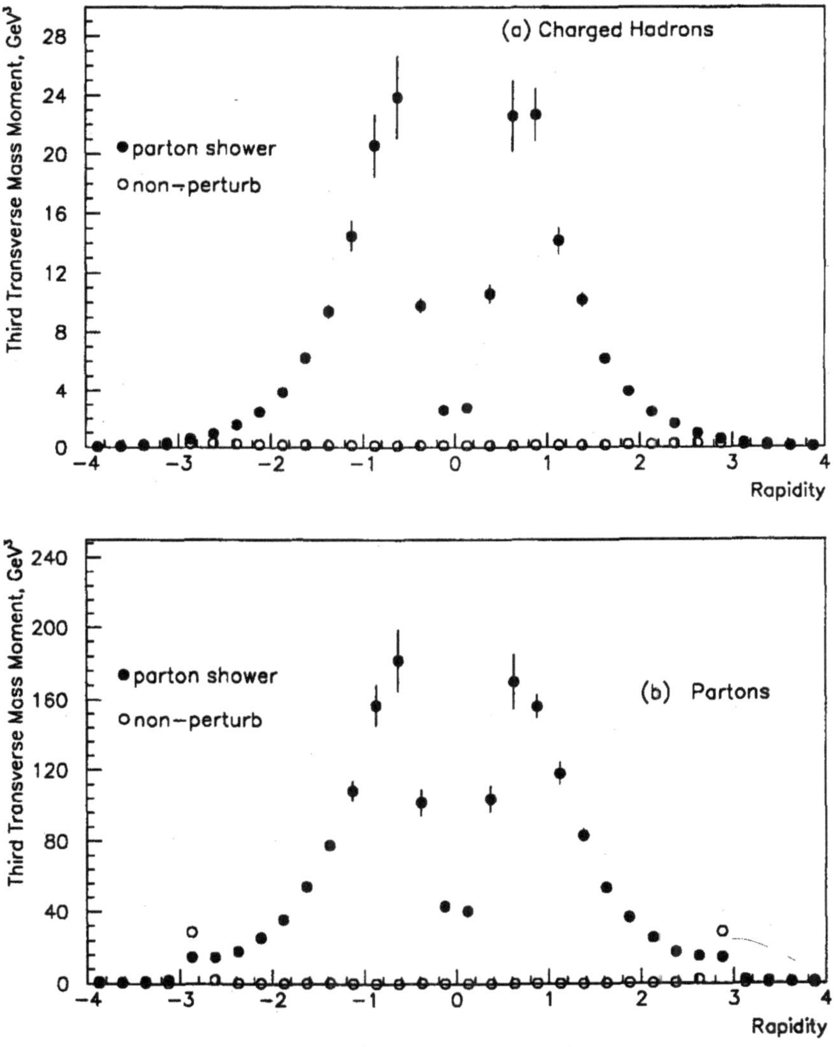

Figure 7: The Third Transverse Mass Moment at Monte Carlo Generator Level for: (a) charged hadrons and (b) partons with and without gluons, showing striking similarity in the behaviour of the moments at the two levels.

It can be seen from equation 4 that, apart from the slowly varying quadratic factor, the density on the $m_t - y$ plane falls as $1/m_t$ and is independent of y. For any given y there is a kinematic limit on the maximum value of M_t which, along with the density distribution, controls the value of the moment. The result is a smooth, structureless variation in the moments with y. The central structure seen in the moments has some origin other than the matrix element probability.

The density in the $m_t - y$ plane above was calculated with respect to the quark direction, while in the analysis the axis is chosen to maximise the event thrust. The corresponding procedure at the parton level causes the axis to align near to the most energetic parton. If the variables are re-evaluated with respect to the thrust axis then the maximum possible value of m_t occurs when the three partons have equal energies, $x_1 = x_2 = x_3$, separated by 120°. The distributions in the $m_t - y$ plane split such that the central dip emerges in m_t as a function of y. The predicted positions of the peaks, namely $|y| = \ln(\tan 60°) = 0.55$, correspond closely with those seen in the differential moments.

The difference in the moments defined with respect to the thrust or sphericity axis can also be seen at the parton level. In an event with two equally energetic partons and a third smaller energy parton (an example of the type of event which populates the central region of the rapidity distribution), the sphericity axis will be along the symmetry axis of the event. The thrust will choose one or the other of the two high energy partons creating the kinematic splitting described above. No corresponding choice occurs with sphericity and therefore no dip structure is produced.

7 Summary and Conclusions

The global moments, standard and normalized, have been presented for charged particles in rapidity space for ALEPH data. Rapidity was measured with respect to the thrust axis. Lund PS reproduced the data well, in agreement with the DELPHI[7] and OPAL[8] results.

The magnitude of the fluctuations in particle density required to match those observed in the data was estimated by comparison with a simple model, and found to correspond to local increases in the density of tracks of a factor 3.5 over a range of 1 - 1.5 units of y. The introduction of differential moments showed that the fluctuations themselves have a strong y dependence, with sharp peaks near $|y| = 0.6$. These moments give detailed information on the fluctuations over the rapidity range. The only source of fluctuations of the required magnitude, found on the considered scale, arose from perturbative QCD effects. When the gluon radiation was switched off, corresponding to the fragmentation of a straight, rather than a kinked, string the structure disappeared.

The total transverse mass of a set of particles in a rapidity bin, whether hadrons or partons, was also studied. The differential moments, the transverse mass moments for hadrons and the transverse mass moments for partons all showed the same structure, confirming that the source of the effect lay at the parton level. Finally, it was demonstrated that the main features of the structure seen in the moments, and hence in the global moments, can be traced back to the first order matrix element level, coupled with kinematic bounds and the use of thrust to determine the event axis.

8 Acknowledgements

This work was completed by David Binnie and I in the ALEPH collaboration. We thank W. Ochs for several useful discussions. We also thank the SL division for the excellent performance of the machine. We also wish to acknowledge the contribution made to ALEPH by the technical staff of CERN and the home institutes.

9 References

1. M. Bengtsson and T. Sjöstrand, Phys. Lett. B 185 (1987) 435; B. Andersson et al., Phys. Rept. 97 (1983) 31
2. D. Decamp et al.,(ALEPH Collaboration), *Phys. Lett.* **B234** (1990) 209
3. D. Decamp et al.,(ALEPH Collaboration), *CERN-PPE/91-121*
4. D. Decamp et al.,(ALEPH Collaboration), Properties of hadronic Z decays and test of QCD generators, *to be published.*
5. A. Bialas and R. Peschanski *Nucl. Phys.* **B273** (1986) 703, also *Phys. Lett.* **B207** (1988) 59
6. W. Braunschweig et al.,(TASSO Collaboration), *Phys. Lett.* **B231** (1989) 548
7. P. Abreu et al.,(DELPHI Collaboration), *Phys. Lett. B*; A De Angelis *CERN-PPE/90-129* Sept 1990
8. M.Z.Akrawy et al.,(OPAL Collaboration), Phys. Lett. B262 (1991) 351

Intermittency analysis in higher dimensions

VOLKER RAAB

Max-Planck Institut für Physik, Werner Heisenberg Institut

Föhringer Ring 6, D-8000 München 40, Germany

and

Physics Department, Technische Universität München

Abstract

Factorial moments are analyzed for one, two and three dimensional subspaces of the three dimensional phase space $(y, \phi, \ln \frac{2p_\perp}{E_{CM}})$. Apart from the influence of the Dalitz decay and photon conversions all moments saturate for more than $\approx 2^5$ divisions per dimension. Some moments in two and three dimensions are found to factorize into products of moments in lower dimensions. The data are compared to a set of models with different methods for calculating multi parton states. The intermittent behaviour, i.e. the rise of the factorial moments with decreasing bin size, is found to be stronger in models with larger numbers of partons. Taking event samples with given minimal values of thrust it is shown that different scales of transverse momentum contribute to the increase of the factorial moments.

1 Introduction

The subject of the analysis of intermittency is the measurement of nonpoissonian fluctuations in the distribution of particles over phase space in particle collision experiments and their dependence on the resolution. Charged particles produced by e^+e^--annihilations at a center of mass energy of 91 GeV measured with the ALEPH detector at LEP are analyzed. The one or more dimensional phase space is prescribed in terms of rapidity (y) with respect to thrust axis, azimuthal angle (ϕ) with $\phi = 0$ oriented parallel to the minor axis and the scaled logarithm of transverse momentum ($\ln \frac{2p_\perp}{E_{CM}}$). Unless otherwise stated, the intervals $y \in [-2, 2]$, $\phi \in [0, 2\pi]$ and $\ln \frac{2p_\perp}{E_{CM}} \in [-5.5, -4]$ were used.

To measure fluctuations of the number of particles n per bin in a region of phase space, the factorial moments of order i and resolution M are introduced:

$$F_i(M) := \frac{< n(n-1)\cdots(n-i+1) >}{< n >^i} \quad . \tag{1}$$

Here the averaging extends over all M subdivisions of a given region of phase space and over all events. A power-law dependence of $F_i(M)$ with respect to M then hints at self-similar processes in the generation of particles. Sometimes the parametrization $M = 2^{md}$ will be used, in which d gives the dimensionality of the pase space.

If inclusive distributions of the variables of phase space are not flat, the $F_i(M)$ will reflect those (non-statistical) fluctuations because of the averaging over the M bins. These contributions due to the inclusive shapes are called geometrical or topological effects. To flatten their densities $n(y)$, Białas, Gazdizicki[2] and Ochs[10] have suggested a transformation of the variables according to

$$\tilde{y}(y) = \frac{\int_{y_{min}}^{y} n(y')\, dy'}{\int_{y_{min}}^{y_{max}} n(y')\, dy'} \quad . \tag{2}$$

It has been argued by Ochs[9] that multidimensional fractal structures may be hidden when projected down to fewer dimensions. Therefore the analysis has also been expanded to two and three dimensional phase space by taking pairs or the triplet from y, ϕ and $\ln \frac{2p_\perp}{E_{CM}}$.

2 The data sample

In this analysis have been used the data that were taken with the ALEPH detector* in the years 1989 and 1990. Only charged particles were included. Tracks were required to have at least five three dimensional time-space-koordinates, to have an angle of at least 20° and a transverse momentum of at least 200 MeV both with respect to the beam axis and to have a distance from the interaction point of not more than 5 cm in the direction along the beam and 2 cm in radial direction. An event was included if it had more than four charged tracks, the total energy of charged particles exceeded 15 GeV and the angle between sphericity axis and beam pipe was at least 35°. These requirements left 50 602 events to be analyzed.

To correct the measured factorial moments for detector effects, monte carlo generators were used to determine correction factors. For this, factorial moments at the generator level (HVFL†) were compared with moments after a detailed detector simulation (GALEPH[13]).

*For a detailled description of the detector see B. Decamp [6].

†This generator is based upon JETSET 7.3, but it is modified in how it treats heavy flavour particles.

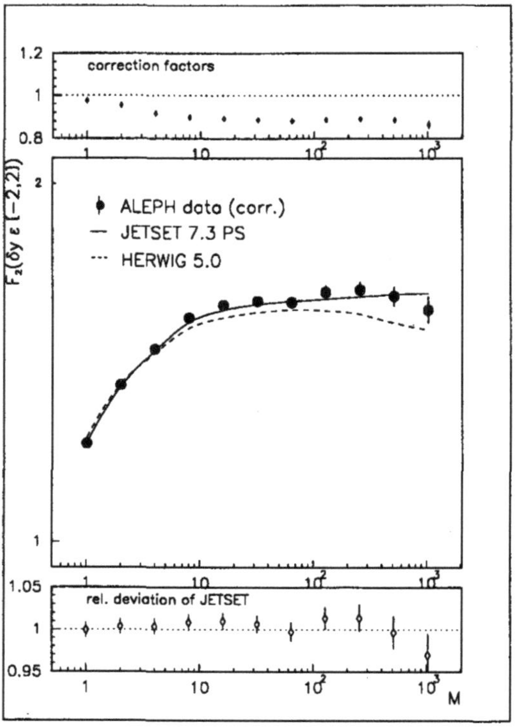

Figure 1: The second moment in $y \in [-2, 2]$, $F_2(\delta y)$, for corrected ALEPH data and for the models JETSET 7.3 PS and HERWIG 5.0 (central part). In the upper box are the correction factors. The lower box indicates the relative deviation between ALEPH and JETSET 7.3.

3 One dimensional analysis

3.1 Factorial moments in rapidity y

The one dimensional factorial moments for a binning in rapidity are measured in three different ways: taking the interval $|y| < 2$ without transformation, the interval $|y| < 5$ without transformation and the interval $|y| < 5$ with transformation according to Białas, Gazdizicki and Ochs. In the center of figure 1 it can be seen that discrepancies between the predictions of the JETSET 7.3[15,1] model and data for the reduced interval $|y| < 2$ are at a level of approximately $1 - 2\%$ (lower box) and are equivalent to the statistical accuracy[‡, §]. The correction factors are shown in the upper box.

‡Error bars are purely statistical in all graphs shown.

§Numerical tables of ALEPH data for the shown factorial moments can be found in V. Raab[12].

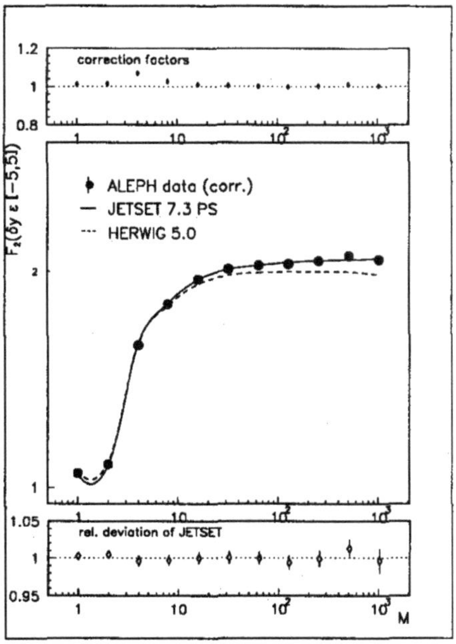

Figure 2: The second factorial moment of $y \in [-5, 5]$, $F_2(\delta y)$, for corrected ALEPH data.

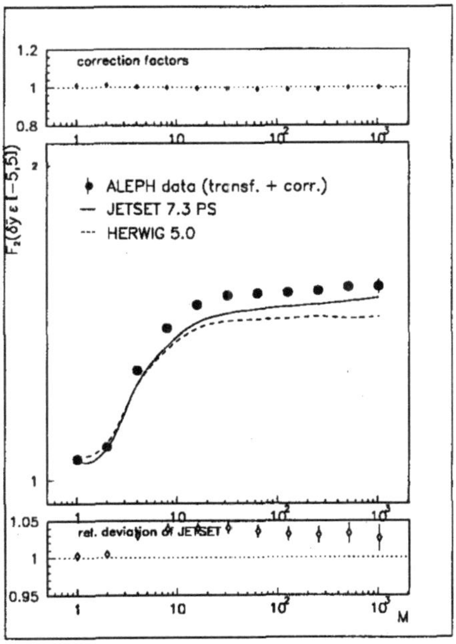

Figure 3: The second factorial moment of $\tilde{y}$, $F_2(\delta\tilde{y})$, for corrected ALEPH data.

For the full interval $|y| < 5$ as shown in figure 2, the absolute value at which the factorial moments level out increases due to geometrical/topological effects. However the agreement between models and data essentially remains the same.

If one then performs a transformation of y into $\tilde{y}$ according to Białas and Gazdzicki or Ochs for the same data sample, the discrepancies clearly increase as seen in figure 3. This was also found by the DELPHI collaboration[4]. The discrepancies may be due to the combination of the cuts on transverse momentum and the orientation of sphericity axis. The most apparent effect of these cuts is a strong distortion of the central region in the inclusive distribution of y. This in turn leads to *different mappings* to transform y into $\tilde{y}$ for data than for Monte Carlo generators. It is rather unlikely that the discrepancies are due to correlations in the range $2 < |y| < 5$, insufficiently described by the models, since the agreement is comparatively good for the whole interval without transformation. Here the given procedure to correct data for detector effects thus seems to be incapable to compensate for all effects.

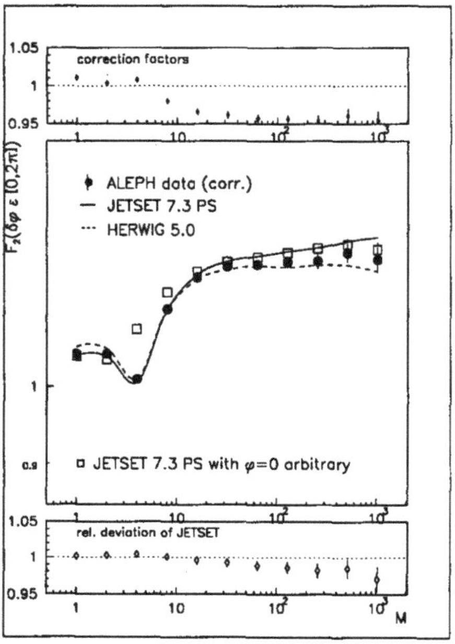

Figure 4: The second factorial moment of ϕ, $F_2(\delta\phi)$. The open squares '□' indicate the factorial moments for JETSET 7.3 when the reference for $\phi = 0$ is chosen at an arbitrary angle.

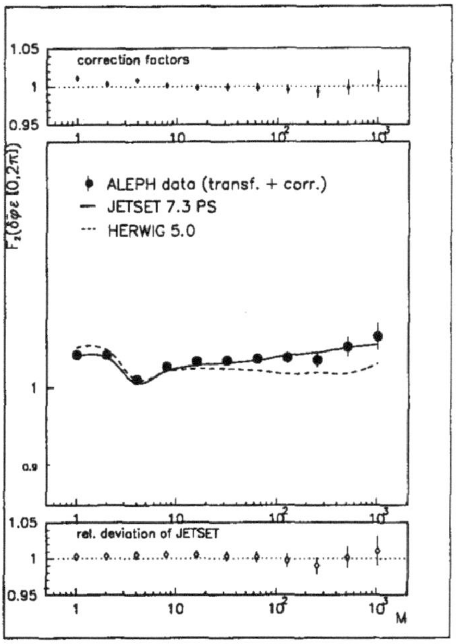

Figure 5: The second factorial moment of $\tilde{\phi}$, $F_2(\delta\tilde{\phi})$, for corrected ALEPH data.

3.2 Factorial moments in azimuthal angle ϕ

If one takes the azimuthal angle ϕ around the thrust axis for the analysis, and chooses $\phi = 0$ along the minor axis, he finds that the absolute values of the factorial moments decrease as shown in figure 4. In addition a dip at four bins due to momentum conservation develops. It is also present in models which contain no correlations at all except energy and momentum conservation.

For JETSET 7.3 PS the agreement between data and Monte Carlo is slightly different for ϕ than for y, with discrepancies sometimes exceeding the level of one standard deviation.

The major contribution to the absolute value seems to originate from the shape of the inclusive distribution which has two peaks as can be seen after application of the transformation into $\tilde{\phi}$ (fig. 5) where, apart from the kinematical dip, the curve is almost flat.

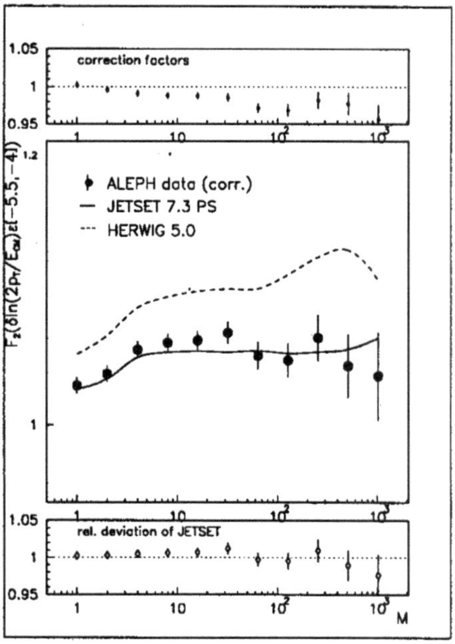

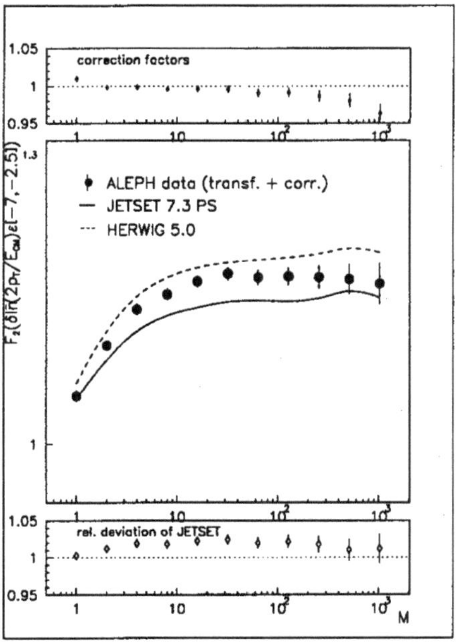

Figure 6: The second factorial moment of $\ln\frac{2p_\perp}{E_{CM}} \in [-5.5,-4]$, $F_2(\delta\ln\frac{2p_\perp}{E_{CM}})$, for corrected ALEPH data.

Figure 7: The second factorial moment of $\widetilde{\ln\frac{2p_\perp}{E_{CM}}}$, $F_2(\delta\widetilde{\ln\frac{2p_\perp}{E_{CM}}})$, with $\ln\frac{2p_\perp}{E_{CM}} \in [-7,-2.5]$ for corrected ALEPH data.

It has been suggested[3] to use an arbitrary coordinate system to force inclusive distributions to be flat and thus spare the geometrical effects. Apart from a smearing of the dip, the moments are very similar, since in *every event*, the shape in ϕ has peaks and valleys. Only the superposition of independently *located* peaks leads to a flat distribution. Thus topological effects are still present. This can be seen in figure 4 by comparing the open squares ('□') where in JETSET 7.3 the orientation for ϕ has been chosen along the minor axis with the line ('—') where $\phi = 0$ is parallel to the beam pipe.

3.3 Factorial moments in transverse momentum $\ln(\frac{2p_\perp}{E_{CM}})$

The factorial moments for the one dimensional analysis in $\ln\frac{2p_\perp}{E_{CM}}$ are shown in figure 6 for the reduced interval $\ln\frac{2p_\perp}{E_{CM}} \in [-5.5,-4]$ and in figure 7 for the transformed variable $\widetilde{\ln\frac{2p_\perp}{E_{CM}}}$ fom the full interval $\ln\frac{2p_\perp}{E_{CM}} \in [-7,-2.5]$.

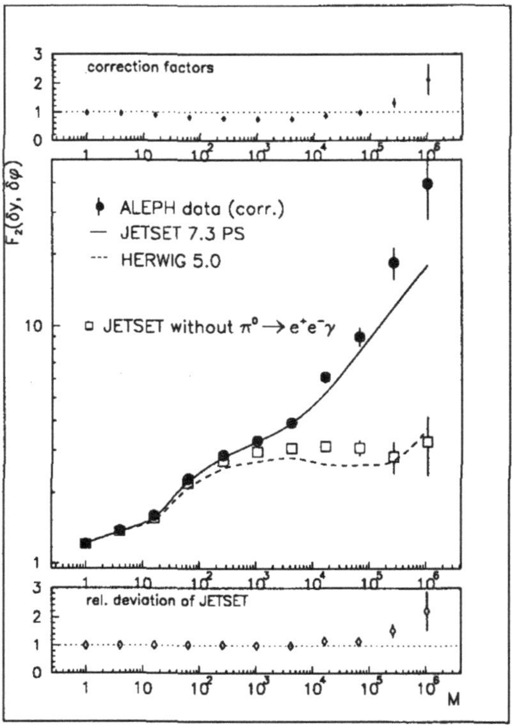

Figure 8: The second factorial moment in y and ϕ, $F_2(\delta y, \delta\phi)$, for corrected ALEPH data and the models JETSET 7.3 PS and HERWIG 5.0. For the open squares ('□') the Dalitz decay $\pi^0 \to e^+e^-$ has been inhibited inside JETSET 7.3 PS.

The fact that the increase of the differences between data and Monte Carlo when using transformed variables is smaller for $F_2(\delta\tilde{\phi})$ and larger for $F_2(\delta \widetilde{\ln \frac{2p_\perp}{E_{CM}}})$ and $F_2(\delta\tilde{y})$, supports the interpretation that the transformation (artificially) increases the discrepancies, since the previously mentioned cuts on sphericity axis and transverse momentum do not have much influence on the distribution in ϕ, but shift the distribution in $\ln \frac{2p_\perp}{E_{CM}}$ and distort the distribution in y.

4 Higher dimensional analysis

4.1 Two dimensional analysis in y and φ

The two dimensional analysis in y and ϕ is shown in figure 8. Here the reduced interval $|y| < 2$ with no transformation have been used. One notices that a linear fit through the data points is not possible because of the two kinks at $M = 16$ and $M = 4096$. The first kink is again due to momentum conservation in ϕ. While the second is caused by the Dalitz decay which produces very collinear e^+e^--pairs.

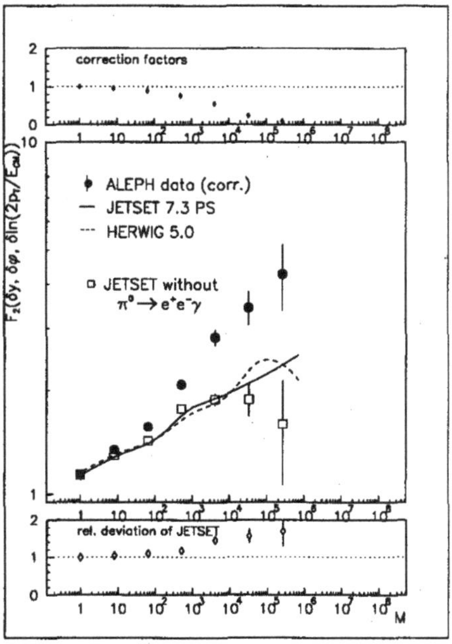

Figure 9: The second factorial moment in y, ϕ and $\ln \frac{2p_\perp}{E_{CM}}$, $F_2(\delta y, \delta\phi, \delta \ln \frac{2p_\perp}{E_{CM}})$, for corrected ALEPH data. For the open squares ('□') the Dalitz decay $\pi^0 \to e^+e^-$ has been inhibited inside JETSET 7.3 PS.

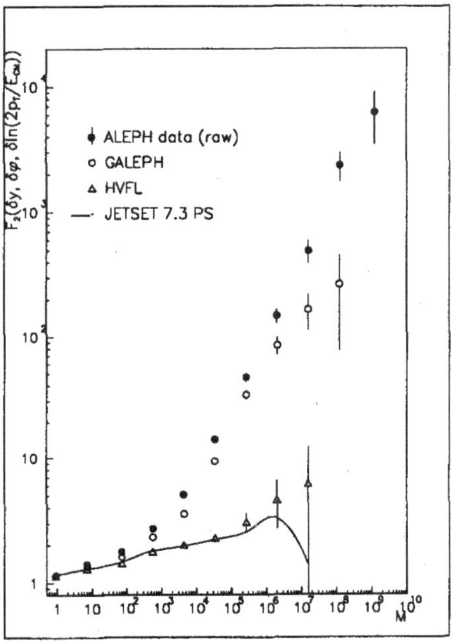

Figure 10: The second factorial moment $F_2(\delta y, \delta\phi, \delta \ln \frac{2p_\perp}{E_{CM}})$ for raw ALEPH data, the detector simulation (GALEPH), the corresponding generator level (HVFL) and JETSET 7.3 PS.

Therefore the approximate accordance of the two slopes below and above $M \approx 4096$ is only by chance. Without the Dalitz decay the moments also show saturation for $M > 4096$ as can be seen from the Monte Carlo study with JETSET 7.3 PS where the Dalitz decays were inhibited ('□' in figure 8). Since in the HERWIG 5.0 model this specific decay is not implemented, deviations from measured data are of approximately the same magnitude in this region.

4.2 Three dimensional analysis in y, ϕ and $\ln \frac{2p_\perp}{E_{CM}}$

Adding $\ln \frac{2p_\perp}{E_{CM}}$ to the previous analysis reveals deviations of up to a factor of 1.7, equivalent to between four and seven standard deviations (fig. 9). One explanation for this may be the presence of strong detector effects (photon conversions in detector materials[3]). This conclusion is demonstrated in figure 10 where raw (i.e. uncorrected) data ('•') are seen to be larger than the predictions of JETSET 7.3

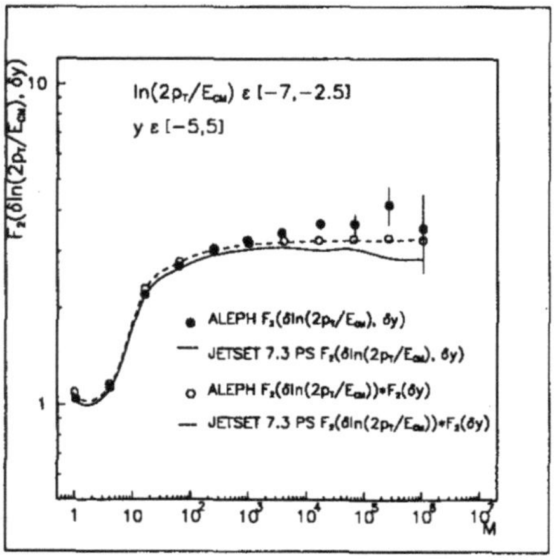

Figure 11: Comparison of the two dimensional factorial moment $F_2(\delta \ln \frac{2p_\perp}{E_{CM}}, \delta y)$ with the product of the one dimensional $F_2(\delta \ln \frac{2p_\perp}{E_{CM}})$ and $F_2(\delta y)$. The complete intervals $y \in [-5, 5]$ and $\ln \frac{2p_\perp}{E_{CM}} \in [-7, -2.5]$ have been used.

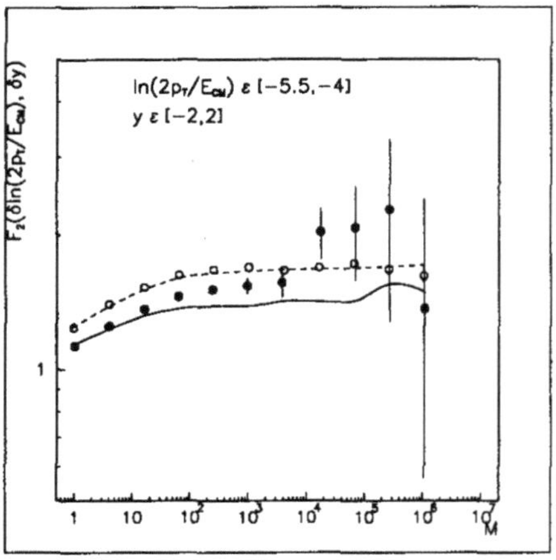

Figure 12: Comparison of the two dimensional factorial moment $F_2(\delta \ln \frac{2p_\perp}{E_{CM}}, \delta y)$ with the product of the one dimensional $F_2(\delta \ln \frac{2p_\perp}{E_{CM}})$ and $F_2(\delta y)$. The reduced intervals $y \in [-2, 2]$ and $\ln \frac{2p_\perp}{E_{CM}} \in [-5.5, -4]$ have been used. (For the legend see fig. 11.)

('—') by up to a factor of 30. The detector simulation ('o') is capable of describing the presence of strong detector effects in general. This is also reflected in the small correction factors in figure 9.

From this analysis it cannot be decided whether the discrepancies originate from insufficiencies in the JETSET 7.3 generator or in the detector simulation GALEPH.

$F_2(\delta y, \delta\phi, \delta \ln \frac{2p_\perp}{E_{CM}})$ also saturates for JETSET 7.3 if the Dalitz decay is inhibited.

4.3 Factorization

Taking a different pair of variables—here $\ln \frac{2p_\perp}{E_{CM}} \in [-7, -2.5]$ and $y \in [-5, 5]$ for $F_2(\delta \ln \frac{2p_\perp}{E_{CM}}, \delta y)$—for the two dimensional analysis, one observes saturation as in the one dimensional cases. This is shown in figure 11, where the full dots ('•') indicate ALEPH data and the line ('—') represents the JETSET 7.3 model. The difference again may be due to conversions.

In addition to calculating the two dimensional moment $F_2(\delta \ln \frac{2p_\perp}{E_{CM}}, \delta y)$ for $M = 2^{2m}$ bins, one can also calculate the product from $F_2(\delta \ln \frac{2p_\perp}{E_{CM}})$ and $F_2(\delta y)$ each determined for $\sqrt{M} = 2^m$ bins. This is indicated in figure 11 by the open circles 'o' (dashed line '- - -') for the ALEPH data (JETSET 7.3 PS model). Comparing the data with the models, one sees that the JETSET 7.3 model factorizes to within one standard deviation and the data to within approximately two.

The two dimensional $F_2(\delta\phi, \delta \ln \frac{2p_\perp}{E_{CM}})$ also factorizes and $F_2(\delta y, \delta\phi, \delta \ln \frac{2p_\perp}{E_{CM}})$ factorizes into the product of $F_2(\delta y, \delta\phi)$ and $F_2(\delta \ln \frac{2p_\perp}{E_{CM}})$ in the three dimensional case (both not shown). $F_2(\delta y, \delta\phi)$ does not factorize (because the Dalitz decay introduces correlations in y *and* ϕ simultaneously).

However the degree of factorization drastically decreases when the reduced intervals $\ln \frac{2p_\perp}{E_{CM}} \in [-5.5, -4]$ and $y \in [-2, 2]$ are used (fig. 12). The main contribution to this difference is a parallel shift which arises from the fact that in reduced intervals the total multiplicity is not poissonian anymore. The moments for $M = 1$ thus deviate from 1 which is counted twice in the product but only once in the two dimensional moment. The variable transformation formula 2 also has disturbing influence on the factorization (without figure).

5 Model studies at the parton level

To investigate contributions to the factorial moments at various stages within the development of a parton cascade, a series of Monte Carlo studies has been performed. ALEPH data were compared with five models available within JETSET 7.3, namely:

- PS: The parton shower model (default). Parameters have been optimized to fit inclusive distributions measured with ALEPH. This gives an average of 7.8 partons before the onset of string fragmentation.

- ERT: Second order matrix elements according to Ellis, Ross, Terrano[7]. This allows up to four partons before string fragmentation.

- MEOPT: Second order matrix elements with optimized renormalization scale. This allows up to 4 partons *and* in addition to ERT the experimental 4-jet rate is reproduced.

- ABEL: Abelian parton shower which means that the triple gluon vertex has been switched off.

- QQBAR: A theory free of gluons, i.e. the primary quark-antiquark-pair produces a straight string which then decays according to the JETSET 7.3-model. Only two partons are allowed.

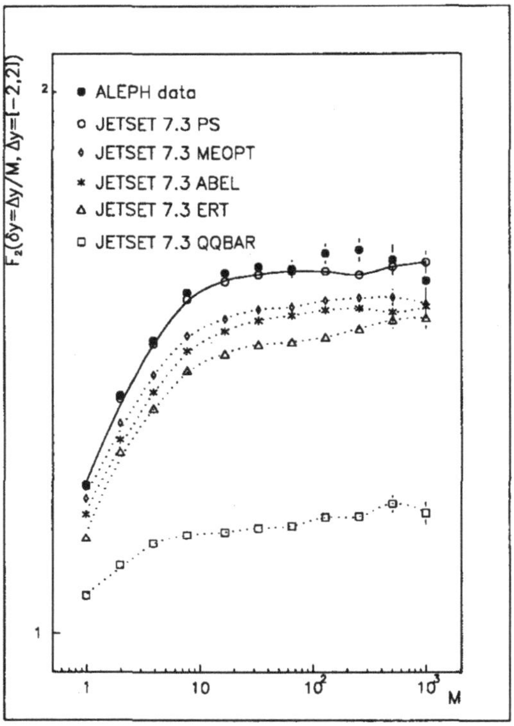

• Corrected ALEPH data.

o Lund model JETSET 7.3 PS ('parton shower').

◇ Second order matrix elements with "optimized" scale ('MEOPT').

* Abelian parton shower without vertices $g \rightarrow gg$ ('ABEL').

△ Second order matrix elements according to Ellis, Ross und Terrano ('ERT').

□ A gluon-free theory. Only fragmentation and decaying hadrons produce higher multiplicity ('QQBAR').

Figure 13: Comparison of different Monte Carlo models with varying amount of built in cascading structure at the parton level during the generation of hadronic events.

For all the models except QQBAR and ABEL there exist sets of parameters which have been fit to ALEPH inclusive distributions[14]. However the factorial moments have not been included in the fit. Therefore different sets of parameters for the various models still could improve agreement between models and data with respect to factorial moments but only at the expense of other observables. In the cases of QQBAR and ABEL the fragmentation parameters have been adjusted to yield the measured average thrust and multiplicity[12].

The results are shown in figure 13. Without a parton cascade ('QQBAR', '□') there is only a very small rise of the factorial moments. This remaining increase can be shown to be due to decays of hadrons, which are still present in identical proportions as in the other models[12].

Allowing the partons to split twice ('ERT', '△'), thus producing up to four partons, gives rise to a large part of the observed increase, but deviations are still

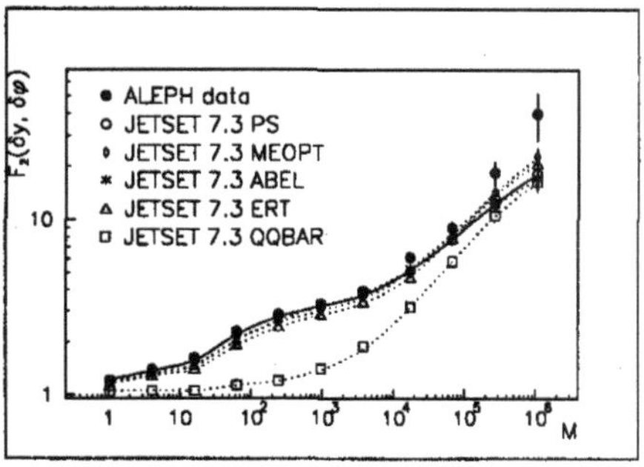

Figure 14: Two dimensional studies with models (JETSET 7.3 PS, MEOPT, ABEL, ERT and QQBAR) compared with ALEPH data.

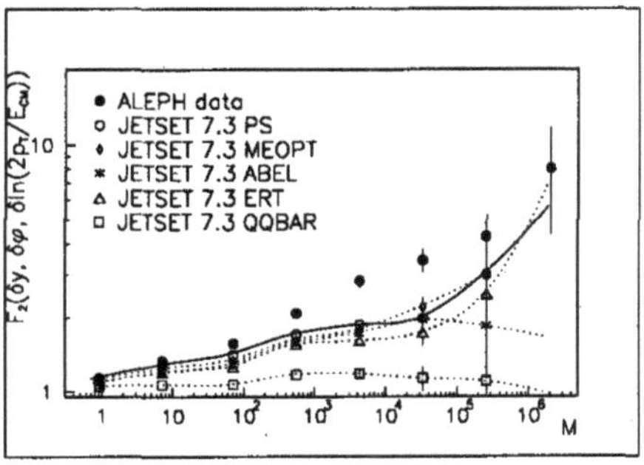

Figure 15: Three dimensional studies with models (JETSET 7.3 PS, MEOPT, ABEL, ERT and QQBAR) compared with ALEPH data.

significantly beyond the statistical accuracy (typically about 25%).

By "optimizing" the renormalization scale and thus having the predicted four-jet-rate to agree with the observed, one can achieve a better description than before. However deviations remain considerably larger than the statistical error (typ. 14%).

The only (known) models which are able to describe e^+e^- data to an accuracy of approximately the statistical spread (deviations of about 1 – 2% for JETSET 7.3), are parton shower models like JETSET 7.3 PS and HERWIG 5.0 (compare figure 1 for HERWIG 5.0).

As mentioned previously, JETSET 7.3 PS with the parameters as used here produces an average of 7.8 partons which are then fragmented. Naming the primary $q\bar{q}$-pair as *first* parton generation, a four parton state as *second* and every doubling as next higher—the invariant mass thus will decrease to less than half from one generation to the next. A multiplicity of 7.8 is then equivalent to an almost fully developed third generation. The good agreement between data and JETSET 7.3 PS is well known, however models still tend to be systematically below measured data for $F_2(\delta y)$ by about one standard deviation seen also by DELPHI[5] and OPAL[11].

One can thus conclude that a multistep process of branching on parton level is important for a rise of factorial moments of the observed extent. Primary (hard) gluons give the largest contribution but they are not sufficient to describe the data alone. Contributions to the increase of the moments decrease with decreasing energy scale.

A different model with an abelian parton shower ('ABEL', '*') with 6.2 partons on average is able to yield factorial moments of the size of optimized matrix elements. From the fact that this rather simple cascading structure, when compared to the

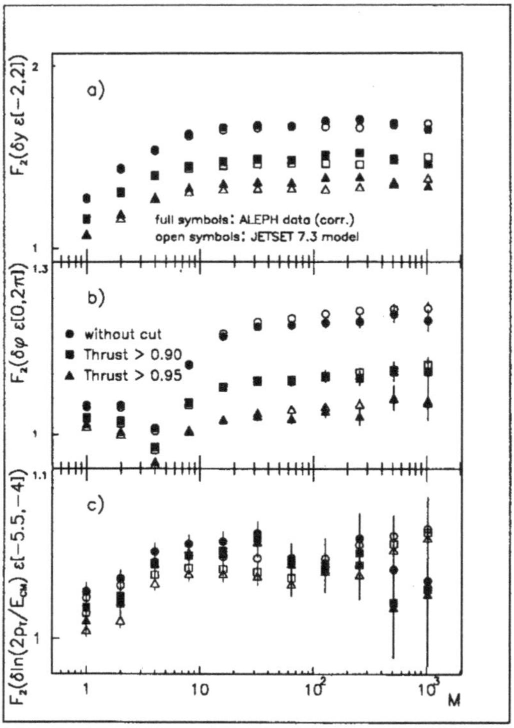

Figure 16: Influence of a thrust cut on the one dimensional factorial moments *(a)* $F_2(\delta y)$, *(b)* $F_2(\delta\phi)$ and *(c)* $F_2(\delta \ln \frac{2p_\perp}{E_{CM}})$ for ALEPH data (full symbols) and the JETSET 7.3 model (open symbols).

rather elaborate low-parton-multiplicity-ERT-model, gives results closer to those that were measured, indicates again that a cascading structure is a useful concept for exhibiting rising moments. Quantitatively the nature of the QCD vertices is important.

The same conclusions are derived from model studies using $F_2(\delta\phi)$. However $F_2(\delta \ln \frac{2p_\perp}{E_{CM}})$ does not reveal sizeable differences between the models. When analyzed in two or three dimensions, figures 14 and 15, the moments still exhibit the same trend—the more cascade the higher the moments—but they approach the ALEPH data more rapidly. Also the factorial moments predicted by the QQBAR-model rise to large values in the region dominated by the Dalitz-Decay ($M > 4^5$).

6 Influence of a cut on thrust

The primary $q\bar{q}$-pair defines the thrust axis of an event. At this time the thrust T is equal to one. Almost each "decay" of a quark or gluon into two new partons

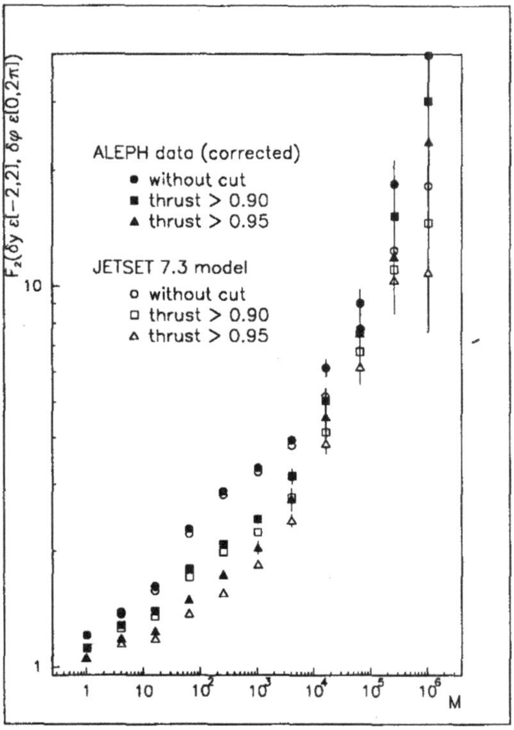

Figure 17: Influence of a thrust cut on the two dimensional factorial moment $F_2(\delta y, \delta\phi)$ for ALEPH data and the JETSET 7.3 model.

decreases T. So the thrust value reflects how often partons have split and how "hard" the single processes have been.

Successive splittings of single partons into pairs are ordered in terms of invariant mass and transverse momentum (see e.g. R. D. Field[8]). Selecting only events with thrust T above a given value thus selects those events with a certain *maximal* transverse momentum, i.e. events which contain cascading only *below* a certain scale of invariant mass.

If the *whole* cascading process contributes to the observed fractal structure, one expects the slope of the factorial moments to decrease when the cut on thrust is tightened . Experimental results clearly support this: In figure 16, for both Monte Carlo Simulations ('o', '□', '△') and ALEPH data ('•', '■', '▲') the moments for fixed M gradually decrease for tighter cuts on thrust T.

The analysis in two dimensions (fig. 17) agrees with the one dimensional cases and exhibits the same tendency of the moments, namely to decrease with tighter cuts, but the differences are larger than one standard deviation only in the region below $M = 4^6$. That confirms the previous conclusion that only this range is dominated by the parton cascade.

However further detailed work is needed here. The influence of geometrical/topological effects of the cuts need more investigation, since the inclusive distributions strongly depend on them. Also the interdependence between thrust and the branching structure is not fully understood yet.

7 Summary

In general the data are well-described by Monte Carlo models which are based on the leading logarithm approximation at the parton level (HERWIG 5.0 and in particular JETSET 7.3 PS). This was not the case in the 3D analysis, but there it was shown that detector effects have a major contribution to the moments. Insufficiencies in *either* the model *or* the detector simulation are responsible for these discrepancies.

The variable transformation to produce flat inclusive distributions can eliminate geometrical/topological effects and thus give an estimate of their magnitude. For this it is important to align the events along a coordinate system determined by the single event. However the transformation seems to have some distorting influence on some of the moments (especially on those whose inclusive distributions differ on generator and detector level).

A multidimensional analysis in (y, ϕ) or $(y, \phi, \ln \frac{2p_\perp}{E_{CM}})$ does not show saturation for small bins. It could be shown that the region of $m \gtrsim 5$ bins is dominated by the Dalitz decay. The three dimensional analysis revealed the importance of detector effects. Kinematical effects are still visible at $m = 2$ in two or three dimensions. A linear fit through all data points is not possible.

Studies in multiple dimensions showed that $F_2(\delta \ln \frac{2p_\perp}{E_{CM}})$. factorizes from the other variables for JETSET 7.3 PS. This factorization is sensitive to the size of the total intervals, to the variable transformations and to multidimensional correlations like Dalitz decays or conversions. Therefore it is less pronounced for ALEPH data.

By the use of different models it was shown that the absolute values of the factorial moments depend on the multiplicity of the parton state. The description becomes better with more parton branching. Only JETSET 7.3 PS with an average of 7.8 partons was able to describe data at the level of statistical accuracy.

Using a cut on thrust, it could be shown that a *range* of scales of invariant mass (transverse momentum) contributes to an increase of factorial moments. This was confirmed on both one and two dimensional moments.

Overall the conclusion is that a *multistep process at parton level* is responsible for the increase of the factorial moments in the range of $m < 5$ and the region above $m \approx 5$ is dominated by the Dalitz decay and detector effects.

References

[1] M. Bengtsson and Torbjörn Sjöstrand. The lund monte carlo for jet fragmentation and e^+e^- physics: jetset version 6.3 — an update. *Comput.Phys.Commun.*, **43**:367, (1987).

[2] A. Białas and M. Gazdzicki. A new variable to study intermittency. *Phys. Lett.*, **B252**:483–486, (1990).

[3] CELLO Collaboration. Fractal dimensions from a 3-dimensional intermittency analysis in e^+e^- annihilation. (1990). DESY 90-114.

[4] Alessandro de Angelis (DELPHI) et al. Fluctuations in hadronic final states from the decay of the Z^0. This volume.

[5] Alessandro de Angelis (DELPHI) et al. A study for intermittency in hadronic Z^0 decays. *Phys. Lett.*, **B247**:137–147, June 1990. CERN EP 90-78.

[6] B. Decamp (ALEPH-Collaboration). ALEPH: a detector for electron-positron annihilations at LEP. *Nucl.Instrum.Meth.*, **A294**, (1990).

[7] R. K. Ellis, D. A. Ross, and A. E. Terrano. The perturbative calculation of jet structure in e^+e^- annihilation. *Nucl. Phys.*, **B178**:421, (1981). CALT 68-785.

[8] Richard D. Field. *Applications of Perturbative QCD.* Addison Wesley, (1989).

[9] Wolfgang Ochs. The importance of phase space dimension in the intermittency analysis of multi hadron production. *Phys. Lett.*, **B247**:101–106, (1990).

[10] Wolfgang Ochs. Scaling and scalebreaking phenomena in multiparticle production. September 1990. Invited Talk presented at the XXth Int. Symp. on Multiparticle Dynamics, Gut Holmecke, near Dortmund, MPI-PAE/PHt 78/90.

[11] OPAL Collaboration. Intermittency in hadronic decays of the Z^0. *Phys. Lett.*, **B262**:351–361, (1991). CERN-PPE-91-37.

[12] Volker Raab. *Intermittenz bei* 91 *GeV in der Reaktion* $e^+e^- \longrightarrow$ *Hadronen.* Diploma thesis, Technische Universität München, September 1991. to be published.

[13] F. Ranjard (ALEPH-Collaboration) et al. GALEPH user guide. (1988). ALEPH Note 119.

[14] Gerald Rudolph. Properties of hadronic Z decays and test of QCD generators. to be published.

[15] Torbjörn Sjöstrand. The lund monte carlo for e^+e^- jet physics. *Lund: Univ.*, June 1982. LUTP 82-7.

CELLO results on multiparticle production

Oliver Podobrin

Deutsches Elektronen-Synchrotron, DESY

Notkestraße 85, W-2000 Hamburg 52, F.R.G.

Bitnet: F14POD@DHHDESY3

Abstract

In this talk results from CELLO on multiparticle production in e^+e^- annihilation are presented. The three-dimensional distribution of charged particles, and its one- and two-dimensional projections, are extensively studied through the bin-size dependence of their factorial moments. The data are found to have an excellent description by standard Monte Carlo models. This implies an explanation of the observed intermittency behaviour in terms of known physics. The study is extended by an analysis of two-particle rapidity correlations.

1 Introduction

During recent years the study of multiparticle production has received important contributions from e^+e^- experiments [1,2]. Owing to the clean initial state, these experiments are very well suited for this type of analysis and have for the first time made it possible to identify unambiguously the nature of the intermittency behaviour [1,2,3,4,5]. It has been in e^+e^- annihilation reactions where fluctuations in three-dimensional phase space, at scales ranging from the centre of mass energy down to the electron mass, have been observed and identified as known physical processes, making it definitely clear that the intermittency behaviour is not a new phenomenon [1,3,4]. These findings also affect the interpretation of results obtained from other collision processes, where the experimental and theoretical situation is less clear and consequently wild speculations about the origin of the intermittency behaviour have made their appearance [6].

The main interest in these studies is focussed on the experimentally observable properties of hadron distributions and their explanation in terms of QCD, the gauge theory of the strong interactions. A serious problem is faced here: as long as the theory is solved only perturbatively and hence restricted to high Q^2 processes, it leaves the description of hadron formation to phenomenological Monte Carlo models. These models are therefore of vital importance, to relate the experimentally accessible information to the fundamental theory.

Various such models exist [7], and can all be divided into four steps as it is depicted in Fig. 1. e^+e^- annihilation into $q\bar{q}$ pairs proceeds via virtual γ/Z° ex-

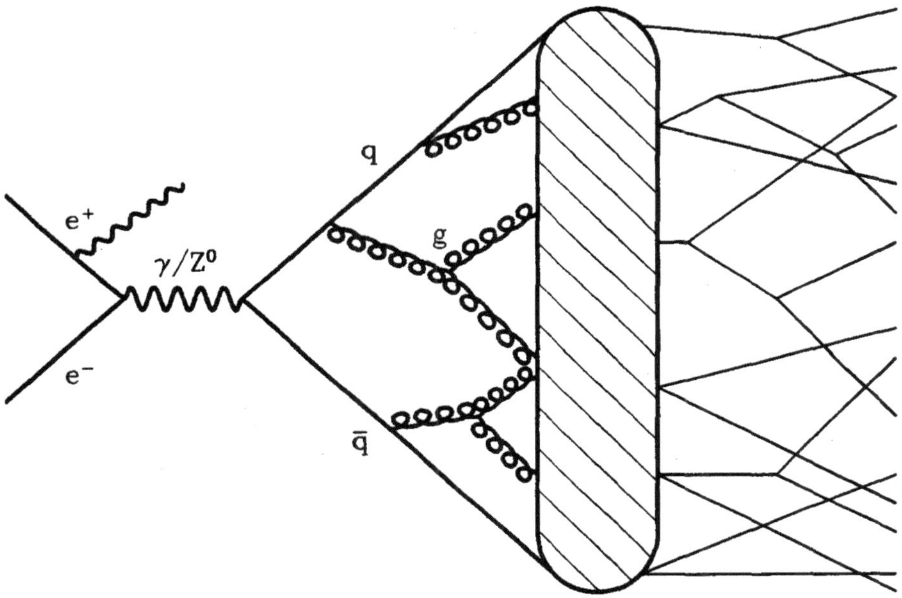

Figure 1: Schematic view of an e^+e^- annihilation event.

change, with possible photon radiation off the initial state. This first step is purely electro-weak and therefore calculable to high precision. In the second phase the leading quarks emit gluons, which eventually split into pairs of quarks or gluons. The simulation of this process relies on approximations to QCD, of which three different types are used in the corresponding models: the $O(\alpha_S^2)$ matrix element model (ME) generates at most four-parton states, while the parton shower model (PS) and the colour dipole model (CD) may generate multi-parton states. In the third step the parton state is transformed into a system of colour-singlet hadrons. This process is governed by low momentum transfer, implying large values of α_S, which makes a perturbative description impossible, and therefore requires the use of fragmentation models. Among these the string- and cluster models are most successful. In the final step matrix elements or simple phase space decays are used to transform primary hadrons into stable particles. These decays proceed through measured decay channels with their respective branching ratios.

The study of particle correlations by means of factorial moments, as proposed by Bialas and Peschanski [8], turned out to be very sensitive to fine details of the hadronization process [4,5]. For this reason, these analyses not only put stringent constraints on Monte Carlo models, in addition they test our present understanding of fragmentation, implying the possibility to discover new phenomena.

The outline of this talk is as follows: the next section contains a description of the experiment, event selection and background reactions. The section closes with a definition of variables and a description of variable transformations. In Sect. 3 the analysis of factorial moments is presented along with detailed Monte Carlo studies. Following this, two-particle rapidity correlations are presented in Sect. 4 and finally the main results are summarized and conclusions are given.

2 Experimental aspects

2.1 CELLO experiment

The analysis presented here is based on data taken with the CELLO detector at the PETRA e^+e^- storage ring. The data has been taken in 1986 at 35 GeV centre of mass energy and corresponds to a time integrated luminosity of 86 pb^{-1}. The detector system is described in detail in [9]. In the present analysis information from the central tracking device and the lead-liquid-argon calorimeter is used to measure energy and momentum of charged and neutral particles.

Multi-hadronic events are identified by a standard selection procedure, the details of which may be found in [1]. 18543 events passed the above selection and are the basis for the present analysis. The remaining background of approximately 3% consists mainly of beam-gas-, two-photon- and τ-pair events. The latter are of particular importance for the intermittency analysis, inasmuch as the τ's decay into very narrow jets of low multiplicity. This issue will be discussed in detail in Sect. 3.

The selection criteria for charged and neutral particles are also given in [1], and are chosen to ensure good momentum resolution and to suppress spurious track candidates generated by the track finding algorithm but not corresponding to a real particle. Note that in this analysis neutral particles are solely used to improve the reconstruction of the event axis.

It is obvious that a very accurate simulation of the detector performance is required to allow the analysis of particle production in small regions of phase space. In this context the experience CELLO has gained through many two-photon analyses [10] turned out to be very helpful. These analyses typically deal with detection efficiencies of a few per mille and are therefore ideal to test and improve the detector simulation.

2.2 Variable definition

It is the aim of this analysis to investigate the correlation structure of particle production in three-dimensional phase space and in the corresponding one- and two-dimensional projections. For this purpose the Lorentz invariant phase space element $dLIPS$ of a single particle is decomposed as follows;

$$dLIPS = \frac{1}{(2\pi)^3} \cdot \frac{1}{4} \cdot dy \cdot d\phi \cdot dp_\perp^2 \,, \tag{1}$$

where the variables $y = 0.5 \cdot \ln[(E + p_{||})/(E - p_{||})]$, ϕ and $p_\perp^2$ must be evaluated with respect to a suitably chosen reference axis. Since intermittency studies are mainly concerned with the soft hadronization process, it is natural to take the event axis as reference, which eliminates fluctuations due to the angular distribution of the hard scattering process $e^+e^- \to q\bar{q}$. This axis is obtained as the first eigenvector of the *linearized* sphericity tensor;

$$S_{lin}^{ab} = \frac{\sum_i p_i^a p_i^b / |\vec{p_i}|}{\sum_i |\vec{p_i}|} \,, \tag{2}$$

which is determined by the charged and neutral particles in an event. Usually the orientation of the sphericity axis is arbitrary. For this analysis it has been aligned with the direction of the most energetic jet. By means of this definition positive rapidity values indicate particles belonging to the most energetic jet, and negative values signify particles from the less energetic jets. In an analogous way the origin of the ϕ coordinate, i.e. $\phi \equiv 0$, is taken to be the second eigenvector of the sphericity tensor (Eq. 2). With these definitions the corresponding inclusive distributions contain information about "trivial" fluctuations, which are present even in individual events due to limited statistics or hard gluon radiation. In Fig. 2 the inclusive y, ϕ, $p_\perp^2$ and charged multiplicity distributions are shown and compared to the Jetset 7.2 PS+BE* Monte Carlo simulation. The perfect reproduction of these distributions by the Monte Carlo is a necessary precondition for the subsequent intermittency analysis. In this context it is important to note that the Jetset 7.2 Monte Carlo [11] is run with default parameters, except for the inclusion of Bose-Einstein correlations [4,5]. It is clearly visible in Fig. 2 that the inclusive distributions are not constant. Their variations would modify the factorial moments at the corresponding resolution scales and therefore fake intermittent behaviour, even in case of complete absence of genuine particle correlations. However, it is possible to unfold these variations from the factorial moments by an appropriate variable transformation, which will be explained in the following section.

---

*In what follows BE is used to indicate that Bose-Einstein correlations are included in the Monte Carlo simulation.

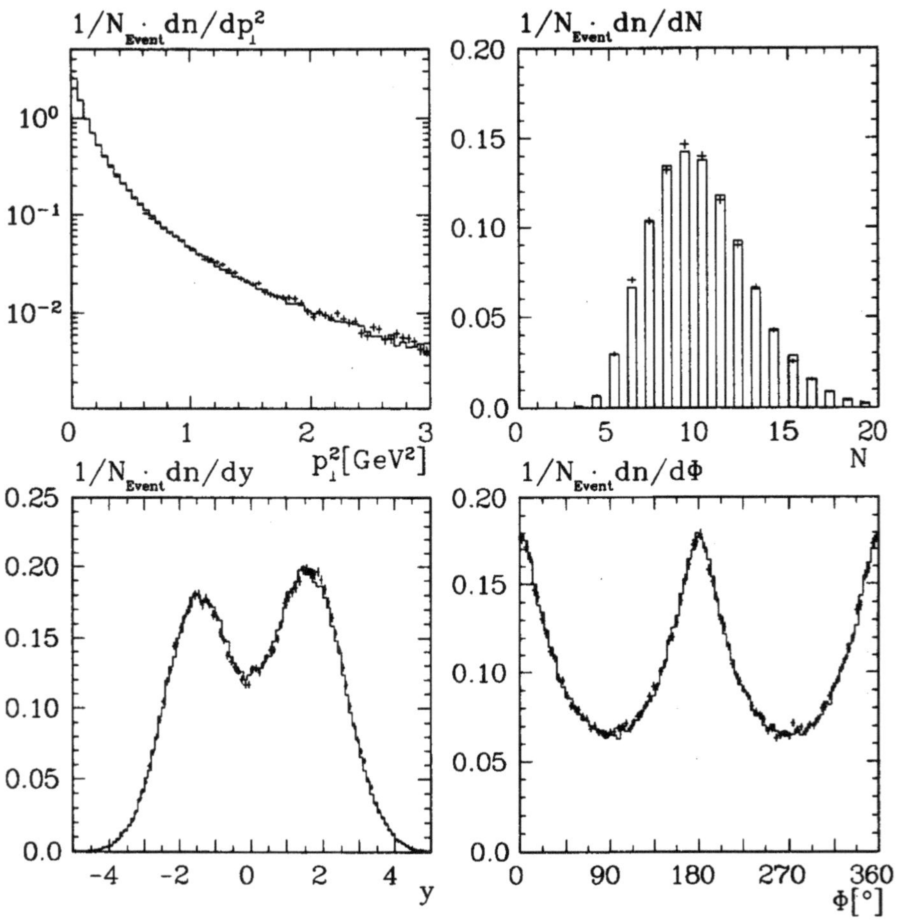

Figure 2: Inclusive y, ϕ, $p_\perp^2$ and charged multiplicity distributions (crosses) compared to the Jetset 7.2 PS+BE simulation (histograms).

2.3 Variable transformation

Different procedures have been proposed to unfold effects due to non-constant inclusive distributions. In [4,5] we have presented one- and two-dimensional analyses without corrections, and three-dimensional analyses according to the procedures advocated by Ochs [12] and by Bialas and Gazdzicki [13]. The latter procedure has also been used for the one-, two- and three-dimensional analyses presented here. It consists of a variable transformation $x \to \tilde{x}$, such that the distribution in $\tilde{x}$ is constant. This is achieved by the transformation;

$$\tilde{x}(x) = \frac{\int_{x_{min}}^{x} \rho(x')dx'}{\int_{x_{min}}^{x_{max}} \rho(x')dx'} \,, \qquad x = y, \phi, p_\perp^2 \,, \tag{3}$$

where $\rho(x)$ are the inclusive density functions. This method is applicable in any dimension, where the correlations between the different variables are either neglected [12] or taken into account [13]; for a discussion of these methods cf. [5]. Since the computational realization of the method proposed in [13] is non-trivial, a short description of our algorithm is given in Appendix A.

3 Factorial moment analysis

The factorial moments $\langle F_m^q \rangle$ of rank q are computed according to the following formula;

$$\langle F_m^q \rangle = \left\langle \left(\frac{m}{\langle N \rangle} \right)^q \cdot \frac{1}{m} \cdot \sum_{k=1}^{m} n_k(n_k - 1) \cdots (n_k - q + 1) \right\rangle \,, \tag{4}$$

where $\langle N \rangle$ is the average particle multiplicity. The subscript m signifies the total number of bins obtained after B successive bisections in each dimension; i.e. $m = 2^{D_T \cdot B}$, where $D_T = 1, 2, 3$ is the topological dimension and n_k denotes the particle content of bin k.

Eq. 4 implies two averages: the average over all events, as indicated by the brackets $\langle\rangle$ and the average over all m bins. Note that other definitions of factorial moments have been given. However, in case of constant inclusive distributions, as obtained from Eq. 3, the different definitions are equivalent. For the present one-, two-, and three-dimensional analyses the complete phase space has been considered and has been transformed according to the prescription given in [13]. In Fig. 4 the resulting factorial moments $\langle F_m^2 \rangle - \langle F_m^5 \rangle$ are displayed and compared to the Jetset 7.2 PS+BE model. It is clearly seen that the slope of the factorial moments increases as D_T is increased from one to three. This behaviour is expected for a system with a

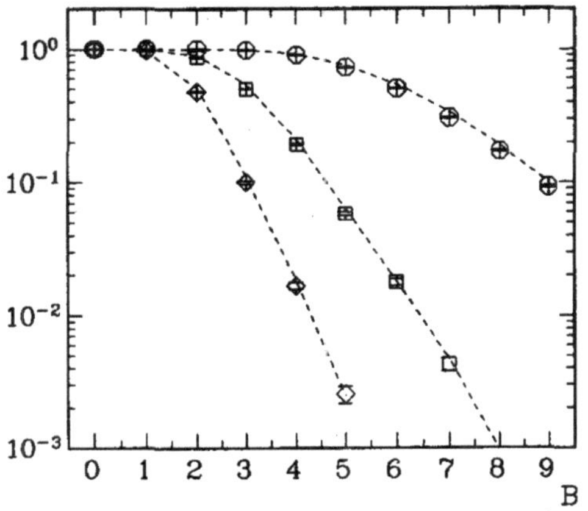

Figure 3: Fraction of events contributing to F^2 in one dimension (circles), two dimensions (squares) and three dimensions (diamonds) compared to the Jetset 7.2 PS+BE model (dashed lines).

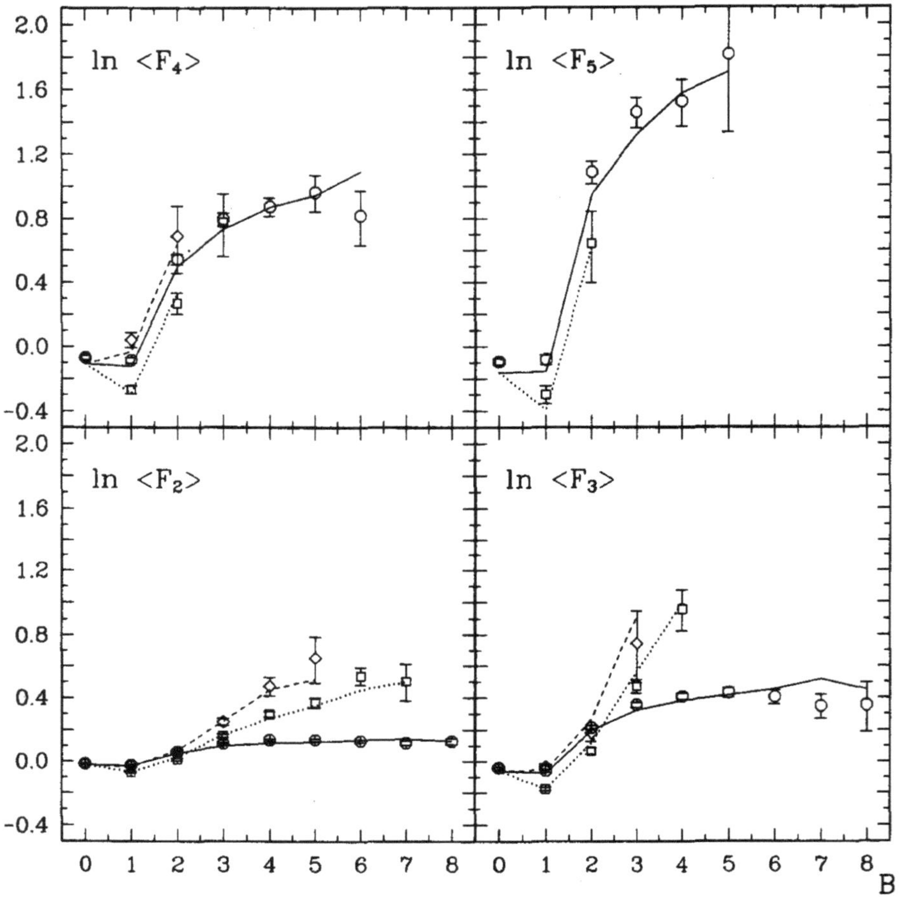

Figure 4: Factorial moments $\langle F_m^2 \rangle - \langle F_m^5 \rangle$ for increasing number of bins $m = 2^{D_T \cdot B}$. The open symbols represent the CELLO data, while the curves show the Jetset 7.2 PS+BE model. $D_T = 1$; y analysis – open circles and solid lines: $D_T = 2$; y, ϕ analysis – open squares and dotted lines: $D_T = 3$; $y, \phi, p_\perp^2$ analysis – open diamonds and dashed lines.

truely three-dimensional correlation function [14]. It is evident from Fig. 4 that the Monte Carlo provides a consistent description of the data in all dimensions. In Fig. 3 the fraction of events, contributing to the measurement of factorial moments, is seen to decrease rapidly as the resolution increases and finally drops below one per mille. This behaviour is precisely followed by the Monte Carlo. At this point it should be mentioned that factorial moment data are only plotted if at least ten events contributed, in order to have a reliable error estimate. On the basis of the excellent Monte Carlo description it is fair to say that the intermittency behaviour is understood and hence that the Monte Carlo is appropriate to analyse the complex structure of the factorial moments.

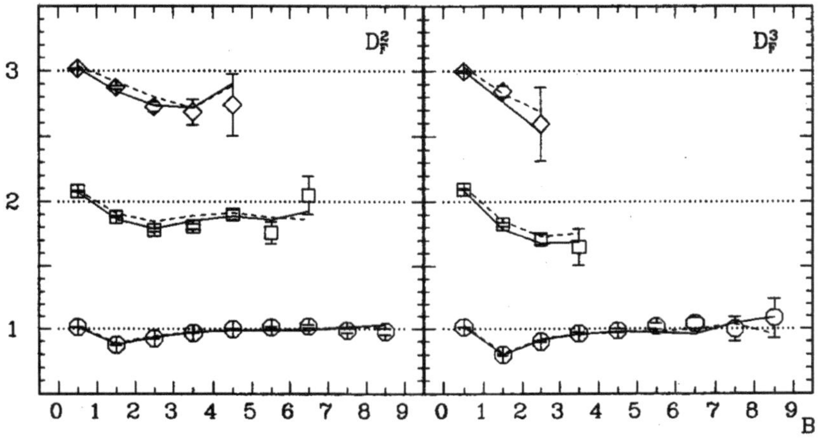

Figure 5: Fractal dimensions D_F^2 and D_F^3 for increasing number of bins $m = 2^{D_T \cdot B}$. The open symbols represent the CELLO data, while the solid and dashed lines show the Jetset 7.2 PS model with and without Bose-Einstein correlations.

The local slopes α_m^q of the factorial moments can be related to a *fractal dimension* D_F^q [1], which is, of course, resolution dependent and in general also depends on the rank q.

$$D_F^q = D_T \cdot \left(1 - \frac{\alpha_m^q}{q-1}\right) \tag{5}$$

$$\alpha_m^q = \frac{\partial \ln\langle F_m^q \rangle}{\partial \ln m}$$

The D_F^q's are presented in Fig. 5. Apparently the topological dimension is not reached in the two- and three-dimensional analyses, even with ultrafine resolution. From this it follows that particle correlations exist at the corresponding scales. In Fig. 5 the Monte Carlo is seen to follow precisely the data, where the inclusion of Bose-Einstein correlations improves the agreement. In the following, the observed complex behaviour will be explained by means of Monte Carlo studies and detailed analyses of the data. But first we will discuss the covariances between data points at different resolution.

3.1 Covariances

The covariances V^q_{lm} between factorial moments F^q_m and F^q_l of the same rank q are defined by the following matrix;

$$V^q_{lm} = \langle F^q_l \cdot F^q_m \rangle - \langle F^q_l \rangle \cdot \langle F^q_m \rangle \,, \tag{6}$$

and the F^q_m are given by Eq. 4. From this matrix the error on the mean value $\langle F^q_m \rangle$ is obtained as $\frac{1}{N_{event}} \cdot \sqrt{V^q_{mm}}$; these are the errors displayed on the figures. The covariance matrix has also been used to propagate the errors on $\langle F^q_m \rangle$ to the corresponding fractal dimension, cf. Eq. 5. Moreover the covariance matrix is interesting in itself: at first it can be transformed into a correlation matrix $C^q_{lm} = V^q_{lm} / \sqrt{V^q_{ll} \cdot V^q_{mm}}$, and second its inverse ${V^q_{lm}}^{-1}$ is predicted to be tri-diagonal in the α-model [15]. In Fig. 6 the correlation matrices C^2 from the one-, two-, and three-dimensional analyses of F^2 are shown. Large positive correlations are observed between all data points. These have to be taken into account when e.g. data and Monte Carlo are compared. The inverse covariance matrix from the analysis of F^2 in one dimension is presented

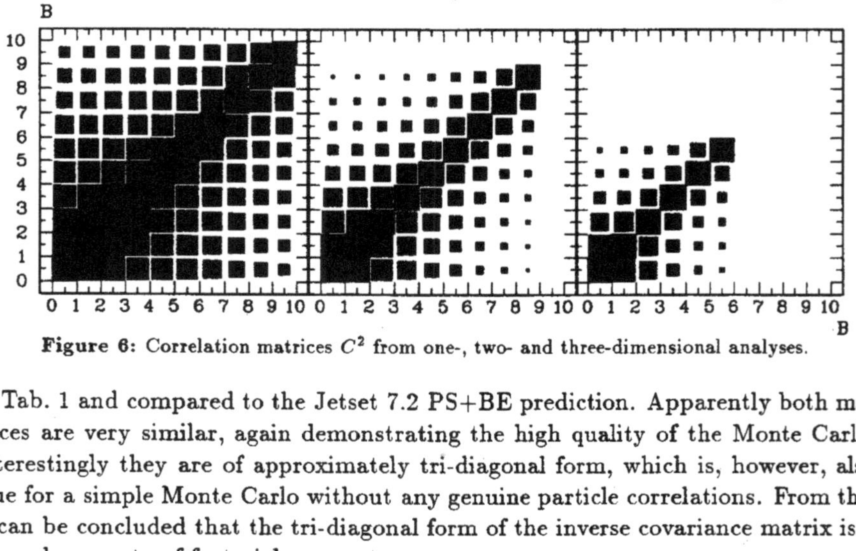

Figure 6: Correlation matrices C^2 from one-, two- and three-dimensional analyses.

in Tab. 1 and compared to the Jetset 7.2 PS+BE prediction. Apparently both matrices are very similar, again demonstrating the high quality of the Monte Carlo. Interestingly they are of approximately tri-diagonal form, which is, however, also true for a simple Monte Carlo without any genuine particle correlations. From this it can be concluded that the tri-diagonal form of the inverse covariance matrix is a general property of factorial moments.

3.2 Factorial moments revisited

In e^+e^- annihilation different event topologies occur, e.g. due to the emission of hard gluons. It has been pointed out already in [1,5] that the transformation (Eq. 3) is based on the *event average* and therefore induces fluctuations in the total event sample, where two- and three-jet events are mixed. To avoid this effect, the Luclus

$$\begin{pmatrix} 1.00 \\ -1.01 & 1.33 \\ 0.04 & -0.27 & 0.38 \\ 0.01 & -0.02 & -0.17 & 0.26 \\ 0.00 & 0.00 & 0.00 & -0.10 & 0.15 \\ 0.00 & 0.00 & 0.00 & 0.00 & -0.05 & 0.07 \\ 0.00 & 0.00 & 0.00 & 0.00 & 0.00 & -0.02 & 0.04 \\ 0.00 & 0.00 & 0.00 & 0.00 & 0.00 & 0.00 & -0.01 & 0.02 \\ 0.00 & 0.00 & 0.00 & 0.00 & 0.00 & 0.00 & 0.00 & -0.01 & 0.01 \end{pmatrix} \begin{pmatrix} 1.00 \\ -1.00 & 1.30 \\ 0.04 & -0.26 & 0.37 \\ 0.00 & -0.01 & -0.16 & 0.25 \\ 0.00 & 0.00 & 0.00 & -0.09 & 0.14 \\ 0.00 & 0.00 & 0.00 & 0.00 & -0.05 & 0.07 \\ 0.00 & 0.00 & 0.00 & 0.00 & 0.00 & -0.02 & 0.04 \\ 0.00 & 0.00 & 0.00 & 0.00 & 0.00 & 0.00 & -0.01 & 0.02 \\ 0.00 & 0.00 & 0.00 & 0.00 & 0.00 & 0.00 & 0.00 & -0.01 & 0.01 \end{pmatrix}$$

Table 1: Inverse covariance matrices: data (left) and Jetset 7.2 PS+BE (right). The matrix elements are normalized to ${V_{11}^{q}}^{-1}$.

algorithm [11] with a resolution parameter $D_{join} = 7\% \cdot W_{vis}$ has been used to define two-jet events; 50% of the data passed this selection. In Fig. 8 the corresponding factorial moments and fractal dimensions are presented. It is clearly seen that this two-jet sample shows fewer fluctuations than the total event sample, as is expected. Another interesting observation is the occurrence of negative slopes, especially in the two-dimensional analysis. This is interpreted as anti-correlations due to local $p_\perp$ conservation during the hadronization process. The Lund model, which contains explicit $p_\perp$ conservation due to the tunneling mechanism, provides a good description of this effect.

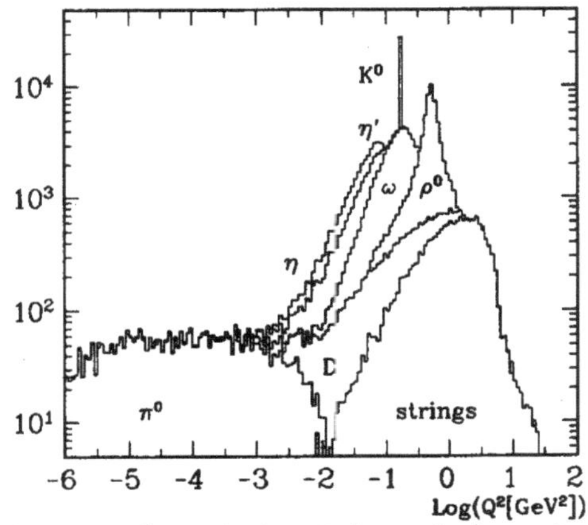

Figure 7: Q^2 distribution of e^+e^- and $\pi^+\pi^-$ pairs from resonance decays.

The analysis of like-sign charged particles will provide more information on the importance of Bose-Einstein correlations since the effect of resonance decays is reduced. The corresponding results are shown in Fig. 9. Obviously the Monte Carlo including Bose-Einstein correlations provides a very good description of the data. And furthermore the difference between both Monte Carlo versions is much more pronounced than in the analysis of all charged particles (cf. Figs. 4, 5). This makes it clear that Bose-Einstein correlations have to be taken into account to describe the factorial moments. The importance of resonance decays is demonstrated in Fig. 10, where the dramatic change in the fractal dimension due to π^0-Dalitz decays is shown. It should be noted that the Dalitz decays are only visible in higher dimensional analyses and moreover they occur at scales where the finite detector resolution becomes important and thus makes it impossible to see the full strength of the effect. However, as shown in Fig. 7, there are many more resonance decays at larger correlation scales, which are

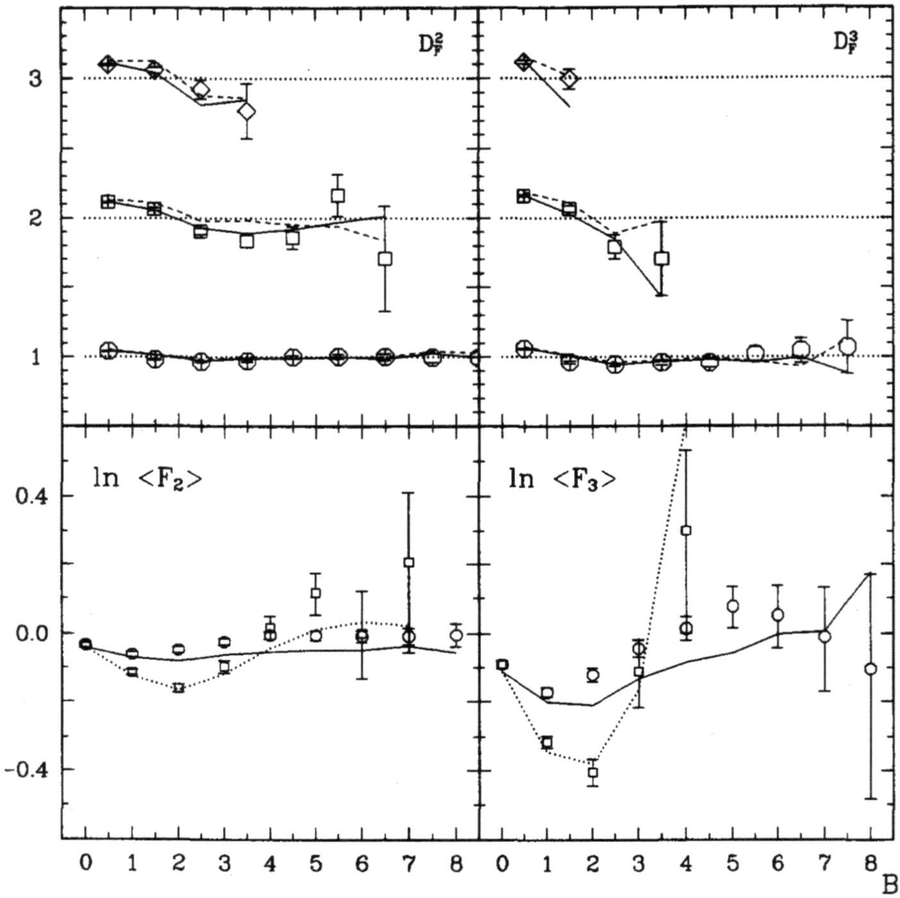

Figure 8: Factorial moments and fractal dimensions for two-jet events. Lower plots: $\langle F_m^2 \rangle$ and $\langle F_m^3 \rangle$. $D_T = 1$; y analysis – open circles (data) and solid lines (Jetset 7.2 PS+BE): $D_T = 2$; y, ϕ analysis – open squares (data) and dotted lines (Jetset 7.2 PS+BE). Upper plots: D_F^2 and D_F^3. The open symbols show the CELLO data while the solid and dashed lines correspond to the Jetset 7.2 PS model with and without Bose-Einstein correlations.

easily accessible by the experiment. Under these circumstances it is important to note that CELLO has measured a variety of inclusive particle cross sections [20] to be in good agreement with the Lund model.

Following a suggestion by Ochs [16] the multiplicity dependence of factorial moments has been analysed. For this purpose the total event sample has been divided into three parts with mean multiplicities $\langle N \rangle = 6.8, 9.9, 13.5$. The results

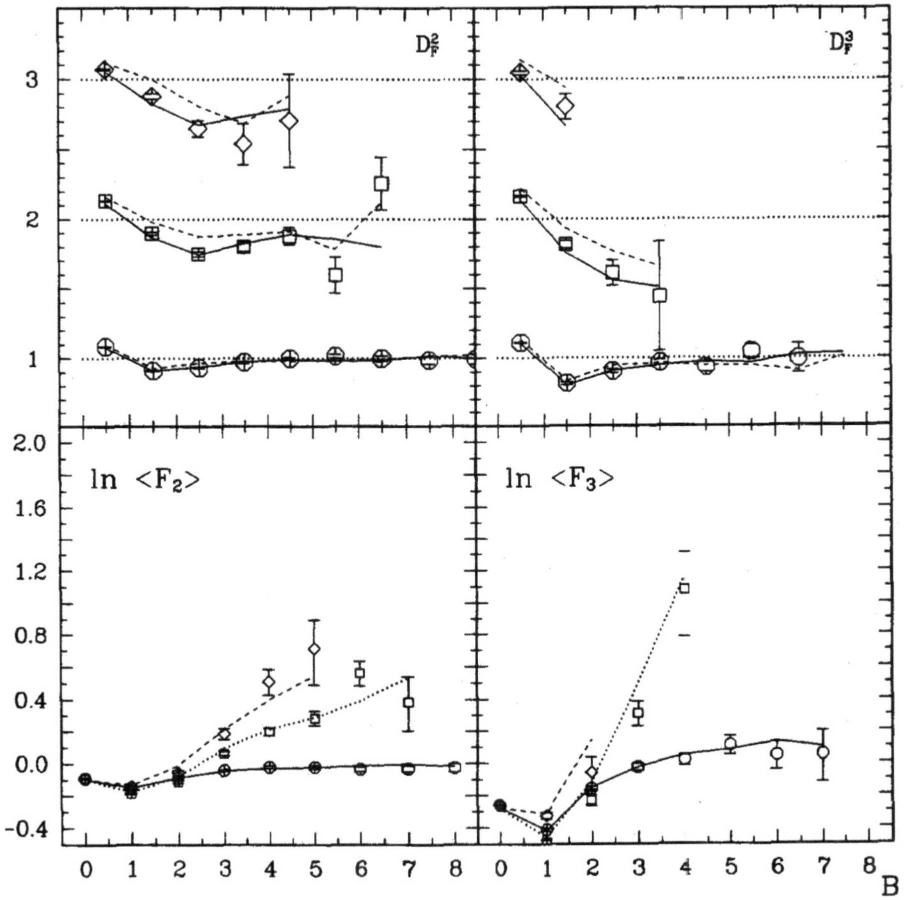

Figure 9: Factorial moments and fractal dimensions for like-sign charged particles. Lower plots: $\langle F_m^2 \rangle$ and $\langle F_m^3 \rangle$. $D_T = 1$; y analysis – open circles (data) and solid lines (Jetset 7.2 PS+BE): $D_T = 2$; y, ϕ analysis – open squares (data) and dotted lines (Jetset 7.2 PS+BE). Upper plots: D_F^2 and D_F^3. The open symbols show the CELLO data while the solid and dashed lines correspond to the Jetset 7.2 PS model with and without Bose-Einstein correlations.

from the one- and two-dimensional analyses are plotted in Fig. 12, where it is clearly seen that the factorial moments increase as the mean multiplicity increases. The Jetset 7.2 PS+BE is seen to reproduce the data in general quite well, however the slopes of F^2 in the one-dimensional analysis are 50% larger in the data. The origin of this discrepancy appears to be twofold: after residual τ-pair events have been removed from the data sample, by requiring at least one hemisphere with more than four particles, the discrepancy is reduced from 50 to 20%.

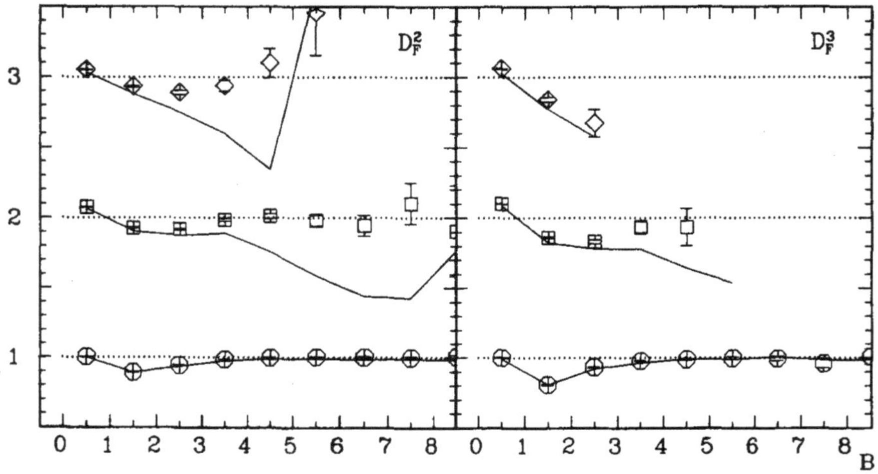

Figure 10: Fractal dimensions D_F^2 and D_F^3. The solid lines correspond to the default Jetset 7.3 PS simulation. The open symbols show the same model, neglecting e^+e^- pairs from π^0-Dalitz decays and $\pi^+\pi^-$ pairs from $\eta \to \pi^+\pi^-\pi^0$ and $\eta' \to \pi^+\pi^-\eta$ decays.

Another reason is the underestimation of primary pion production in the Lund model, as has been pointed out in [17]. In this context the $p_\perp$ conservation mechanism is also important, an assumption which is supported by a comparison of Monte Carlo models (Fig. 11), where e.g. the cluster and string models are seen to behave differently.

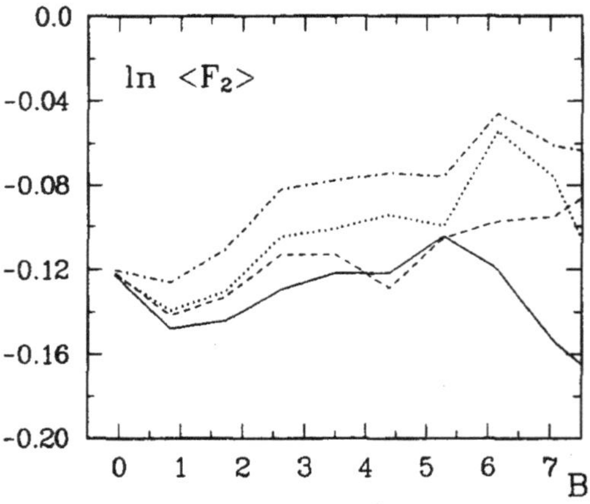

Figure 11: One-dimensional F^2 analysis of low multiplicity events ($N = 2-8$). Jetset 7.3 PS (solid line), Jetset 7.3 ME (dotted line), Ariadne 3.1 CD (dashed line) and Herwig 5.0 PS (dash-dotted line).

Finally we wish to emphasize that the factorial moments of the total event sample are dominated by hard gluon radiation [21] as can be seen in Fig. 13, where a comparison between Jetset 7.3 PS, Jetset 7.3 ME [11], Ariadne 3.1 CD [18] and Herwig 5.0 PS† [19] reveals no significant differences. This is an important fact, since these models differ in the treatment of the perturbative QCD phase (parton shower, dipole radiation or matrix element) and moreover use different hadronization schemes (cluster or strings).

† Herwig has been modified to invoke Jetset for particle decays.

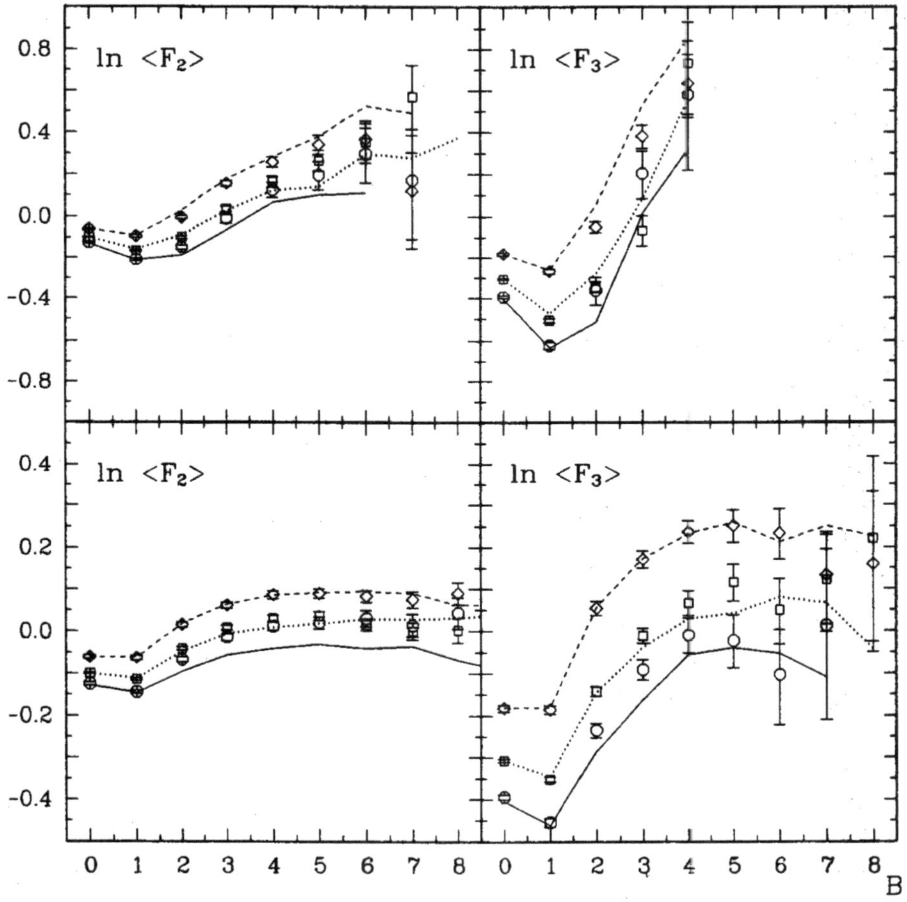

Figure 12: Factorial moments $\langle F_m^2 \rangle$ and $\langle F_m^3 \rangle$ in one dimension (lower plots) and two dimensions (upper plots). The open symbols represent the CELLO data, while the curves show the Jetset 7.2 PS+BE model: $\langle N \rangle = 6.8$; open circles and solid lines: $\langle N \rangle = 9.9$; open squares and dotted lines: $\langle N \rangle = 13.5$; open diamonds and dashed lines.

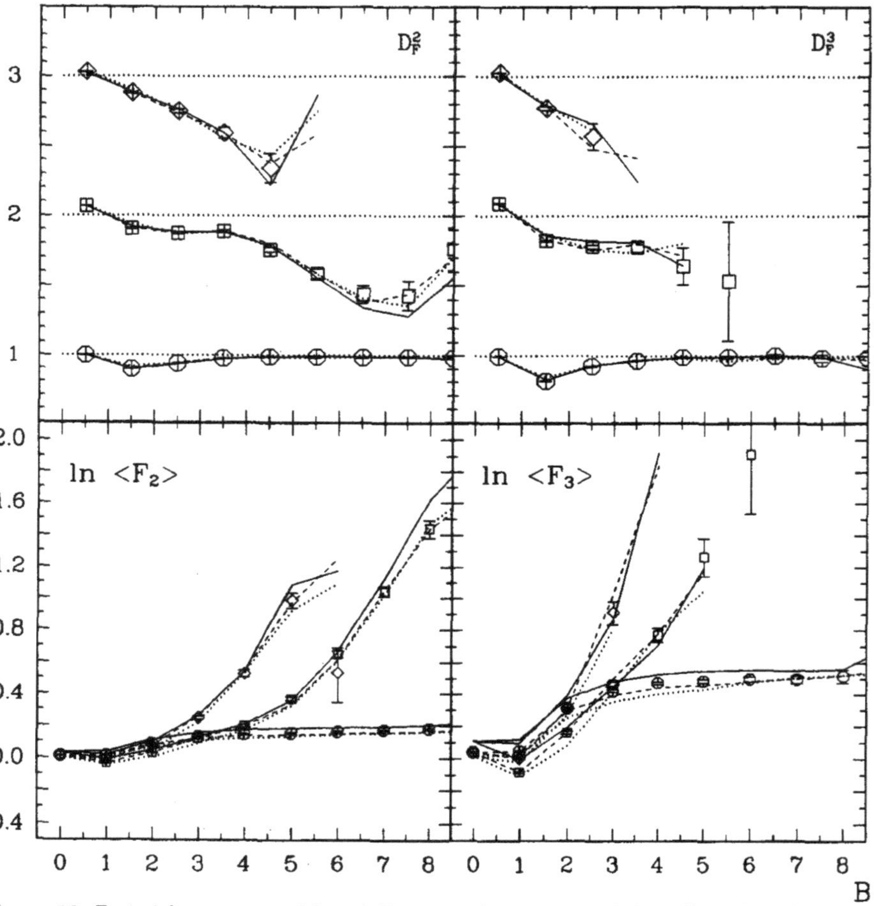

Figure 13: Factorial moments and fractal dimensions in one, two and three dimensions for various Monte Carlo models: Jetset 7.3 PS (open symbols), Jetset 7.3 ME (dotted lines), Ariadne 3.1 CD (dashed lines) and Herwig 5.0 PS (solid lines).

4 Two-particle rapidity correlations

In the previous discussion of intermittency, the relevance of particle correlations has been pointed out. In this context the analysis of two-particle rapidity correlations can provide further information on the underlying physics. The rapidity correlation functions are defined as follows [22],

$$\rho_1(y_1) = \left\langle \sum_i \delta(y_1 - y_i) \right\rangle \tag{7}$$

$$\rho_2(y_1, y_2) = \left\langle \sum_{i \neq j} \delta(y_1 - y_i) \cdot \delta(y_2 - y_j) \right\rangle \tag{8}$$

Here we present results in terms of the reduced cumulant

$$k_2(y_1, y_2) = \frac{\rho_2(y_1, y_2)}{\rho_1(y_1) \cdot \rho_1(y_2)} - 1 \,, \tag{9}$$

apparently $k_2(y_1, y_2)$ vanishes in the case where the two variables become statistically independent. In Fig. 14 the CELLO data on $k_2(y_1, y_2)$ are plotted for the total

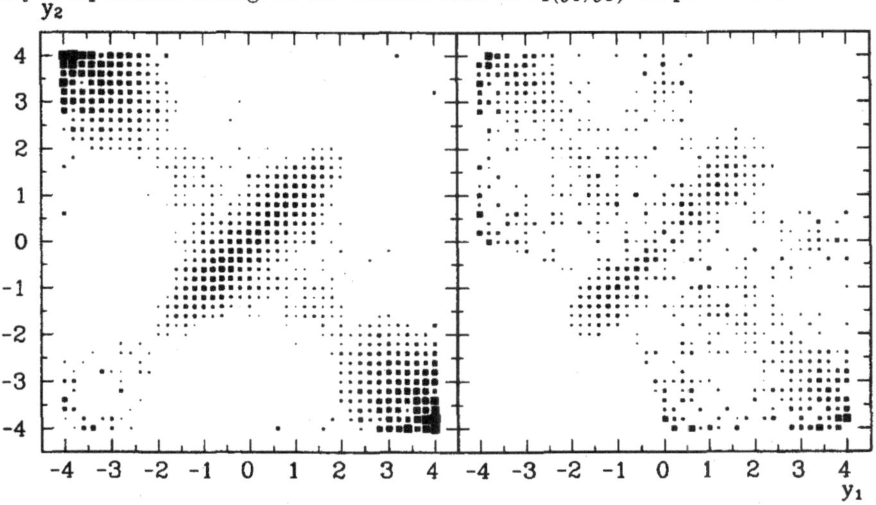

Figure 14: Reduced cumulant $k_2(y_1, y_2)$: total event sample (left) and two-jet events (right).

event sample and for selected two-jet events. Strong correlations with a range of one unit in rapidity are observed. These are seen to be much weaker in the two-jet sample, which indicates their origin due to hard gluon radiation. In addition strong long range correlations due to energy-momentum conservation are observed. The corresponding projections along the main diagonal are displayed in Fig. 15 and

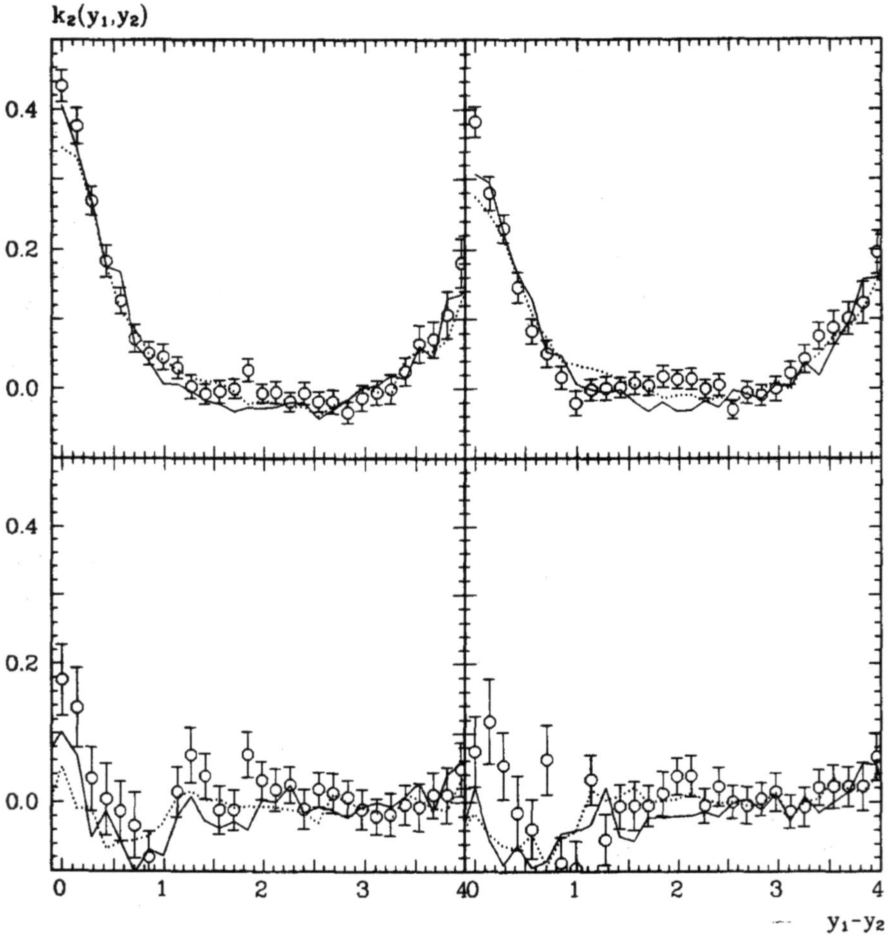

Figure 15: Reduced cumulant $k_2(y_1, y_2)$ averaged over $|y_1 - y_2| \leq 0.5$ for the entire data sample (upper plots) and for selected two-jet events (lower plots). The left plots correspond to negative values of y_1 and y_2 and the right plots to positive values. The data (open circles) are compared to the Jetset 7.2 PS model with (solid lines) and without (dotted lines) Bose-Einstein correlations.

compared to the Lund model, which is seen to reproduce the data quite well. As in the factorial moment analysis the Monte Carlo including Bose-Einstein correlations gives a better description of the data. It is also seen that the correlations are larger for $y \leq 0$, i.e. in the event hemisphere opposite of the most energetic jet, which is likely to contain additional particles from gluon jets.

5 Summary and conclusions

In summary we have presented a comprehensive analysis of multiparticle production via factorial moments in one, two and three dimensions and two-particle rapidity correlations. These studies made it clear that the intermittency behaviour in e^+e^- annihilation is not due to new physics. The effect of several known phenomena on the factorial moments has been demonstrated and explained via *differential* analyses of the data and detailed Monte Carlo studies.

A An algorithm for 3-D intermittency analyses

Consider a three-dimensional intermittency analysis. The problem is to find a decomposition of the three-dimensional phase space into boxes with equal particle content. After e.g. $B = 6$ bisections in every dimension the total number of boxes is $2^{3 \cdot 6} = 262.144$. Given six coordinates to define the position of a box in phase space, a total of $\approx 1.57 \cdot 10^6$ coordinates must be known to enable the assignment of a particle to its box. Once these coordinates are defined, the actual assignment requires us to compare three particle coordinates with six coordinates of 262.144 boxes. The idea to solve this problem is as follows: the information needed to perform a three-dimensional intermittency analysis consists of three phase space variables, e.g. $y, \phi, p_\perp^2$ and the event number for all particles. This data is stored in arrays and is kept in the main memory of the computer. In the first step the y array is sorted in ascending order, providing an index array which is used to trace the remaining variables ($\phi, p_\perp^2$ and the event number) to the elements in the sorted y array. Then the y array is divided into the desired number of sub-arrays containing equal numbers of particles. Particles belonging to the same y sub-array are given a common index, replacing the old y values. In the second step, for every y sub-array separately, the corresponding ϕ array is sorted in ascending order, again using an index array to trace the remaining variables ($p_\perp^2$ and event number) to the sorted ϕ sequence. As before the desired division is made and particles belonging to the same ϕ sub-array are assigned a common index, which replaces the old ϕ values. In a similar way the $p_\perp^2$ values are replaced by appropriate indices. At the end of the procedure the position of each particle in the transformed phase space is defined by three indices and its event number. This information is readily transformed into factorial moments. The complete procedure requires one minute of CPU time (including I/O) on a IBM 9021 for 500.000 particles with six different phase space decompositions.

Acknowledgements

It is a pleasure to thank the organizers of this workshop for the pleasant atmosphere and the royal hospitality experienced at Ringberg castle. I want to take this opportunity to thank my colleagues in CELLO and in particular M. Feindt for helpful cooperation during this analysis.

REFERENCES

[1] CELLO Coll. H.-J. Behrend et al.: *Fractal dimensions from a three-dimensional intermittency analysis in e^+e^- annihilation*, Phys. Lett. B 256 (1991) 97

[2] TASSO Coll. W. Braunschweig et al.: Phys. Lett. B 231 (1989) 548
DELPHI Coll. P. Abreu et al.: Phys. Lett. B 247 (1990) 137
OPAL Coll. M.Z. Akrawy et al.: Phys. Lett. B 262 (1991) 351
ALEPH Coll. D. Decamp et al.: preprint CERN-PPE/91-121 (1991)

[3] M. Feindt: *Intermittency and two-particle correlations in e^+e^- annihilation*, to appear in Procs. XXV. Int. Conf. on High Energy Physics, Singapore, 1990

[4] O. Podobrin: *Intermittency, fractal dimensions and correlation phenomena in e^+e^- annihilation at 35 GeV*, Procs. XX. Int. Symp. on Multiparticle Dynamics, eds. R. Baier and D. Wegener (World Scientific, Singapore, 1991) p. 417

[5] O. Podobrin: *Intermittency and Bose-Einstein correlations in e^+e^- annihilation*, preprint DESY 91-028, to appear in Procs. XXVI. Rencontre de Moriond on High Energy Hadronic Interactions, Les Arcs, 1991

[6] See the following articles for a survey: W. Kittel and R. Peschanski: Procs. EPS-HEP '89, Nucl. Phys. B (Proc. Suppl.) 16 (1990) p. 445
A. Bialas: loc. cit. [4] p. 669

[7] T. Sjöstrand: Int. J. Mod. Phys. A 3 (1988) 751

[8] A. Bialas and R. Peschanski: Nucl. Phys. B 273 (1986) 703

[9] CELLO Coll. H.-J. Behrend et al.: Phys. Scr. 23 (1981) 610

[10] M. Feindt: *A review of two-photon physics*, preprint CERN-PPE/91-90, to appear in Procs. XXVI. Rencontre de Moriond on High Energy Hadronic Interactions, Les Arcs, 1991

[11] B. Andersson, G. Gustafson, G. Ingelman and T. Sjöstrand: Phys. Rep. 97 (1983) 31
T. Sjöstrand and M. Bengtsson: Comp. Phys. Commun. 43 (1987) 367

[12] W. Ochs: Z. Phys. C – Particles and Fields 50 (1991) 339

[13] A. Bialas and M. Gazdzicki: Phys. Lett. B 252 (1990) 483

[14] W. Ochs: Phys. Lett. B 247 (1990) 101

[15] G. Ekspong, R. Peschanski and J. Wosiek: Phys. Lett. B 251 (1990) 455

[16] W. Ochs: loc. cit. [4] p. 434

[17] G. Gustafson: *Fractal structures and Intermittency in QCD*, Lund preprint LU TP 90-9 (1990)

[18] G. Gustafson and U. Pettersson: Nucl. Phys. B 306 (1988) 746
L. Lönnblad: Lund preprint LU TP 89-10 (1989)

[19] G. Marchesini and B.R. Webber: Nucl. Phys. B 310 (1988) 461

[20] Cross section measurements of $K^0_S, K^{*\pm}, \Lambda, \gamma, \pi^0, \eta$ and $D^{*\pm}$ are presented in the following papers: CELLO Coll. H.-J. Behrend et al.: Z. Phys. C – Particles and Fields 46 (1990) 397 and Z. Phys. C – Particles and Fields 47 (1990) 1
O. Podobrin: Procs XXV. Rencontre de Moriond on High Energy Hadronic Interactions, ed. J. Tran Thanh Van (Editions Frontières, Gif-sur-Yvette, 1990) p. 217

[21] B. Andersson, G. Gustafson, A. Nilsson and C. Sjögren: Z. Phys. C – Particles and Fields 49 (1991) 79

[22] P. Carruthers and I. Sarcevic: Phys. Rev. Lett. 63 (1989) 1562

MULTIFRACTAL ANALYSIS OF UA1 MINIMUM BIAS DATA

UA1–Minimum Bias Collaboration

H. DIBON

Institut für Hochenergiephysik der Österr. Akademie der Wissenschaften, Nikolsdorfergasse 18, 1050 Wien, Austria

Abstract

We investigate UA1 minimum bias data at $\sqrt{s} = 630$ GeV in order to find multifractal structures in pseudorapidity η distributions. Such structures are found for relatively large bin sizes $(1.5, 0.75, 0.375)$, while for smaller bins they break down. A possible method to reduce the statistical fluctuations in the η - distribution is discussed and it is found that a linear scaling behaviour in terms of the G_q - moments can be established by the enrichement of the event sample with more spiky events. We compare the results with the Monte Carlo models Gencl and Pythia 4.8. In the multifractal spectral function $f(\alpha_q)$ no agreement between data and models is observed.

1 Introduction

Studies using the concept of intermittency [1] have turned attention in soft hadron interactions from the analysis of the data in terms of distributions averaged over many events, to the investigation of fluctuations of observables within one event. The purpose of this new approach is to obtain more information on processes, governing the evolution from the quark to the hadron level, which so far are explained only by conjecture. For instance, cascading or branching processes that could evolve into self - similar patterns as do geometrical objects, recently well known by the name of fractals or multifractals [2,3].

In this context, an approach which works in analogy with the theory of multifractals has been proposed [4]. A few expressions from this theory have to be defined [2]: consider a distribution given in the form of a histogram $H(\eta)$. In order to describe the bin to bin fluctuations of such a histogram the moments

$$G_q = \sum_{j=1}^{M} p_j{}^q, \; p_j = n_j/n, \; \sum_{i=1}^{M} p_j = 1 \tag{1}$$

are used. n_j is the number of entries in the bin j and n is the total number of entries in $H(\eta)$. p_j is the probability to find n_j entries in bin j. The sum is extended over non - empty bins, because q extends over the whole spectrum of real numbers. In the case where the number of bins in $H(\eta)$ is increased, and the bin to bin fluctuations follow a self-similar pattern, a scaling behaviour of the moments, G_q, would result *.

$$G_q \propto (\delta\eta)^{\tau_q} \qquad \text{for} \qquad \delta\eta \to 0 \tag{2}$$

A measure of the fluctuations of $H(\eta)$ is found by mapping them onto a smooth curve, so that the multifractal spectral function $f(\alpha_q)$ is obtained. $f(\alpha_q)$ can be derived from a generalized exponent τ_q by Legendre transformations using the Lipschitz - Hölder exponent α_q.

$$\alpha_q = \frac{d\tau_q}{dq} \tag{3}$$

$$f(\alpha_q) = q\alpha_q - \tau_q \tag{4}$$

The generalized dimensions D_q are given by

$$D_q = \frac{\tau_q}{q-1} = \frac{1}{q-1} \lim_{\delta\eta \to 0} \frac{\ln G_q}{\ln \delta\eta} \tag{5}$$

We have the following definitions:

$D_0 = f(\alpha_0)\ldots$ fractal dimension

$D_1 = f(\alpha_1) = \alpha_1 \ldots$ information dim.

$D_2 = 2\alpha_2 - f(\alpha_2)\ldots$ correlation dim.

⋮

According to this theory, the shape of $f(\alpha_q)$ in the case of self-similar behaviour is downward concave in the form

$$\frac{df(\alpha_q)}{d\alpha_q} = q \quad \text{and} \quad \frac{d^2 f(\alpha_q)}{d\alpha_q^2} < 0 \tag{6}$$

This is shown in fig.1. The $q > 0$ side represents the maxima the $q < 0$ side the minima of $H(\eta)$. $f(\alpha_q)$ provides an accurate quantitative description of these fluctuations, since it is a function of an infinite set of moments. A comparison in $f(\alpha_q)$ for data and models is therefore possible in a very exact way.

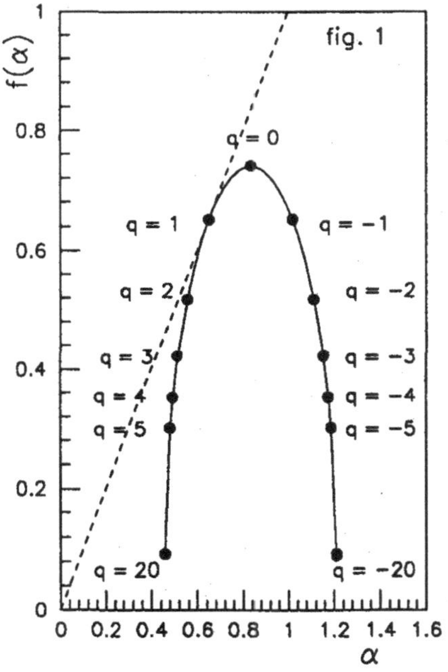

2 Application to Particle Physics

The one to one application of these formulae to particle physics is not possible, since this would require events with an infinite number of particles in the final state. We have to work with approximations, or slightly modified concepts of multifractal theories

*the two-fold scaling Cantor set [2] is an example of such a process

because the number of particles in events are limited. For example, the limit $\delta\eta \to 0$ cannot be achieved. If $\delta\eta$ is small enough only bins containing one particle remain. A further decrease of $\delta\eta$ gives no further information since the moments G_q contain only non-empty bins. Furthermore in this case any scaling property has been lost, since

$$G_q = n^{1-q} = const. \tag{7}$$

Therefore if there is multifractal behaviour in the data, it has to be investigated at rather large $\delta\eta$ (depending on n). In addition to this we deal with events having rather low multiplisities in the final state. In order to compensate, to some extent, for these small event multiplicities the average is taken over many events according to the following formulae [4]:

$$\langle \ln G_q(\mu,\nu)\rangle = \frac{1}{N}\sum_{i=1}^{N} \ln \sum_{j=1}^{M} p_{ij}^q \tag{8}$$

$$\langle \tau_q(\mu,\nu)\rangle = \frac{\langle \ln G_q(\mu,\nu)\rangle_{\mu_2} - \langle G_q(\mu,\nu)\rangle_{\mu_1}}{\ln \delta\eta_{\mu_2} - \ln \delta\eta_{\mu_1}} = -\frac{\Delta\langle \ln G_q(\mu,\nu)\rangle}{\Delta\mu \ln 2} \tag{9}$$

$$\langle \alpha_q(\mu,\nu)\rangle = \frac{d\langle \tau_q(\mu,\nu)\rangle}{dq} \tag{10}$$

$$\langle f(\alpha_q(\mu,\nu))\rangle = q\langle_q(\mu,\nu)\rangle - \langle \tau_q(\mu,\nu)\rangle \tag{11}$$

$$D_q(\mu,\nu) = \frac{\langle \tau_q(\mu,\nu)\rangle}{q-1} \tag{12}$$

$N\ldots$ number of events

$M = 2^{\mu}\ldots$ number of nonempty bins

$n = 2^{\nu}\ldots$ number of tracks

The purpose of this investigation, by means of (8),(9),(10),(11) and (12), is to find an answer to the following questions:

- can one find fractal or multifractal properties in the data?
- is there a self - similar behaviour of dynamical origin in the data and how universal (independent of observables) are the results?
- how much do models like Pythia 4.8[5] or Gencl [6] reproduce the fluctuations in the data?

3 The Data Sample

UA1 minimum bias data from the 1985 data taking run at $\sqrt{s} = 630$ GeV was used for the multifractal analysis. The same sample, of ~ 160000 events, was used in intermittency analysis [7] and also in a previous multifractal analysis [8].

The data comes from the UA1 central driftchambers (CD) and this is subject to corrections. A map of the geometrical acceptance properties of this detector is available in the form of tables, which are divided into 32 subdivisions in η, into 16 intervals in ϕ and 5 intervals in p_t [9]. Visual inspection of these tables enabled the selection of two η- intervals $1.5 \leq |\eta| \leq 3.0$ where the correctionfactors ϵ are close to 1 in order to keep systematic uncertainties small.

In the course of the intermittency investigations two additional systematic problems in the CD were discovered, that turned out to be of major importance for these kind of investigations [7]:

- a strong multiplicity dependence in different regions in the azimuth angle ϕ has been found in the central region $-1.5 \leq |\eta| \leq 1.5$ which has not been taken into account in the tables. Fig. 2 shows that this defect is neglegible in our η region. The ratio

$$R = \frac{\langle n \rangle_{45° \leq |\phi| \leq 135°}}{\langle n \rangle_{-45° \leq \phi \leq 45°}}$$

  is much smaller for the central interval than for the one used here. Therefore this lack of correction in the tables is unimportant here.

- a further problem was the fact that for low momentum tracks a splitting occured when they traversed from one CD module to the adjacent one. This problem was solved by a software procedure called the ghostfinder [7].

In order to find the multiplicities used in this analysis the data was processed by an analysis program which used the acceptance tables and the ghostfinder.

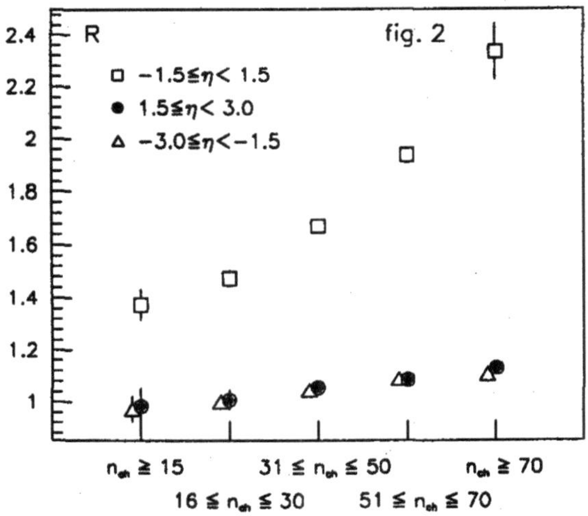

4 Discussion of the Results

4.1 Multifractality

A multifractal analysis was performed by applying formulae(8) to (12) to the data.

In fig. 3 the behaviour of $\langle \ln G_q \rangle$ versus μ is displayed. Due to the power of q a dispersion is observed which is strongest for $\mu = 1$ ($\mu = 1, 2, 3, 4$ stands for subdivision into $2, 4, 8, 16$ bins of the union of $1.5 \leq | \eta | \leq 3.0$). For $\mu > 1$ the dispersion decreases and saturates for higher μ. This saturation is the result of having more events with bins only containing one entry, which contribute to the G - moments, for higher values of μ . According to (7) the scaling behaviour becomes contaminated with such non - scaling events for increasing numbers of subdivisions.

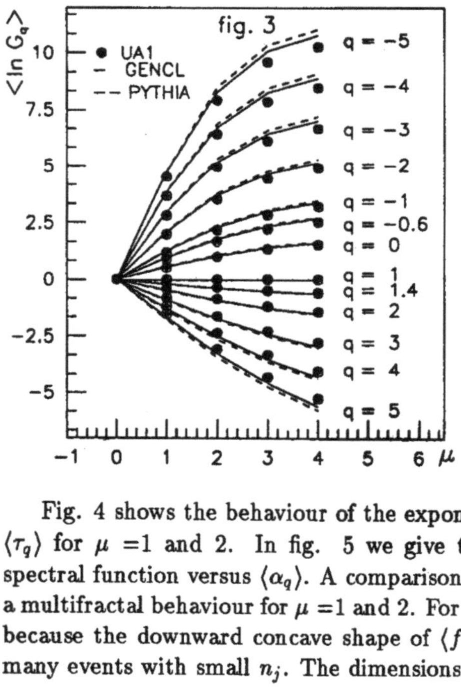

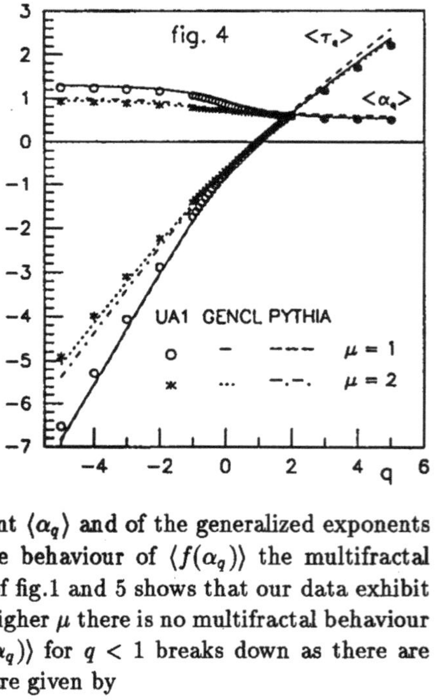

Fig. 4 shows the behaviour of the exponent $\langle \alpha_q \rangle$ and of the generalized exponents $\langle \tau_q \rangle$ for μ =1 and 2. In fig. 5 we give the behaviour of $\langle f(\alpha_q) \rangle$ the multifractal spectral function versus $\langle \alpha_q \rangle$. A comparison of fig.1 and 5 shows that our data exhibit a multifractal behaviour for μ =1 and 2. For higher μ there is no multifractal behaviour because the downward concave shape of $\langle f(\alpha_q) \rangle$ for $q < 1$ breaks down as there are many events with small n_j. The dimensions are given by

$\mu = 1$:	$\mu = 2$:
$D_0 = 0.7676 \pm 0.0011$	$D_0 = 0.6782 \pm 0.00097$
$D_1 = 0.6778 \pm 0.001$	$D_1 = 0.6448 \pm 0.00093$
$D_2 = 0.6219 \pm 0.001$	$D_2 = 0.6108 \pm 0.0009$

Therefore we find a different $\langle f(\alpha_q) \rangle$ for increasing μ. The spectral function moves the $45°$ tangent downward with increasing μ.

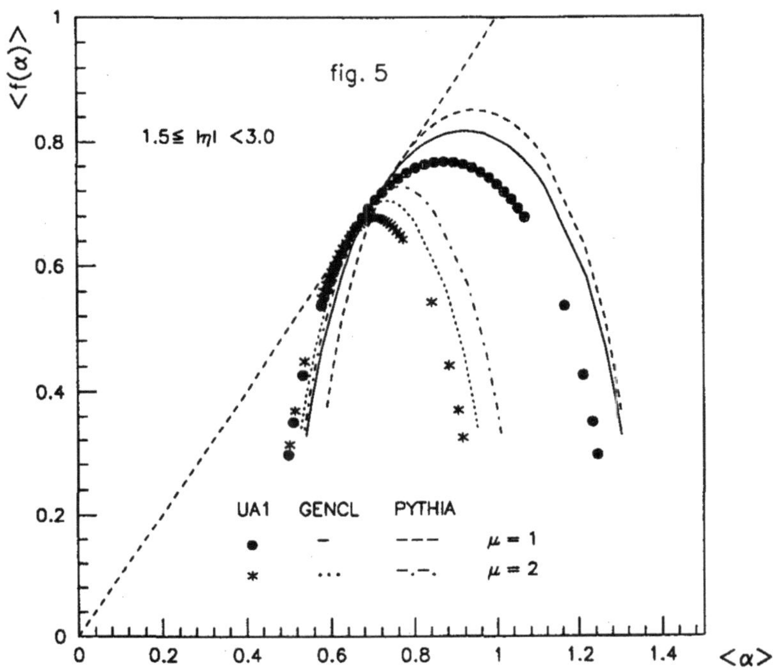
fig. 5
1.5≤ |η| <3.0
UA1 GENCL PYTHIA
μ = 1
μ = 2
<f(α)>
<α>
0
0.2
0.4
0.6
0.8
1
1.2
1.4

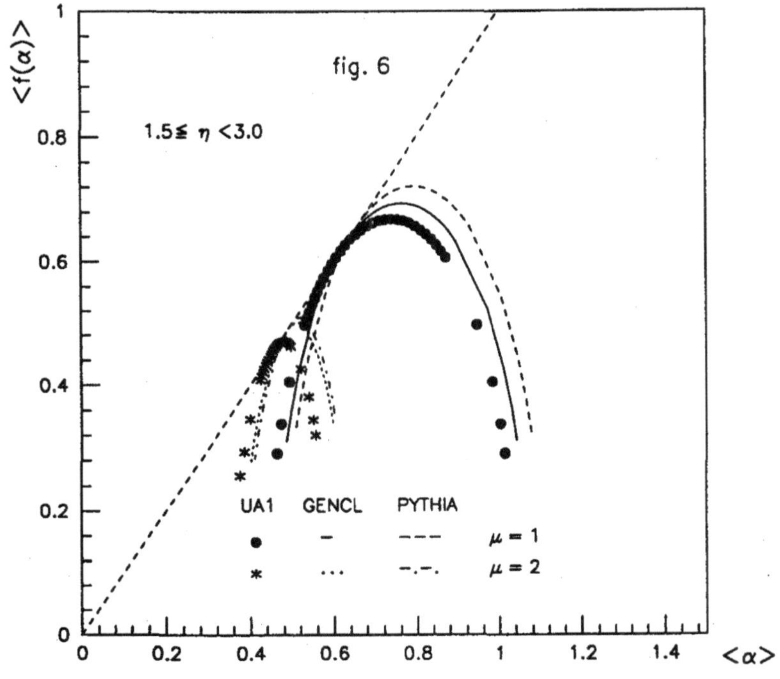
fig. 6
1.5≤ η <3.0
UA1 GENCL PYTHIA
μ = 1
μ = 2
<f(α)>
<α>
0
0.2
0.4
0.6
0.8
1
1.2
1.4

In fig. 6 we show $\langle f(\alpha_q)\rangle$ for $1.5 \leq \eta \leq 3.$ and find a very similar behaviour as in fig. 5. Only the dimensions are decreased:

$\mu = 1:$	$\mu = 2:$
$D_0 = 0.6690 \pm 0.0013$	$D_0 = 0.4719 \pm 0.0011$
$D_1 = 0.6071 \pm 0.0012$	$D_1 = 0.4586 \pm 0.0011$
$D_2 = 0.5648 \pm 0.0011$	$D_2 = 0.4416 \pm 0.0011$

We find a dependence of $\langle f(\alpha_q)\rangle$ on μ and $\Delta\eta$.

In fig.7 we present $\langle f(\alpha_q)\rangle$ for fixed multiplicites $\nu = 2, 3, 4$ for $\mu = 1$ and we find a dependence on ν as well.

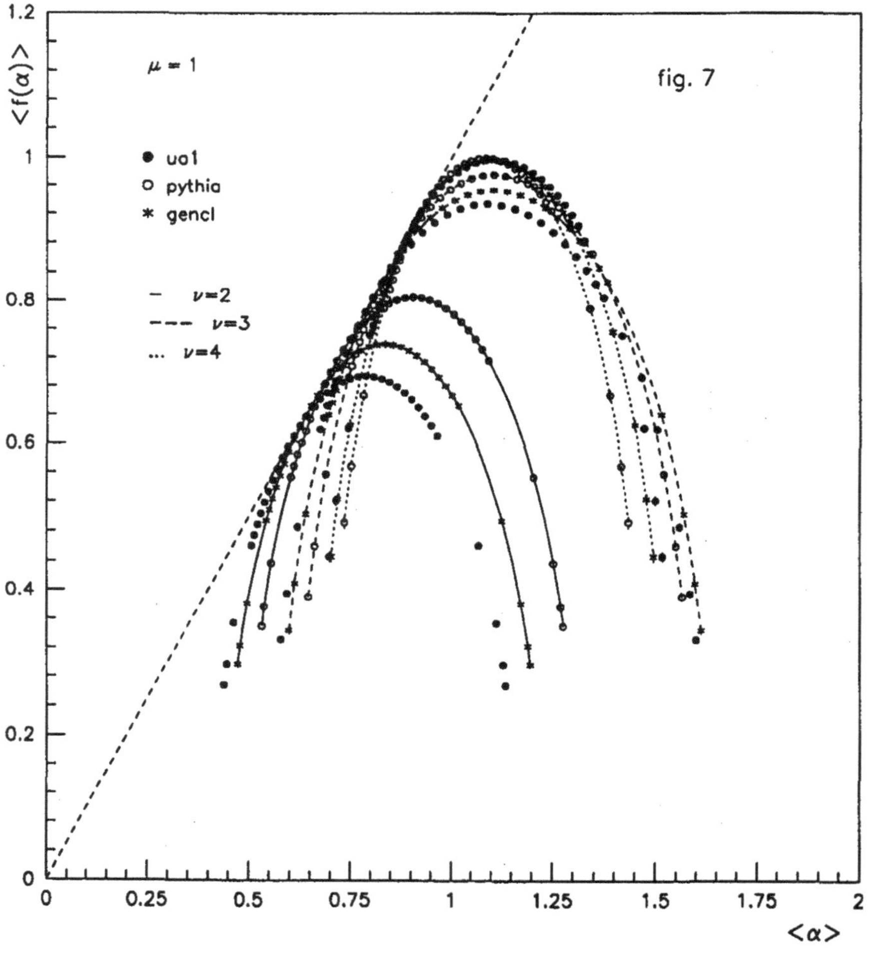

Fig.8 presents data also at $\sqrt{s} = 200$ and 900 GeV (1985 UA1 ramping run) together with the $\sqrt{s} = 630$ GeV data. A clear dependence on s is visible.

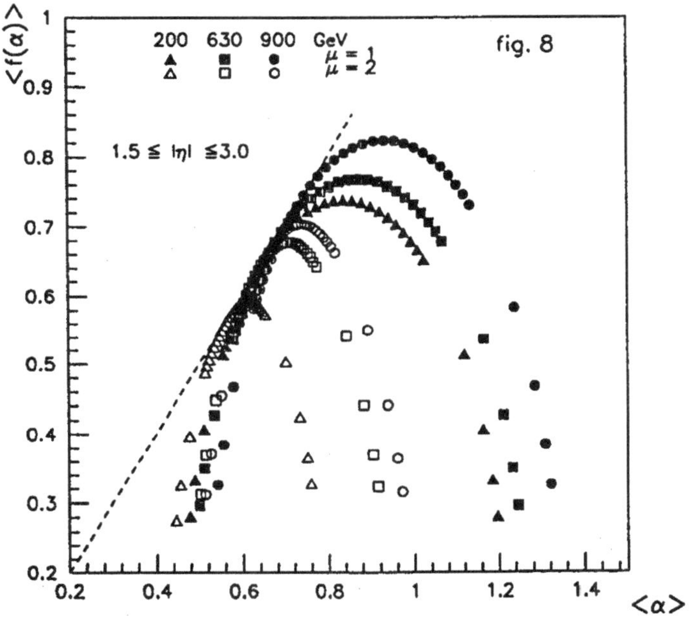

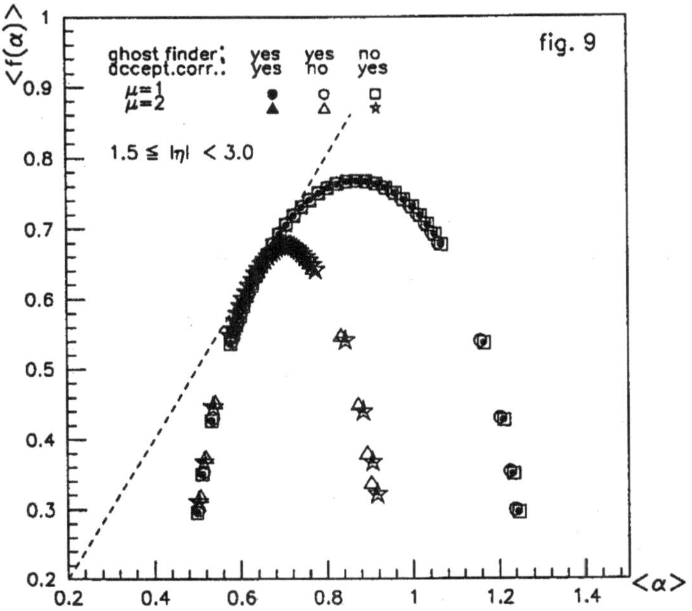

The UA1 data shows a dependence of the spectral function $\langle f(\alpha_q)\rangle$ on the variables s, η and n [10].

In figs.3 - 7 we show also a comparison to Pythia 4.8 and Gencl. We observe a clear difference between data and models; the difference between data and Gencl is less than that between data and Pythia 4.8. From fig.7 we observe that this difference between models and data comes mainly from low multiplicities where the disagreement is greater than that for high multiplicities.

We repeated the analysis without using acceptance tables or ghost finder, to see whether the discrepancies between the models and the data originated from systematic effects. The small variation of the data points in figure 9 indicates that the differences cannot be explained by systematic effects.

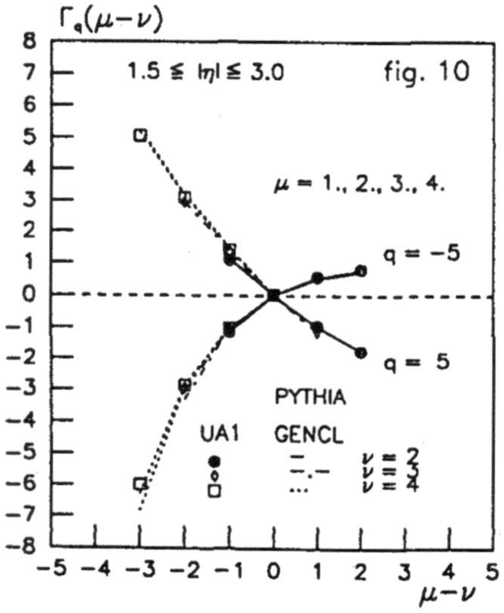

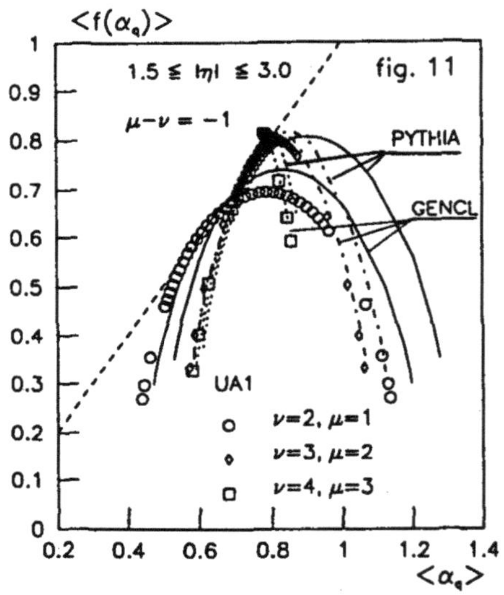

4.2 Universality

Due to the s, η, n dependence [11] universal features are discussed which can be interpreted and derived from self - similar behaviour in the process of multiplicity splitting occurring when the number of bins is increased. In terms of the universality function [11].

$$\Gamma_q(\mu-\nu) = \ln\langle G_q(\mu,\nu)\rangle - \ln\langle G_q(\nu,\nu)\rangle \tag{13}$$

a behaviour is expected which would depend only on $\mu - \nu$ and q.

Fig. 10 shows Γ_q versus $\mu - \nu$ for $q = \pm 5$ where the deviations from universality are larger than for $| q |< 5$.

All data points are close to universal lines indicating universal behaviour. The comparison of data and models shows no great differences. From the previous figures one would expect that they should be larger.

It is also predidted [11] that $\langle f(\alpha_q)\rangle$ versus α coincides for $\mu - \nu =$ constant. Fig.11

demonstrates that this quantity does not display clear universal properties and $\langle f(\alpha_q)\rangle$ is more sensitive to universality effects and differences when comparing with models than $\Gamma_q(\mu-\nu)$ versus $\mu-\nu$. Again there is a larger discrepancy between data and models for small multiplicities.

4.3 Reduction of Statistical Fluctuations

For $\mu > 2$ the multifractal behaviour of the data and models is made unclear by the increasing amount of non - empty bins containing only one track, and causes the saturation effect with increasing μ, seen in fig.3. According to the scaling hypothesis one would expect one slope for each q in this figure to be independent of the value of μ. A procedure has been developed [10] to remove statistical fluctuations by selecting events which contribute more tracks to the bins.

The moments G_q vary in two domains:

1. $0 \leq G_q \leq 1$ for $q > 1$
2. $1 \leq G_q \leq \infty$ for $q < 1$

The first domain describes the peaks in the bins of $H(\eta)$, and the other domain the minima. In order to enrich spiky events and suppress deep non-empty minima the following event selection critera can be applied.

$$\ln G_q > Z\langle \ln G_q\rangle \qquad \text{for} \qquad q > 1$$

$$\ln G_q < Z\langle \ln G_q\rangle \qquad \text{for} \qquad q < 1$$

The factor Z has to be adjusted in order to find straight line behaviour in $\langle G_q\rangle$ versus μ.

To show the linearity of $\langle \ln G_q\rangle$ more clearly a plot similar to the one of [12] has been suggested [13] in the form $\langle \ln G_q\rangle$ versus $-\langle \ln G_2\rangle$.We have applied, for $\mu = 4$, the following cut procedure [10]

$$\ln G_3 \geq Z\langle \ln G_3\rangle$$

$$\ln G_{-2} \leq Z\langle \ln G_{-2}\rangle$$

For Z = 0.7, the behaviour seen in fig.12 agrees well with a straight line. The corresponding multifractal spectrum is shown in fig.13. Here one observes a multifractal behaviour only for $\mu = 1$ and 2 again, although in fig.12 straight line behaviour extends up to $\mu = 4$. The break - down for $\mu > 2$ is probably due to the low multiplicities in our η intervals which is given by $\langle n\rangle \sim 4$. If more than 4 bins are required there are many events that have only one particle in the non - empty bins, and therefore there is strong contamination from non-scaling events. Thus, multifractal behaviour breaks down due to the low multiplicity of events.

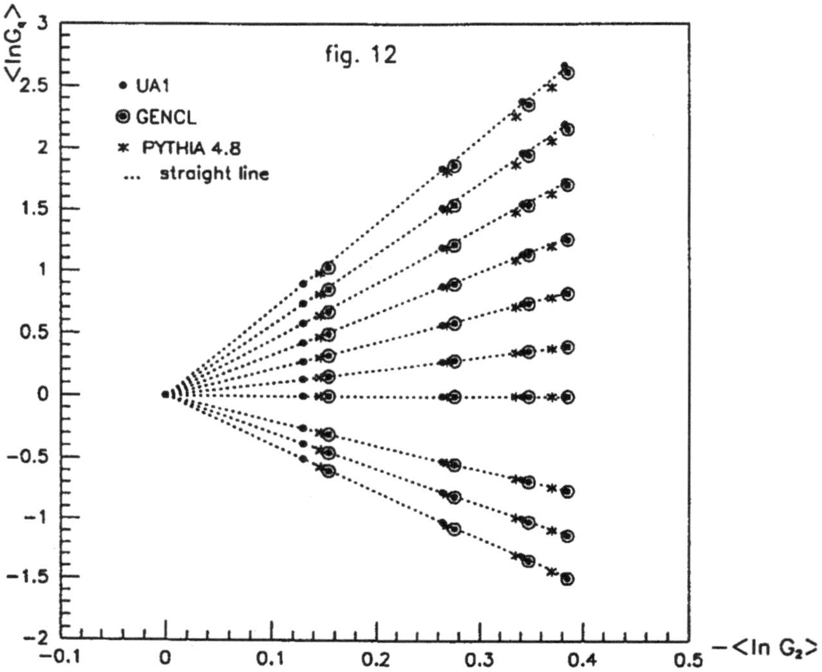
fig. 12
• UA1
GENCL
* PYTHIA 4.8
... straight line
<lnG_q>
-<ln G_2>
3
2.5
2
1.5
1
0.5
0
-0.5
-1
-1.5
-2
-0.1
0.1
0.2
0.3
0.4
0.5

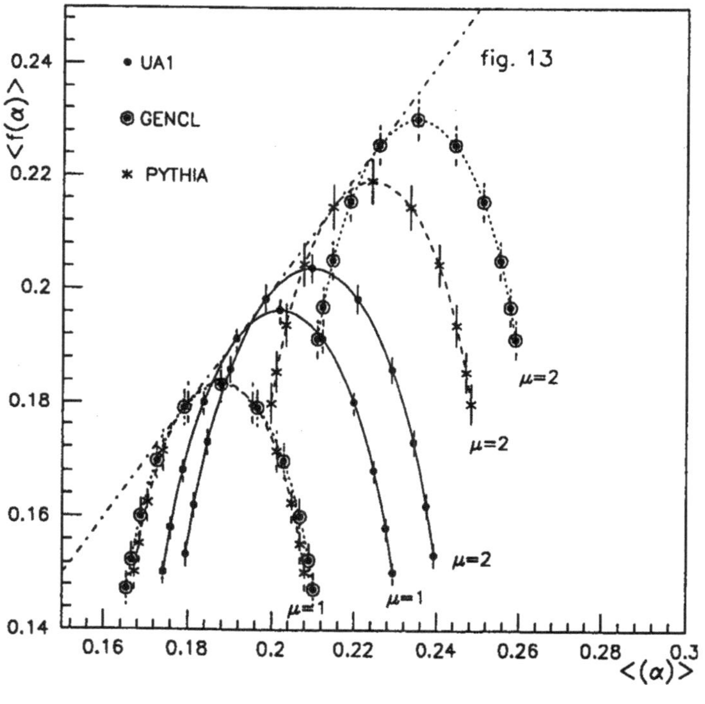
fig. 13
• UA1
GENCL
* PYTHIA
<f(α)>
<(α)>
μ=2
μ=2
μ=2
μ=1
μ=1
0.24
0.22
0.2
0.18
0.16
0.14
0.16
0.18
0.2
0.22
0.24
0.26
0.28
0.3

The shapes of the multifractal spectra in figure 13 for $\mu = 1$ and 2 are much closer together compared with those in fig.5 and 6. This is the result of the approximate straight line behaviour which should result in only one $f(\alpha)$ independently of μ. It can be seen that $f(\alpha)$ is a much more sensitive quantity than for instance the G - moments.

Similar statements can be made for the models Pythia and Gencl. In fig.12 these models also approximate straight lines, although there are small differences to the data. In fig.13 these differences are exaggerated.

Finally, it is concluded that the use of such a cut procedure removes most of the saturation in $\langle \ln G_q \rangle$ observed in fig.4. The expected independence of $f(\alpha)$ on μ is suggested by the closeness of $f(\alpha)$ for $\mu = 1$ and 2. In order to test this, for further increases in μ, one would probably need data with higher average multiplicities in the considered η intervals.

5 Summary

The following remarks answer the previously raised questions:

- We find a multifractal, $f(\alpha)$, compatible with that predicted by the theory of multifractals, in both the original data and in the sample of events obtained using the cut procedure described (see below). However, this multifractality is present only for relatively large bin-sizes $\mu = 1$ and 2, and fails for $\mu > 2$. The reason for this failure is probably due to low average multiplicites ($\langle n \rangle \sim 4$) in the considered η regions.

- A cut procedure was applied in order to remove the saturation effect in the $\ln G_q$ moments in fig.4 to obtain a linear scaling in $\ln G_q$ versus μ. By choosing the value for the adjustment parameter Z = 0.7 a good approximation of the straight line behaviour is achieved(fig.12). This is also confirmed by the proximity of the $f(\alpha)$ functions for $\mu = 1$ and 2 (fig.13). For $\mu > 2$ the multifractality of $f(\alpha)$ breaks down due to lack of average multiplicities in the η intervals. Therefore this method is shown to be a possible solution to remove the saturation in $\ln G_q$ versus μ.

  For relatively large binsizes, such as $\mu = 1, 2$, the expected self-similarity is observed and is dependent on the observables s, η, n. The test of the universal behaviour is inconclusive because the coincidence of all $f(\alpha)$ for constant $\mu - \nu$ (fig.11) is not observed in the universal function $\Gamma_q(\mu - \nu)$. The $\Gamma_q(\mu - \nu)$ is insensitive to this.

- With the exception of the universality functions $\Gamma_q(\mu - \nu)$ (fig.10) the models Pythia 4.8 and Gencl do not reproduce the data. This is demonstrated most cleary in the multifractal spectral functions (fig. 5,6,7,11,13) which appear to be the most sensitive measure used.

All data and model points contain statistical error bars which are in most cases very small and therefore not visible in many figures. We have also demonstrated that the systematic errors due to inefficiencies of the UA1 central detector are small compared with the differences between models and data (fig.9), and so do not account for these differences.

6 Acknowledgements

We are very grateful to R.C.Hwa, C.B.Chiu and W.Florkowski for many suggestions and the exchange of information which was essential in order to obtain the results which are presented. We are also grateful to M.Markytan, C.E.Wulz and P.Lipa for support and many discussions and to R.Fruehwirth for checking the error calculation.

References

[1] A.Bialas and R.Peschanski, *Nucl.Phys.* **B273** (1986) 703; B308(1988) 708.

[2] J.Feder, Fractals (PlenumPress, N.Y. and London, 1988)

[3] G.Paladin and A.Vulpiani, *Phys.Rep.* **156** (1987) 147.
J.L.McCauley, *Phys.Rep.* **186** (1990) 225-266.
B.Mandelbrot, The Fractal Geometry of Nature (Freeman,New York,1982)

[4] R.C.Hwa, *Phys.Rev.* **D41** (1990) 1456.
C.B.Chiu and R.C.Hwa, *Phys.Rev.* **D43** (1991) 100.
C.B.Chiu,K.Fialkowski and R.C.Hwa, *Mod.Phys.Lett.A*,(1991) [OITS-441]

[5] G.J.Alner et al., *Nucl.Phys.* **B291** (1987) 445.

[6] H.U.Bengtsson and T.Sjöstrand, *Comp.Phys.Com.* **46** (1987) 43.

[7] C.Albajar et al., *Nucl.Phys.* **B345** (1990) 1-21.
P.Lipa and B.Buschbeck, *Phys.Lett.* **B223** (1989) 465.
P.Lipa, thesis, University of Vienna (1990), HEPHY-PUB542/90.
B.Buschbeck, P.Lipa , HEPHY - PUB 535/90, in Proceedings of the Santa Fe Workshop on Intermittency in High Energy Collisions, Santa Fe, March 1990, edited by F.Cooper,R.C.Hwa and I.Sarcevic (World Scientific, Singapore 1991) p.96.

[8] H.Dibon and M.Markytan, UA1/TN90 - 11, in Proceedings of the Santa Fe Workshop on Intermittency in High Energy Collisions, Santa Fe, March 1990, edited by F.Cooper, R.C.Hwa and I.Sarcevic (World Scientific, Singapore 1991) p.109.
H.Dibon and M.Markytan, HEPHY - PUB 541/90, in Proceedings of the XX. International Symposium on Multiparticle Dynamics, Gut - Holmecke / Germany, September 10 - 14, 1990, p.449.
M.Markytan and H.Dibon, HEPHY - PUB 547/90, in Proceedings of the XXVth Int. Conference on High Energy Physics Singapore, 2. - 8. August 1990, to be published.

[9] M.Pimiä et al., UA1/TN83-20 (1983).
G.Ciapetti et al., UA1/TN84-55(1984).

[10] C.B.Chiu and R.C.Hwa OITS 466,May 1991.

[11] W.Florkowski and R.C.Hwa, *Phys. Rev.* **D43**,(1991) 1548.

[12] W.Ochs, *Phys.Lett.* **B247** (1990) 101.

[13] R.C.Hwa, private communication.

ON THE MEASUREMENT OF CORRELATIONS AND INTERMITTENCY

P. LIPA
Department of Physics, University of Arizona
Tucson AZ 85721, USA

Abstract

We report on a considerable improvement of the usual factorial moment method used in the study of intermittency by means of integrals over the correlation functions in a strip domain. An algorithm for the actual computation of the so-called strip integrals directly from experimental data is explained in detail. Furthermore the contribution of detector dependent two particle correlations such as Dalitz pairs, γ conversions or double counted tracks to factorial moments is estimated.

1. Introduction

The proposal of Białas and Peschanski to look for intermittency in high energy particle collisions [1] triggered a thorough revival of interest in correlation phenomena both by experimentalists and theorists. Intermittency means that the correlation functions[1] (or more precisely the inclusive density functions) grow as a power law if the distance between their arguments is decreased, i.e. they exhibit a power law like singularity at equal values of their arguments. Such a singularity is characteristic of selfsimilar stochastic processes (cascading phenomena) or phase transitions and is often translated into the language of fractals and fractal dimensions [2,3] in order to stress the selfsimilarity aspect of the phenomenon.

[1]We use the term correlation function as generic name for both, the inclusive densities and the cumulant functions and specify the particular type when needed.

Unfortunately, the direct measurement of correlation functions is limited to orders $q \leq 3$ only and the achieved experimental resolution[2] is not sufficient to study a possible pole structure. Białas and Peschanski circumvented this problem by using factorial moments rather than the correlation functions directly. This trick allows to measure correlations up to order $q = 6$ with bin sizes as small as $\delta y \approx 0.1-0.05$ at the price of losing information about the shape of the correlation functions along the direction of constant coordinate difference. Moreover, by using moments one can study correlations not only in one kinematic variable (e.g. rapidity y or azimuthal angle ϕ), but also in higher dimensional spaces such as (y, ϕ), (y, ϕ, p_t) or any other combined set of variables.

To date, many experimental data (see [4] for recent reviews) confirm the existence of a pole in the correlation functions, especially if one considers two and three dimensional phase space. However, the experimental results may still be improved in at least two ways: first, the factorial moments method may be improved to gather the maximum possible information from a given experiment by extending the integration domain over the correlation functions at a given bin size; secondly, detector dependent biases such as contributions from narrow resonances, Dalitz pairs or γ conversions have to be better understood and accounted for in the interpretation of the data. These subjects are addressed in sections 2 and 3.

2. The Correlation Integrals

We start from the central objects in the study of correlations, the inclusive q-particle (or better q-point) density functions ρ_q, in dependence of some kinematic variables y (for simplicity of notation we have rapidity in mind, but the extension to any other set of variables is straightforward). In multiparticle physics these densities are given by

$$\rho_q(y_1, \ldots, y_q) = \frac{1}{\sigma_I} \frac{d^q \sigma_{\text{incl}}}{dy_1\, dy_2 \ldots dy_q} , \tag{1}$$

with σ_I denoting the inelastic cross section and σ_{incl} the inclusive cross section (we do not distinguish here between particles of different kind, but rather treat them generically as charged pions as most experiments do).

The actual experimental procedure to determine ρ_q would be the following [5,6]: take from the first event which has say n identical particles with rapidities y_i^{ev} $(i = 1, \ldots, n)$ a set of q particles, say $\{y_{i_1}^{\text{ev}}, y_{i_2}^{\text{ev}}, \ldots, y_{i_q}^{\text{ev}}\}$, well considering the order in which they are drawn, and add 1 to the inclusive density at the position[3]

[2]Most experiments measure two particle correlation functions in rapidity by means of histograms with binsize $\delta y \approx 0.3 - 0.5$.

[3]The q-fold product of δ-functions is an idealization valid at infinite resolution capabilities of the detector. In real world experiments they have to be replaced by a product of narrow gaussians with width σ_y, the experimental error on the measurement of the y position of the particle. We do not discuss the finite resolution problem here.

$(y_{i_1}^{ev}, y_{i_2}^{ev}, \ldots, y_{i_q}^{ev})$. This is repeated for all $q!\binom{n}{q} = n(n-1)\ldots(n-q+1) \equiv n^{[q]}$ possible q-tuples and finally the average over the whole event sample $\langle\ \rangle$ is taken:

$$\rho_q(y_1,\ldots,y_q) = \left\langle \sum_{i_1\neq i_2\neq\ldots\neq i_q}^{n} \delta(y_1 - y_{i_1}^{ev})\,\delta(y_2 - y_{i_2}^{ev}) \cdots \delta(y_q - y_{i_q}^{ev}) \right\rangle . \quad (2)$$

From these definitions one can easily deduce that the ρ_q are invariant with respect to permutation of their arguments, i.e. they are symmetric with respect to the axis $y_1 = y_2 = \cdots = y_p$.

Obviously, if we do not care about the positions of the individual particles in a region Δy, but rather count all possible ordered p-tuples of particles within Δy we get

$$\int_\Omega dy_1 \ldots dy_q\, \rho_q(y_1,\ldots,y_q) = \left\langle n^{[q]} \right\rangle_\Omega , \quad (3)$$

where the integration domain Ω is just a q-dimensional cube with edge length Δy. Thus, knowledge of factorial moments $\left\langle n^{[q]} \right\rangle_\Omega$ in all possible domains Ω is equivalent to knowing the entire correlation function ρ_q. At this point we recall that for uncorrelated particles the inclusive density, by definition, factorizes $\rho_q = \rho_1(y_1)\,\rho_1(y_2)\ldots\rho_1(y_q)$ and we get

$$\int_\Omega dy_1 \ldots dy_q\, \rho_1(y_1)\,\rho_1(y_2)\ldots\rho_1(y_q) = \left[\int_{\Delta y} dy\, \rho_1(y)\right]^q = \langle n \rangle_{\Delta y}^q \equiv \bar{N}^q , \quad (4)$$

the factorial moments of the Poisson distribution (≡ purely "statistical fluctuations" [1]).

In the case of intermittency the correlation functions exhibit a power law like singularity along the symmetry axis $y_1 = y_2 = \cdots = y_p$. The key idea of Białas and Peschanski was (apart from proposing the very existence of such power laws in multiparticle production) to measure bin-averaged and normalized factorial moments, which were subsequently recognized as correlation function integrals by Carruthers et al. [7] and Capella et al. [8], linking the intermittency idea to a well established formalism. In this formulation it is obvious that the main advantage of measuring integrals instead of the correlation functions directly is a considerable increase of p-tuple counts in the observation region and consequently a reduction of the statistical errors due to finite event numbers and small multiplicities per event.

To be specific, let us exhibit the translation of the bin-averaged normalized factorial moments as proposed by Białas and Peschanski [1] into the language of correlation functions. Currently two different versions are used for data analysis: the so-called "horizontal" factorial moments

$$F_q^h(\delta y) \equiv \frac{\frac{1}{M}\sum_{m=1}^{M} \left\langle n_m^{[q]} \right\rangle}{\left(\bar{N}/M\right)^q} = \frac{\frac{1}{M}\int_{\Omega_B} \prod_i dy_i\, \rho_q(y_1 \ldots y_q)}{\left(\bar{N}/M\right)^q} \quad (5)$$

and the "vertical" factorial moments

$$F_q^v \equiv \frac{1}{M}\sum_{m=1}^{M}\frac{\langle n_m^{[q]}\rangle}{\langle n_m\rangle^q} = \frac{1}{M}\sum_{m=1}^{M}\frac{\int_{\Omega_m}\prod_i dy_i\,\rho_q(y_1\ldots y_q)}{\int_{\Omega_m}\prod_i dy_i\,\rho_1(y_1)\cdots\rho_1(y_q)}\,. \tag{6}$$

For the computation of these moments a binning of the original region Δy into M subintervals of size δy is introduced and the number of particles in the mth bin, n_m, is counted; the integration domain $\Omega_B = \sum_{m=1}^{M}\Omega_m$ thus consists of M q-dimensional boxes Ω_m of edge length δy. Finally the bin size δy is decreased to the smallest possible values bounded either by the finite resolution of the detector or (in particular for higher orders) by vanishing number of observed q-tuples in the integration domain. Note that in the "vertical" definition the ratio of the integrals over *one* box has to be taken *before* the average over M boxes is done.

The distinction between horizontal and vertical moments becomes important in the case of nonconstant single particle density $\rho_1(y)$ and crucial if ρ_1 itself has a singularity (as e.g. the p_t distribution at $p_t \to 0$); in the latter case the horizontal moments also exhibit a singularity at $\delta p_t \to 0$ even if no correlations are present[4] (i.e. ρ_q factorizes) whereas the vertical moments are constantly equal to 1, due to the "local" normalization.

As an illustration, the integration domain $\sum \Omega_m$ for the two point correlation function $\rho_2(y_1, y_2)$ is shown in figure 1; it consists of M quadratic boxes with edge length δy. A point in the mth box corresponds to a pair (y_1, y_2) whose distance $|y_1 - y_2|$ is smaller than δy *and* which lie in the *same* bin with label m. On the other hand there are many pairs whose distance is less than δy which *do not* lie in the same but in *adjacent* bins (one such point is exhibited as a cross in Fig. 1) and which therefore do not contribute to factorial moments Eq. 5 and 6.

There is no a priori reason for excluding pairs with $|y_1 - y_2| < \delta y$, but which happen to lie in adjacent bins, in an experimental investigation of a possible pole structure at $y_1 = y_2$. Carruthers pointed out [12] that if we wish to extract the maximum possible information from a given experiment we ought to include them in our counting procedure. For a given δy we thus increase our integration domain — and the number of contributing pairs — of the two-particle correlation function by a factor of ≈ 2 and reduce the statistical errors by a factor $\sqrt{2}$). The integration domain would then be a strip[5] Ω_S rather than a chain of boxes as shown in Fig. 2. Thus we name this counting procedure strip or tube correlation integral [12] or shortly correlation integral if there is no danger of confusion.

---

[4]This may lead to confusion in the interpretation of horizontal moment data [9] and we thus prefer the vertical normalization because it removes the contribution from the "shape" of ρ_1.

[5]Note that for a given δy the strip domain of ρ_2 is about twice as large as the boxes domain. To obtain closer correspondence of the strip integrals to the factorial moments Eqs. 5–6 one should compare them at equal integration domain areas rather than at equal δy. See also [10,11,13] for a numerical comparison of these two integration procedures.

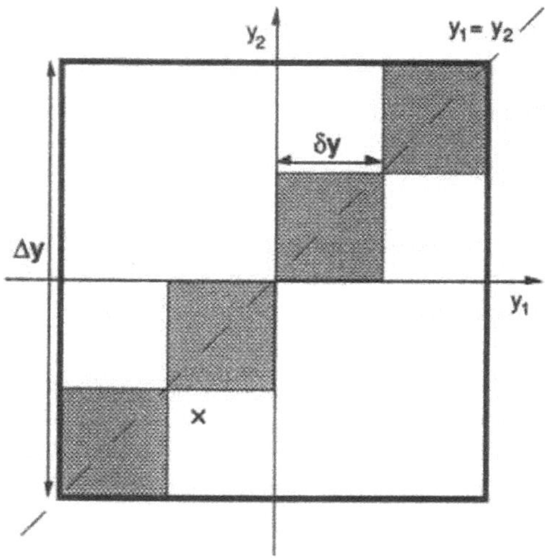

Figure 1: The integration domain $\Omega_B = \sum_m \Omega_m$ of $\rho_2(y_1, y_2)$ for the bin-averaged factorial moments [10] as defined by Białas and Peschanski .

The strip integration pendant to the normalized factorial moments Eq. 5–6 for general order q is [12]

$$C_q^{h,v}(\delta y) \equiv \frac{1}{\mathrm{Norm}_{h,v}} \int_{\Omega_S} \prod_i dy_i \, \rho_q(y_1 \dots y_q) \, . \tag{7}$$

For the horizontal normalization we get in the case of an 1-dimensional kinematic variable y

$$\mathrm{Norm}_h = \bar{N} \times \text{ Strip Volume}/(\Delta y)^q = \bar{N} \left(q\, \Delta y (\delta y)^{q-1} - (q-1)(\delta y)^q \right) / (\Delta y)^q \, , \tag{8}$$

with $\bar{N}$ denoting the average multiplicity in the full Δy region and Ω_S indicating a q-dimensional strip similar to fig. 2. The vertical normalization is simply

$$\mathrm{Norm}_v = \int_{\Omega_S} \prod_i dy_i \, \rho_1(y_1) \cdots \rho_1(y_q) \; ; \tag{9}$$

note, however, that this integral does not factorize into the q-fold product of 1-dimensional integrals over δy as in Eq. 4 because of the strip integration domain Ω_S.

The actual determination of Eq. 7 from experimental data is not as straightforward as one might think. We found that for vertically normalized strip integrals

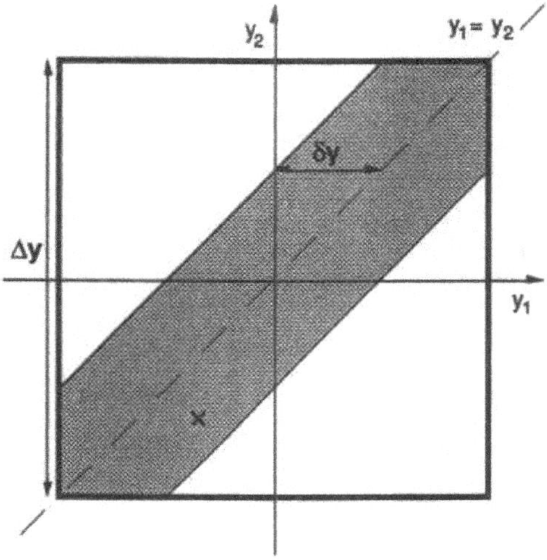

Figure 2: The integration domain of $\rho_2(y_1, y_2)$ for the strip correlation integral as proposed by Carruthers [12].

the following procedure is a very close approximation[6] to Eq. 7: first we choose an analysis domain Δy and a strip width δy for which we want to evaluate $C_q(\delta y)$. Then we take from each event (which has say n particles with positions y_i^{ev} in Δy) a q-tuple $\{y_{i_1}^{ev}, y_{i_2}^{ev}, \ldots, y_{i_q}^{ev}\}$ with $i_1 < i_2 < \cdots < i_q$ and check for each of the $q(q-1)/2$ possible pairs (i_{k_1}, i_{k_2}) if their distance is smaller than δy; if the latter is true for *all* $q(q-1)/2$ pairs we add 1 to the correlation integral. This is repeated for all $\binom{n}{q}$ q-tuples and finally the average over the whole event sample is taken:

$$C_q^{h,v}(\delta y) = \frac{1}{\text{Norm}_{h,v}} \left\langle q! \sum_{i_1 < \cdots < i_q} \prod_{q(q-1)/2} \Theta\left(\delta y - |y_{i_{k_1}}^{ev} - y_{i_{k_2}}^{ev}|\right) \right\rangle , \tag{10}$$

where $\Theta(x)$ is the Heaviside unit step function. Note that Eq. 10 is, apart from the normalization and the event average, equivalent to the correlation integrals proposed by Grassberger and Procaccia [15] and is extensively used in the study of selfsimilarity and fractal behavior in chaotic dynamical systems. Dremin proposed an application of a different kind of correlation integral to single event analyses in high energy physics [16].

[6]The following counting procedure corresponds exactly to Eq. 7 only in the case $q = 2$; for higher orders the exact counting algorithm is not symmetric with change of position of the particles within a q-tuple so that all $q!$ permutations inside the q-tuples actually have to be tested. This increases CPU time by a factor of order $q!$ for an insignificant gain of integration area which we consider

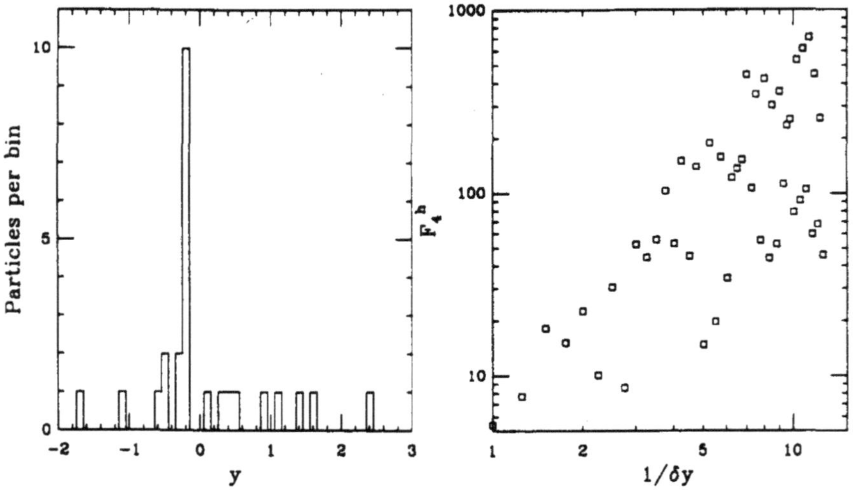

Figure 3: The celebrated NA22 spike event [14,10] and its 4th factorial moment in dependence of δy. Note the strong fluctuations of the factorial moments due to the relative position of the spike to the bin borders.

From the computational point of view Eq. 10 is a hard problem[7]. In fact it is equivalent to the traveling salesman problem and thus NP-complete [17], i.e. the CPU time increases faster than a polynomial with q and n and no faster algorithm can be given. We found that a multidimensional analysis for 160000 UA1 events for $q \leq 5$ takes about 3 CPU hours on an IBM 4381 mainframe[8].

For the vertical normalization Eq. 9 we need to integrate over the q dimensional product of the single particle density representing the case of completely uncorrelated particles in each event. In order to evaluate this integral directly from data we suggest to generate uncorrelated events by mixing particles from different (obviously uncorrelated) events according to following procedure: make a pool of the positions y of as many particles as possible (limited by computer memory) to represent the density $\rho_1(y)$. Then choose for each normalization "event" a multiplicity n according to a poissonian[9] with average multiplicity $\bar{N}_{\mathrm{Norm}}$, and take

not worthwhile. The differences between the exact and the approximate counting procedure will be discussed in an forthcoming publication.

[7]if y is multidimensional. For 1 dimensional analyses an algorithm with CPU time consumption equivalent to the factorial moment algorithm can be given, using the fact that the particles can be ordered along a line.

[8]An efficient procedure for treatment of events with a considerable higher number of particles (e.g. heavy ion collisions) will be given elsewhere.

[9]Any other multiplicity distribution induces (long range) correlations in the normalization events and is not equivalent to Eq. 9.

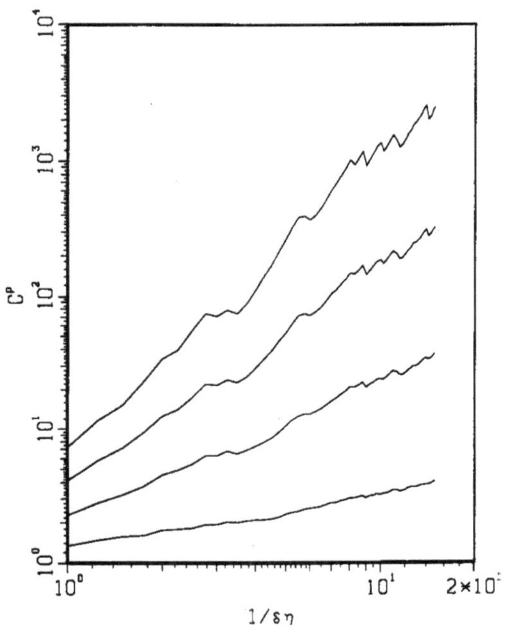

Figure 4: The strip correlation integrals Eq. 10 for q = 2–5 for the NA22 spike event (Fig. 3). The strip integrals make a single event analysis meaningful.

randomly n particles with equal probability from the pool, imposing the constraint that all particles should stem from different events. Finally use the same counting procedure as for the numerator of Eq. 10 for these uncorrelated events.

Unfortunately, the uncorrelated "events" used for the normalization are generally more vulnerable to the empty bin effect [18] than the correlated events (whose multiplicity distribution is in general broader than the poissonian). This is why one should choose $\bar{N}_{\text{Norm}}$ considerably larger than the average multiplicity $\bar{N}$ in Δy of the real events. Of course one has then to rescale the normalization by the factor $(\bar{N}/\bar{N}_{\text{Norm}})^q$.

Apart from higher computational complexity the strip approach has many practical advantages over the usual measured factorial moments. The most important one is that this algorithm does not introduce an artificial partition into bins (or boxes in higher dimensional analysis) and therefore does not lead to ambiguities in the definition of boxes (which are sometimes chosen as squares, rectangles, sectors of circles,...); moreover, artificial fluctuations stemming from the accidental splitting of clusters by bin borders are removed. As an illustration we show in Fig. 3 the

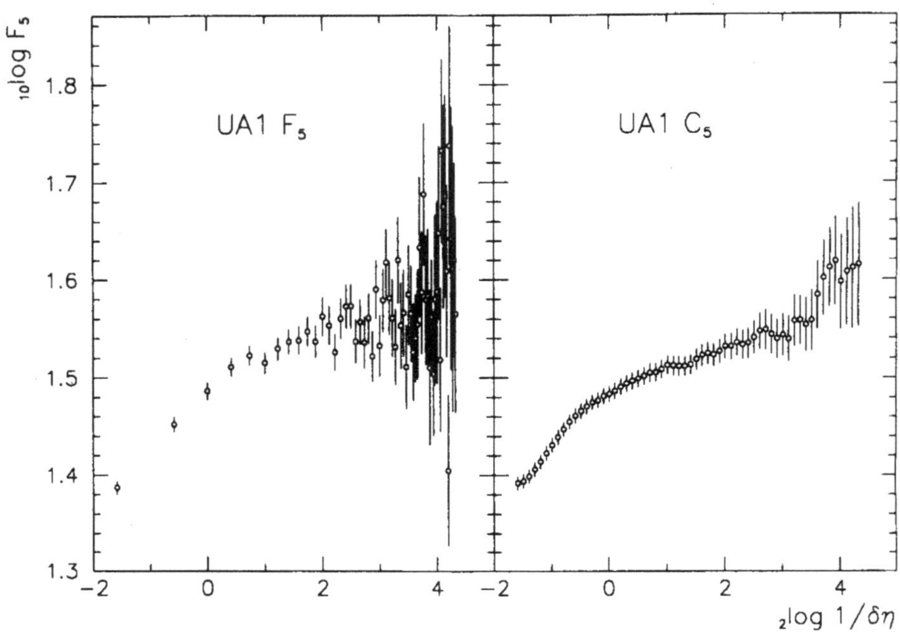

Figure 5: The strip correlation integral C_5 compared to the factorial moment F_5 for one dimensional UA1 data [19]. In the strip version the errors are significantly reduced.

famous NA22 spike event [14] and its factorial moment F_4^h [10] computed according to Eq. 5 (no error bars are shown because it is just one event). Note the strong fluctuations of F_4^h with decreasing bin size, depending on whether the prominent spike lies entirely in one bin or is split across two. Contrastingly we show $C_{2-5}^h(\delta y)$ for this event in Fig. 4. The difference in stability is striking.

Another advantage is the reduction of the statistical error, which is proportional to the square root of the number of contributing q-tuples per event at given δy. Since the strip volume is substantially larger than the M-box volume, the variances should decrease proportionally; this effect may be crucial to extract information for higher orders $q \geq 3$ in ongoing 3 dimensional experimental investigations, which might not be accessible by the usual factorial moment method.

We illustrate this point in Fig. 5 with a comparison of F_5^v with C_5^v in a one dimensional analysis (pseudorapidity η) of UA1 data (see [19] for more details on UA1 data). The strip integral provides clearly more statistical accuracy for high moments at small bin sizes.

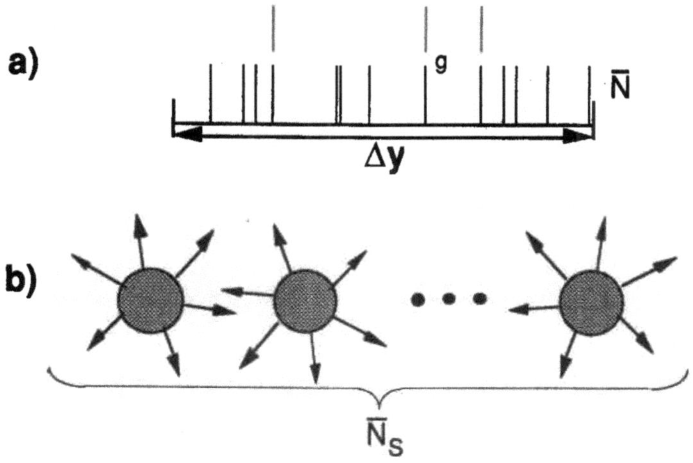

Figure 6: a) The split track model: $\bar{N}$ tracks are generated randomly; with a small probability g each track is split into two parts at the same location. b) The generalization: $\bar{N}_S$ sources (clusters, clans) are produced independently to form one event.

Next we want to mention a minor — but in practice annoying — point: the factorial moment approach requires a division of Δy into an integral number M of bins which means that at best one can sample the correlation functions with a sequence $\delta y/\Delta y = 1, 1/2, 1/3, ...$, leading to huge gaps for large δy and a clumping of data at small δy (see Fig. 5). This is obviously not an optimal representation of the correlation functions. We stress the fact that the strip integration does not restrict the choice of a sequence in δy in any way (which is then conveniently chosen so that the data points appear equidistant in a log–log plot).

Finally we point out that the above (vertical) counting procedure is closely related to the procedure to measure Bose-Einstein correlations, if one restricts the analysis to particles of same charge; in fact, if one chooses y as three or four momentum, the expression Eq. 10 for $q = 2$ is nothing but the integrated version[10] of the well known Bose-Einstein curve apart from different normalization procedures. We strongly suggest a unified study of Bose-Einstein correlations and the intermittency phenomenon...

3. What is the effect of pairs with small distance in δy?

As a warning and tutorial example we discuss here a simple model to see

[10] A differential version of Eq. 10 will be discussed elsewhere.

qualitatively the effect of known narrow two-particle correlations such as Dalitz pairs, γ conversion pairs or simply misidentification of track segments which leads to a double counting of the same particle. All these effects depend strongly on the particular detector setup and track reconstruction procedure so that comparisons between experiments might be questionable.

In order to estimate the size of the contribution of such narrow correlations we start with the simplest possible model which can be extended later to include characteristics of known real world particle decays. We assume that for each event we have $\bar{N}$ uncorrelated particles in Δy following a Poissonian distribution. For that case all normalized factorial moments (both the box and strip versions) Eqs. 5–10 are equal to 1, independent of the bin size δy. We then simulate narrow two-particle correlations by adding a 2nd particle (daughter) with a small probability $g \ll 1$ to each particle (parent) from the previous step at the same[11] position as the parent (see Fig. 6a.). The "coupling strength" g represents the probability that one particle splits into two pieces in a very narrow phase space region, be it a double counted track, a Dalitz or γ conversion pair or any other narrow two particle resonance, and can be estimated by the experimentator for his particular detector set up.

For such simple models one can easily determine their generating functions, and via appropriate derivatives one can then find the normalized cumulants K_q, defined e.g. in [20]. We get for the above case:

$$K_2(\delta y) = \frac{2g}{(1+g)^2} \frac{1}{\bar{N}} \frac{\Delta y}{\delta y} \tag{11}$$

and all higher order cumulants vanish (we introduced only two-particle correlations!). From Eq. 11 we observe 1) that K_2 follows a power law with decreasing bin size with slope 1 (intermittency??), 2) that the strength g of the two-particle correlations does not affect the intrinsic slope 1 but merely the intercept of K_2 at $\delta y = \Delta y$, and 3) that the intercept decreases with $1/\bar{N}$ (see Fig. 7a).

The corresponding normalized factorial moments are then [20]:

$$\begin{aligned} F_2 &= 1 + K_2 \\ F_3 &= 1 + 3K_2 \\ F_4 &= 1 + 6K_2 + 3(K_2)^2 \\ \vdots &\quad \vdots \end{aligned} \tag{12}$$

with K_2 from Eq. 11. *This may look like intermittency*, but on closer inspection one recognizes a *bending upwards* of the factorial moments in the log–log plot (see

[11]This means that the difference in position is smaller than the minimum bin size $\delta y_{\rm min}$ in an analysis. The above model can easily be extended to include deviations σ_0 of the daughter position from the parent position; result for the cumulants is the same as Eq. 11 in the region $\delta y > \sigma_0$, but shows a bending to $K_2 =$ constant in the region $\delta y < \sigma_0$ as expected. The corresponding equations are not shown here because they are more involved but not more instructive.

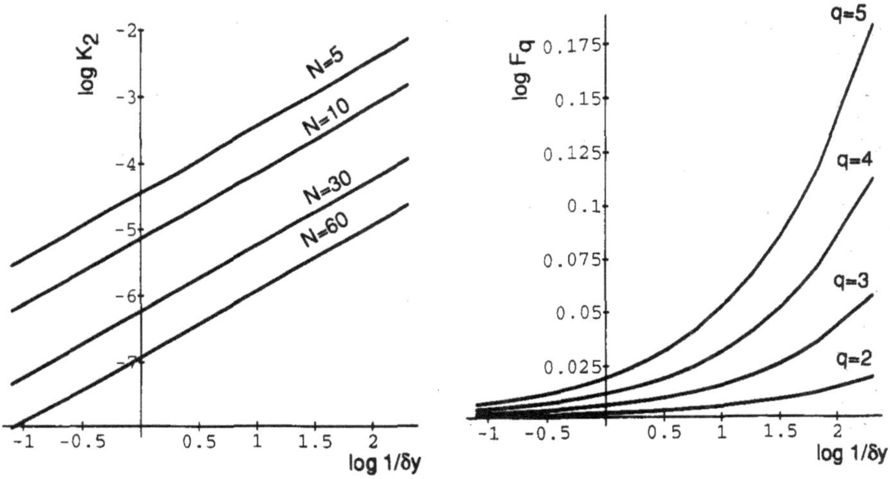

Figure 7: a) The normalized factorial cumulant K_2 for the split track model with very small $g = 0.01$ and different $\bar{N} = 5, 10, 30, 60$; the slopes (=1) do not depend on $\bar{N}$. b) The corresponding factorial moments F_{2-5} (shown for $\bar{N} = 10$) bend upwards in the log-log plot and may fake an intermittency signal.

Fig. 7b for an example with $g = 0.01$, $\bar{N} = 10$ and $\Delta y = 3$). Note that the factorial moments of *all* orders increase, in spite of the fact that we introduced only *two*-particle correlations. Moreover, if one approximates the curved rise of the F_q in Fig. 7b with straight lines (i.e. one expands in the way Fialkowski showed at this workshop [21]) one finds that the "slopes" decrease approximately with $1/\bar{N}$ if the average multiplicity $\bar{N}$ is decreased. Remember that such a behavior is actually observed in experiments [4].

Please note, that we do *not* suggest that all observed intermittency data can be reduced to such two-particle correlations (see e.g. the cumulant decomposition in [20] for a proof that this is indeed not the case), but we stress that their contribution ought to be carefully studied.

Finally we want to show the relevance of the use of (factorial) cumulants in the analysis of intermittency by explaining the generalization of the above model. Suppose an event can be decomposed into $\bar{N}_S$ identical "clusters" or sources (see Fig. 6b), where a source is defined by nonvanishing cumulant correlation functions, and suppose that these sources are produced in an uncorrelated way (i.e. their production mechanism is a poisson process). The generating function G_{tot} of the multiplicity distribution in any bin can then be written as

$$G_{\text{tot}}(z) = G_{\text{poisson}}\left(G_{\text{source}}(z)\right) , \tag{13}$$

with $G_{\rm source}(z)$ denoting the generating function of a single source. The (factorial) cumulants $K_q = (d/dz)^q \log G(z)|_{z=1}$ are then proportional to the factorial moments of a single source:

$$K_{q,\rm tot} = \frac{1}{\bar{N}_S^{q-1}} F_{q,\rm source} \; . \tag{14}$$

This means that if a single source has a fractal structure (i.e. a power law dependence of $F_{q,\rm source}$ on the bin size) the cumulants of the composed events show the same power law as the single source. In other words, if one plots the cumulants $K_{q,\rm tot}$ against δy in an log-log plot one actually sees the factorial moment of *one* source, shifted downwards by a factor $(q-1)\log \bar{N}_S$, but retaining the slope of $F_{q,\rm source}$. On the other hand, factorial moments of superimposed fractal sources show a *bending upwards* in the log-log plot by virtue of Eqs. 12.

The two parametric negative binomial distribution $P_{\rm NBD}(n; \bar{n}(\delta y), k(\delta y))$ is a special case of Eq. 13 with $G_{\rm source}(z)$ being the generating function of a logarithmic distribution, representing so-called clans [22] (see e.g. [23] for a review). In this case the cumulants extract the δy dependence of the clans according to

$$K_{q,\rm NBD} = \frac{1}{\bar{c}^{q-1}} F_{q,\rm clan} \; , \tag{15}$$

with $\bar{c} = k \ln(1 + \bar{n}/k)$ as the average number of clans per event.

4. Conclusions

- We suggested that the strip integrals represent a considerable improvement over the usual bin-averaged factorial moment method used in the analysis of intermittency and explained in detail an algorithm to actually perform these integrals with experimental data.

- The strip integrals have a larger integration domain at a given distance δy than the factorial moment counterparts, and thus reduce the squared statistical errors proportionally. This fact becomes crucial in the analysis of higher correlation functions in three dimensional phase space.

- No artificial choice of binning is required for the strip integrals. Therefore clusters are not split at bin borders and spurious fluctuations of the moments are avoided. This fact makes single event analyses meaningful.

- Narrow two-particle correlations such as Dalitz pairs, γ conversions or double counting of tracks (track splitting) may increase the slopes of factorial moments in a detector dependent way and even fake an intermittency signal (even for very small probability of occurrence!).

- In a simple model we tried to estimate their contribution and find qualitatively that they cause the factorial moments to bend upwards in the log-log plot.
- We showed one more example of the usefulness of cumulants for the analysis of intermittency: the cumulants of the total events are proportional to the factorial moments of a single source (cluster, clan), if the latter are independently superimposed.

Acknowledgement: I am indebted to P. Carruthers not only for drawing my attention to the strip integrals but also for the uncountable discussions, lessons and encouragements I received from him. Much of the actual implementation of the strip integrals and in particular of their normalization is due to intensive discussions with B. Buschbeck; my contribution to this subject can not be separated from hers. There is no thought in this report which was not thoroughly discussed with and (mostly) approved by H. Eggers. Finally I want to thank the organizers of the Ringberg workshop for providing such stimulating and inspiring surroundings and for their patience. This work was supported by the US Department of Energy, Divisions of High Energy and Nuclear Physics.

References

[1] A. Bialas and R. Peschanski, *Nucl. Phys.* **B273** (1986) 703; *Nucl. Phys.* **B308** (1988) 857.

[2] P. Grassberger, *Phys. Lett.* **97A** (1983) 227; H.G.E. Hentschel and I. Procaccia, *Physica* **8D** (1983) 435.

[3] P. Lipa and B. Buschbeck, *Phys. Lett.* **B223** (1989) 465.

[4] B. Buschbeck, Invited review at the XXV Rencentres de Moriond, Les Arcs, Savoie, France, March 17-23, 1991; A. Białas, Summary talk at the Ringberg Workshop on Multiparticle Production, Fluctuations and Fractal Structue, Ringberg Castle, Germany, June 25-28, 1991.

[5] Y. L. Klimontovich, *"The Statistical Theory of Non-Equilibrium Processes in a Plasma"*, (Pergamon, Oxford, 1967).

[6] P. Carruthers, *Phys. Rev.* **A43** (1991) 2622.

[7] P. Carruthers and I. Sarcevic, *Phys. Rev. Lett.* **63** (1989) 1562.

[8] A. Capella, K. Fialkowski and A. Krzywicki, *Phys. Lett.* **B230** (1989) 149.

[9] H. J. Behrend et al. (CELLO Coll.), *Phys. Lett.* **B256** (1991) 97.

[10] H.C. Eggers, PhD Thesis, University of Arizona, Tucson, USA (1991).

[11] P. Carruthers, H.C. Eggers, Qiang Gao and I. Sarcevic, *Int. J. Mod. Phys.* **A6** (1991) 3031.

[12] P. Carruthers, *"Galaxy Correlations, Counts and Moments"*, University of Arizona preprint AZPH-TH/91-08, (to appear in *Astrophysical J.* Oct. 10, 1991).

[13] P. Carruthers, *"From Photons to Hadrons to Galaxies – How to Analyze the Texture of Matter Distributions"*, University of Arizona preprint AZPH-TH/91-34, Lectures presented at the Zakopane Summer School, June 5-15, (1991).

[14] M. Adamus et al. (NA22 Coll.), *Phys. Lett.* **B185** (1987) 200.

[15] P. Grassberger and I. Procaccia, *Phys. Rev. Lett.* **50** (1983) 346; *Physica* **D9** (1983) 189.

[16] I. M. Dremin, *Mod. Phys. Lett.* **A3** (1988) 1333; *Mod. Phys. Lett.* **A4** (1989) 1072.

[17] J. E. Hopcroft and J. D. Ullman, *Introduction to Automata Theory, Languages and Computation*, (Addison-Wesley Publishing Company, 1979).

[18] P. Lipa, H. C. Eggers, F. Botterweck and M. Charlet, University of Arizona preprint AZPH-TH/91/14.

[19] C. Albajar et al., (UA1 Coll.), *Nucl. Phys.* **B345** (1990) 1.

[20] P. Carruthers, H. C. Eggers and I. Sarcevic, *Phys. Lett* **B254** (1991) 258.

[21] K. Fialkowski, Talk at the Ringberg Workshop on Multiparticle Production, Fluctuations and Fractal Structue, Ringberg Castle, Germany, June 25-28, 1991.

[22] A. Giovannini and L. Van Hove, *Z. Phys.* **C30** (1986) 391.

[23] P. Carruthers and C.C. Shih, *Int. J. Mod. Phys.* **A2** (1987) 1447.

RECENT RESULTS FROM UA1

UA1-Minimum Bias Collaboration

P. LIPA*, N. NEUMEISTER and D. WESELKA

presented by B. Buschbeck

Institut für Hochenergiephysik der Österr. Akademie der Wissenschaften, Nikolsdorfergasse 18, 1050 Wien, Austria

Abstract

A 2-dimensional intermittency analysis at collider energies shows a much stronger rise of the factorial moments than a 1-dimensional analysis. An almost linear rise of log(FM) with log(1/l) is observed in the whole region which has been analysed. It indicates that intermittency occurs in the higher dimensional phase space.

The higher order Bose Einstein correlations show a steep rise with decreasing Q^2 incompatible with a Gaussian parametrization. It is argued that the correlations from same charged pions may cause a singularity in the correlation function.

This report covers two contributions from our recent analysis of $\sqrt{s}$ = 630 GeV $\bar{p}p$ collisions. In both cases the data sample consists of about 160 000 non single diffractive minimum bias events observed in the UA1 central detector.

1. Preliminary data from a 2-dimensional intermittency analysis

Fig. 1 shows the number of recorded tracks in dependence of η (pseudorapidity) and ϕ (azimuthal angle). The UA1 central detector consists of several pictorial drift chambers [1]. The dips at $\eta = 0$ and at $\phi = \pm\pi$ and 0 have their origin in the diminished detector acceptances in these regions because there the tracks are approximately parallel to the wires and to the magnetic field. On the other hand the regions arround $\phi = \pm\pi/2$ and $-3 \leq \eta \leq 3$ are regions of good

**Now at the University of Arizona, Tucson AZ 85721, USA*

acceptances ($\geq$ 94%) and the observed flat valleys at $\eta \approx 0$ arise because we have choosen η instead of y (rapidity). A 2-dimensional (2-d) intermittency analysis in $\eta - \phi$ has to give special attention to the influence of the landscape observed in Fig. 1.

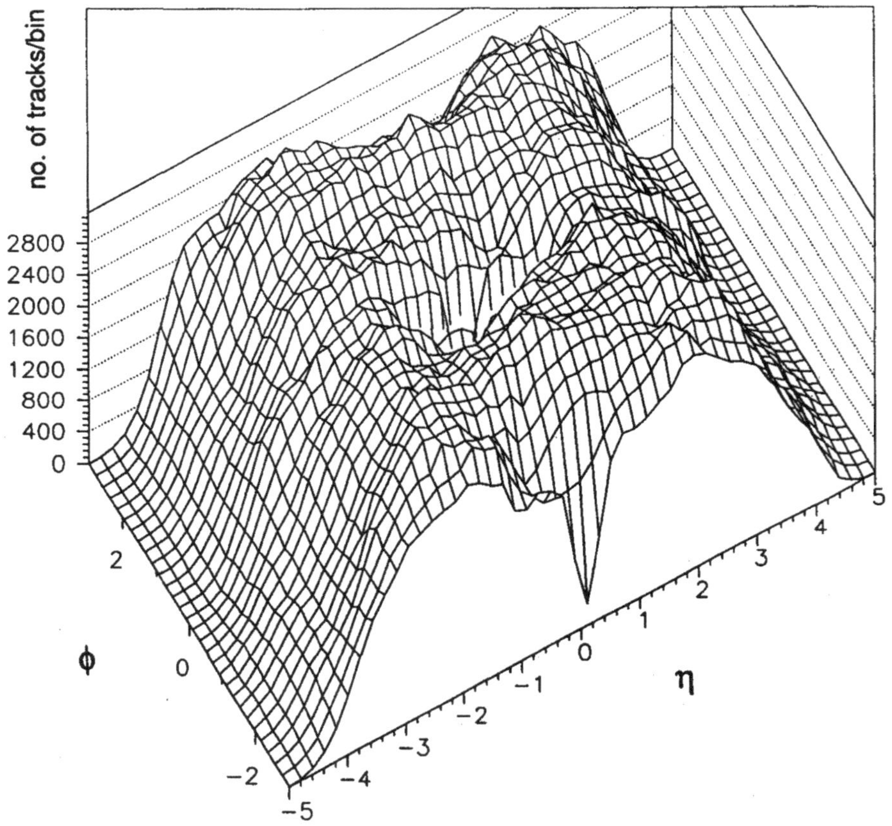

Fig. 1: Number of recorded tracks in the UA1 central detector as a function of η and ϕ.

Therefore we compare in Fig. 2 plots of factorial moments $\langle F^i \rangle$ (Ref. 2) which are obtained with the "horizontal" analysis (1) with those from a "vertical" analysis (2). It has been shown [3] that the latter corrects for non-flat distributions to a large extent. The $\langle F^i \rangle$ from a horizontal analysis are calculated as follows:

$$\langle F^i \rangle_{horiz} = \left\langle \frac{1}{M} \sum_{m=1}^{M} \left(\frac{M}{<n>} \right)^i \cdot (n_m(n_m - 1) \ldots (n_m - i + 1)) \right\rangle \qquad (1)$$

where $\langle\rangle$ denote the averages over the event-sample, n_m is the individual multiplicity in bin m in one event and M is the number of subdivisions of the whole region of investigation $\Delta\eta \cdot \Delta\phi$ with a mean multiplicity $\langle n \rangle$.

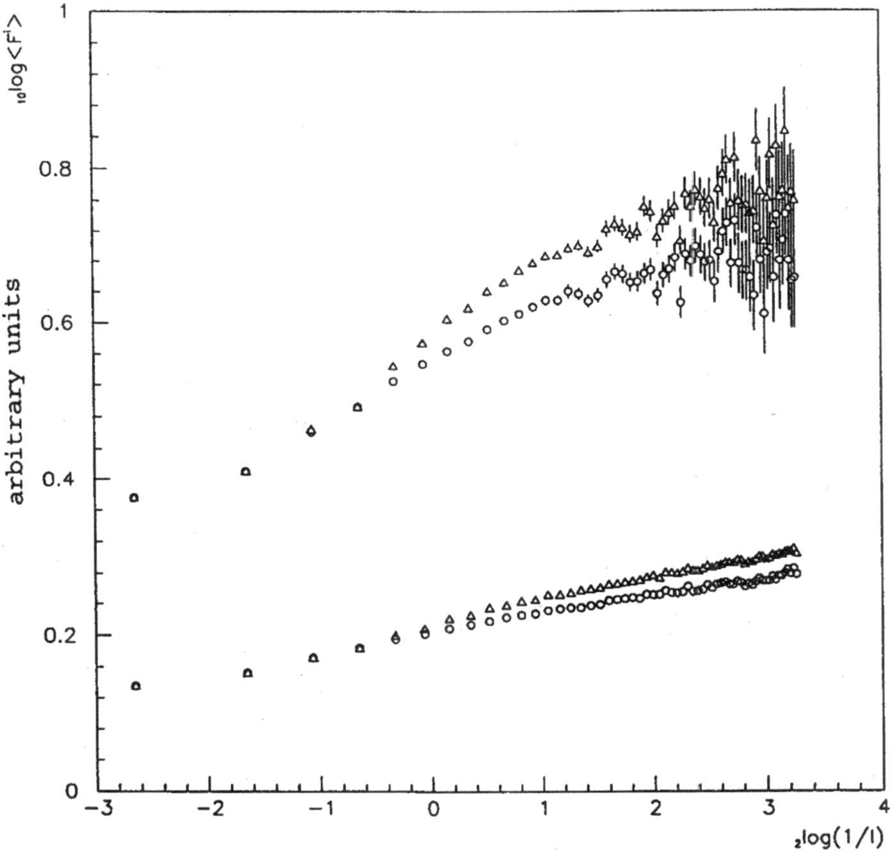

Fig. 2: A comparison of "horizontal" $\langle F^i \rangle$'s (Eqn.(1)), triangles with "vertical" ones (Eqn.(2)), circles, for $i = 2, 3$. The intervall length $\delta\eta, \delta\phi$ is given by l.

In the following, the length of the subintervalls $\delta\eta = \delta\phi$ will be denoted by l, hence their number in the 2-d analysis will be given by $M = \Delta\eta \cdot \Delta\phi / l^2$.

In our case the region has been choosen:

$$\begin{aligned} \eta &= [-\pi, \pi], & \Delta\eta &= 6.28 \\ \phi &= [-\pi, \pi], & \Delta\phi &= 6.28 \end{aligned}$$

The vertical $\langle F^i \rangle$'s are calculated according to

$$\langle F^i \rangle_{vert} = \frac{1}{M} \sum_{m=1}^{M} \frac{\langle n_m(n_m - 1)\dots(n_m - i + 1)\rangle}{\langle n_m \rangle^i} \tag{2}$$

Fig. 2 shows that both representations give similar results:

- for $l \geq 1$ almost no difference is visible between $\langle F^i \rangle_{horiz}$ and $\langle F^i \rangle_{vert}$
- no saturation of $\langle F^2 \rangle$ is observed in both cases in the region $0.1 \leq l \leq 1.0$.
- the differences of slopes in the region $0.1 \leq l \leq 1.0$ are only 20-30% for $i = 2 - 5$.

An important remark should be added here: Our region $\Delta\eta, \Delta\phi$ is quite large and one may ask if the mean values over bins in very different regions of η do make sense. It has been shown in e^+e^- reactions, that the $\langle F^i \rangle$ values obtained in a 1-dimensional analysis vary considerably with rapidity [4]. We want to stress here, that contrary to e^+e^- reactions and also contrary to low energy hadronic reactions [5] we only observe weak dependences on (pseudo) rapidity. In the central interval $\Delta\eta \simeq 6$ a fully developed plateau exists at collider energies and translational invariance is fulfilled to a good approximation. This is demonstrated for the 1-d slope parameters in Fig. 3 taken from [6]. Similarly it is shown in Ref. 7 that the $\langle F^2 \rangle$ values vary only by $\leq 10\%$ in the region considered. Our preliminary 2-dimensional studies in different η-regions show similar results [8].

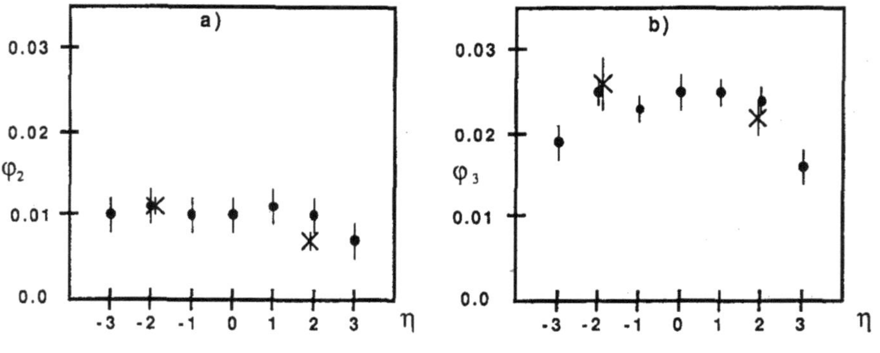

Fig. 3: Dependence of the slope parameters φ_i on pseudorapidity for $i = 2, 3$. The full circles are displayed in the centre of each analysed pseudorapidity interval with an original width of $\Delta\eta = 2$. The crosses are obtained in the region $1.6 \leq |\eta| \leq 2.4$ where the detector acceptance is almost 1.

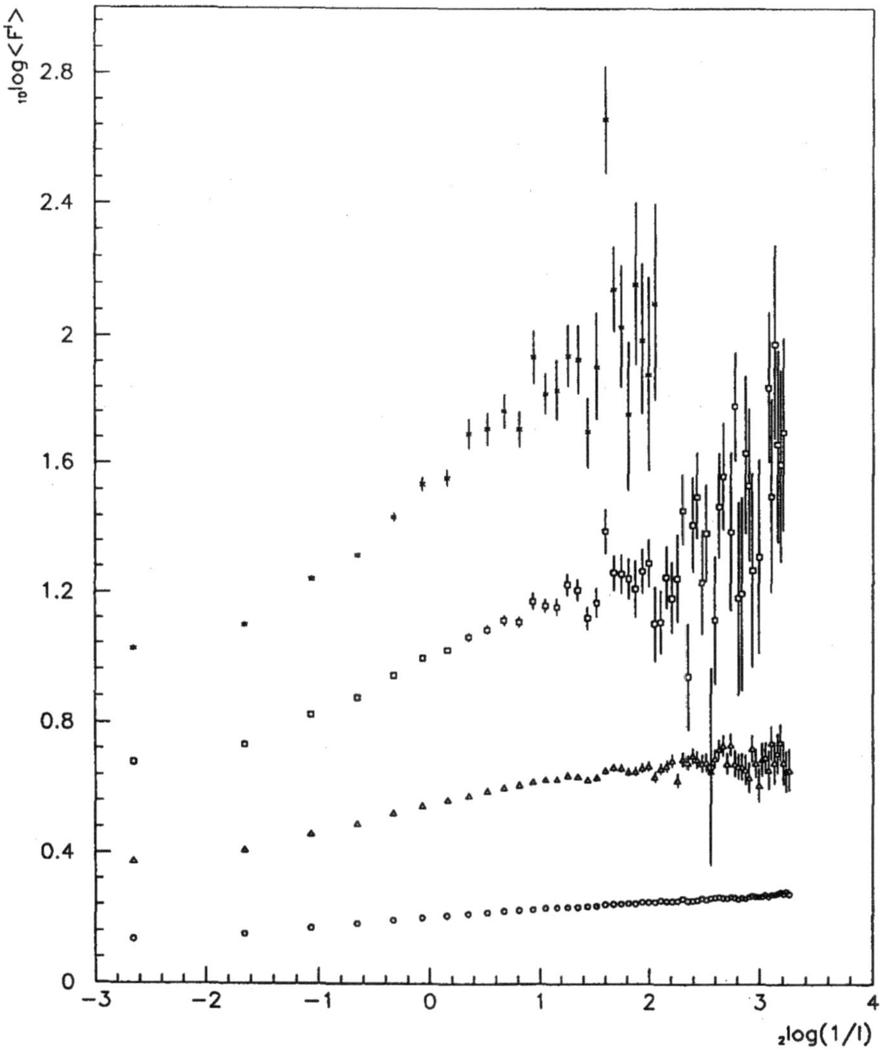

Fig. 4: The variation of the factorial moments with the interval length $\delta\eta = \delta\phi = l$ in a 2-dimensional analysis. Due to the empty bin effect [9] only $\langle F^2\rangle$ and $\langle F^3\rangle$ can be measured reliably ($< 10\%$ error) for $l \geq 0.1$. $\langle F^4\rangle$ needs $l \geq 0.2$ and $\langle F^5\rangle$ needs $l \geq 0.45$.

Fig. 4 shows 2-dimensional plots for $i = 2 - 5$ obtained with the vertical normalization (2) and Fig. 5 shows a comparison of $\langle F^2 \rangle_{vert}$ between 1-d and 2-d plots.

We observe:

1. The strong bending around $\delta\eta = 1$ which shows up in the 1-dimensional plots has almost vanished in the 2-dimensional plots.

2. The slope parameters in $0.1 \leq l \leq 1$ of the 2-dimensional plots are about a factor of 8 higher than those of the 1-dimensional ones. This is also shown in Table 1.

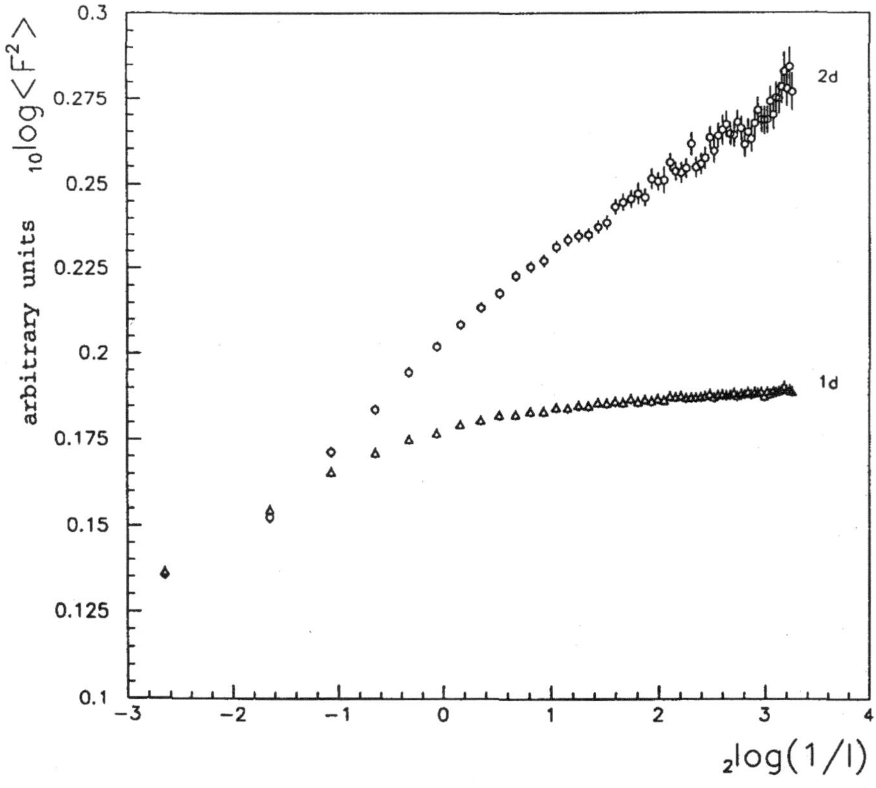

Fig. 5: $\langle F^2 \rangle$ from the 2-dimensional analysis compared to $\langle F_2 \rangle$ from the 1-dimensional analysis. In the region $l < 1$ the slopes differ by a factor of 7.9.

Both observations are in agreement with the findings in other reactions [10–14] and the predictions of W. Ochs [15] and A. Bialas and J. Seixas [16] based on models with intermittency in the 3-dimensional phase space.

Table 1 gives a comparison of 1-dimensional slope parameters with 2-dimensional ones in 2 different regions of l:

(a) for $0.1 \leq l \leq 1$

(b) for $1.0 \leq l \leq 3.2$

Table 1: Comparison of the slope parameters φ_i of an 1-dimensional intermittency analysis with a 2-dimensional one (both vertical). The region of analysis is $\eta = [-\pi, \pi], \phi = [-\pi, \pi]$.

	REGION of FIT	φ_i 1 - d	φ_i 2- d	$R_1^{(i)} = \frac{\varphi_i(2-d)}{\varphi_i(1-d)}$
(a)	$0.1 \leq l \leq 1.0$	$\varphi_2 = 0.009 \pm 0.009$	0.071 ± 0.001	7.9
		$\varphi_3 = 0.022 \pm 0.0014$	0.166 ± 0.007	7.6
		$\varphi_4 = 0.042 \pm 0.003$	0.401 ± 0.030	9.55
		$\varphi_5 = 0.067 \pm 0.006$		
(b)	$1.0 \leq l \leq 3.2$	$\varphi_2 = 0.046 \pm 0.004$	0.104 ± 0.004	2.26
		$\varphi_3 = 0.109 \pm 0.006$	0.283 ± 0.007	2.60
		$\varphi_4 = 0.184 \pm 0.009$	0.523 ± 0.012	2.84
		$\varphi_5 = 0.270 \pm 0.014$	0.810 ± 0.025	3.00
	Linearity tests:	$R_2^{(2)} = 5.11$	$R_2^{(2)} = 1.46$	
	$R_2^{(i)} = \frac{\varphi_i \text{ (region (a))}}{\varphi_i \text{ (region (b))}}$	$R_2^{(3)} = 4.95$	$R_2^{(3)} = 1.70$	
		$R_2^{(4)} = 4.38$	$R_2^{(4)} = 1.30$	
		$R_2^{(5)} = 4.03$		

Due to the large bending in the 1-dimensional plots at $\delta\eta \simeq 1$ which is absent in the 2-dimensional plots, very different slope ratios (last column) are obtained in these 2 regions.

We have estimated our present uncertaincies due to inclusion of γ-conversions in the beam pipe and Dalitz pairs on one side and loss of nearby tracks in the phase space on the other side. Both effects cause systematic errors. A preliminary estimation of possible systematic uncertainties of the 2-d slope parameters in $0.1 \leq l \leq 1$ gives $\pm$ (20 - 30)%.

Fig. 6 shows a comparison of the 2-d data with the LUND model PYTHIA 4.8 (horizontal analysis) [17]. Similar to the 1-dimensional analysis [6] but even more pronounced - discrepancies occure at small l. We want to remark that this Monte Carlo includes resonance production and (multiple) jet production with ratios in agreement with the experiment. Only the Bose Einstein correlations have been introduced in a rather crude way [18]. Therefore it is desirable to study the influence of the latter effect using the data only [19].

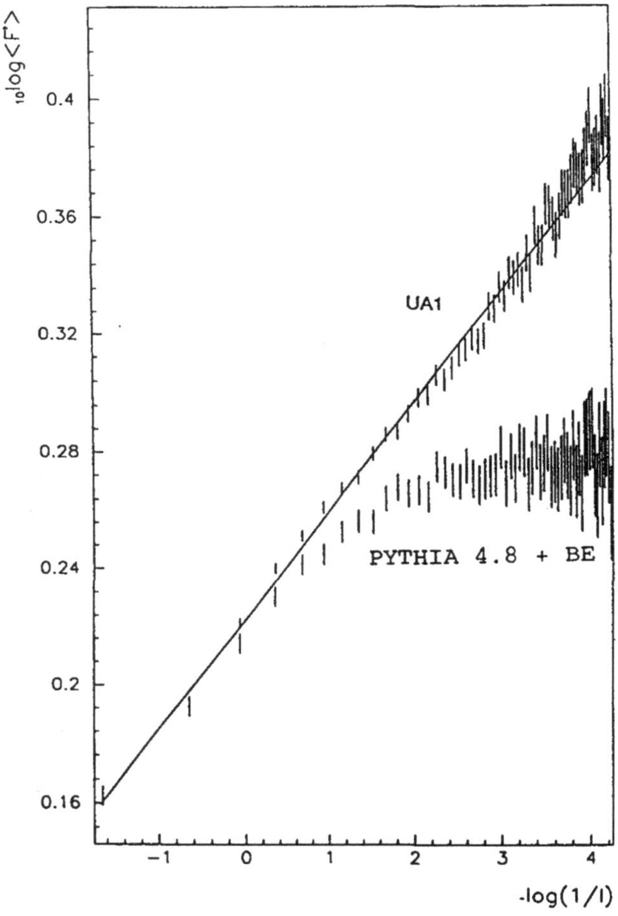

Fig. 6: Comparison of the 2-dimensional $\langle F^2\rangle$ of UA1 data and the PYTHIA+BE (with Bose-Einstein effect included) Monte Carlo model (data points below) [17]. Here the region $\eta = [-1.5, 1.5], \phi = [-\pi, \pi]$ has been chosen.

2. Higher Order Bose-Einstein Correlations and the Possible Influence on Intermittency

Following a suggestion of Ref. 20, the Bose-Einstein correlations up to the 5th order have been measured and compared to theoretical predictions in [21].

The correlation functions

$$R_i\left(Q_{i\pi}^2\right) = N_i\left(Q_{i\pi}^2\right)/Ni^{BG}\left(Q_{i\pi}^2\right) \tag{3}$$

are the ratios of the distributions of like-charged pion-pairs, triplets, ... i-tuples $N_i\left(Q_{i\pi}^2\right)$ with a distribution of reference pairs, triplets ... i-tuples $Ni^{BG}\left(Q_{i\pi}^2\right)$ obtained by event mixing. The analysing variable $Q_{i\pi}^2$ is defined as follows:

$$Q_{i\pi}^2 = q_{12}^2 + q_{13}^2 + \dots q_{(i-1)i}^2 \tag{4}$$

with $q_{ij}^2 = -(p_i - p_j)^2$ where p_i and p_j are the 4-momenta of particle i and j.

The measured correlation functions are steeper than expected. They cannot be fitted by functions of the Gaussian type, neither by the simple functions

$$R_i = 1 + \lambda_i \cdot \exp\left(-r_i^2 Q_{i\pi}^2\right) \tag{5}$$

nor by the more complicated expressions from the quantum optical approach [20] with quadratic terms $Q_{i\pi}^2$:

$$
\begin{aligned}
N^{(2)}/N_2^{BG} = 1 &+ 2\lambda(1-\lambda)e^{-r^2Q_{2\pi}^2} + \lambda^2 e^{-2r^2Q_{2\pi}^2} && (6)\\
N^{(3)}/N_3^{BG} = 1 &+ 6\lambda(1-\lambda)e^{-\frac{1}{3}r^2Q_{3\pi}^2} + 3\lambda^2(3-2\lambda)e^{-\frac{2}{3}r^2Q_{3\pi}^2} + 2\lambda^3 e^{-r^2Q_{3\pi}^2} && (7)\\
N^{(4)}/N_4^{BG} = 1 &+ 12\lambda(1-\lambda)e^{-\frac{1}{6}r^2Q_{4\pi}^2} + 6\lambda^2(7-8\lambda+2\lambda^2)e^{-\frac{1}{3}r^2Q_{4\pi}^2} && \\
&+ 4\lambda^3(11-9\lambda)e^{-\frac{1}{2}r^2Q_{4\pi}^2} + 9\lambda^4 e^{-\frac{2}{3}r^2Q_{4\pi}^2} && (8)\\
N^{(5)}/N_5^{BG} = 1 &+ 20\lambda(1-\lambda)e^{-\frac{1}{10}r^2Q_{5\pi}^2} + 10\lambda^2(13-18\lambda+6\lambda^2)e^{-\frac{1}{5}r^2Q_{5\pi}^2} && \\
&+ 20\lambda^3(16-21\lambda+6\lambda^2)e^{-\frac{3}{10}r^2Q_{5\pi}^2} && \\
&+ 5\lambda^4(53-44\lambda)e^{-\frac{2}{5}r^2Q_{5\pi}^2} + 44\lambda^5 e^{-\frac{1}{2}r^2Q_{5\pi}^2} && (9)
\end{aligned}
$$

Contrarily, simple exponentials

$$R_i = 1 + \lambda_i \exp\left(-r_i Q_{i\pi}\right) \tag{10}$$

fit the data well and equally good fits are obtained with the expressions (6) to (9) where the Gaussian form is replaced in all orders by an exponential form.

This is shown in Fig. 7. Since we loose pairs in the detector due to the limited resolution, the measured correlation functions at the smallest Q^2 have to be regarded as lower limits: A power law behaviour cannot be excluded. The contributions of the same charged pairs and opposite charged pairs to the

2-particle-correlation function dependent on $Q^2 = q_{12}^2$ is displayed in Fig. 8. It shows that a possible singularity could arise from the same charged pairs only. This leads to the suspicion that in very small phase space regions the Bose-Einstein effect will dominate the intermittency effect [22].

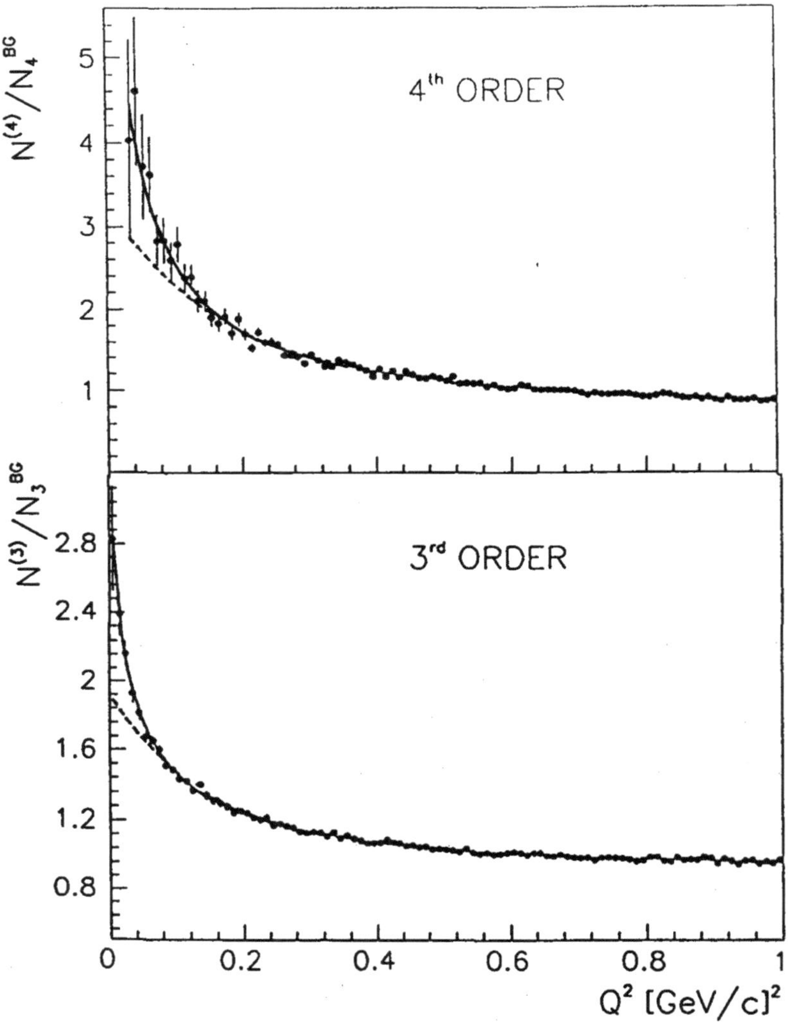

Fig. 7: Bose Einstein ratio for three and four particles. The dashed line represents a fit with terms containing $A \cdot \exp(-ar^2Q^2)$ given by Eqn.(7) and (8), the solid line represents the fit with exponential terms $A \cdot \exp(-arQ)$.

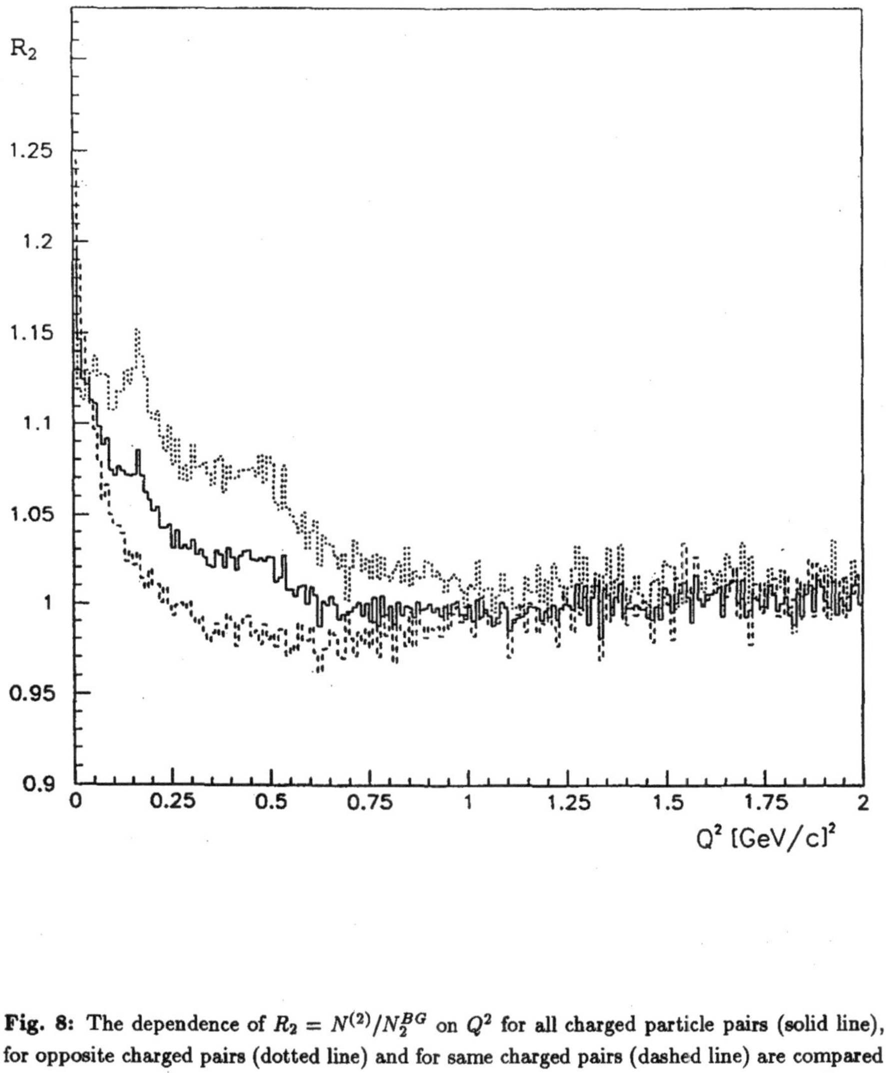

Fig. 8: The dependence of $R_2 = N^{(2)}/N_2^{BG}$ on Q^2 for all charged particle pairs (solid line), for opposite charged pairs (dotted line) and for same charged pairs (dashed line) are compared to each others. In each case the ratios have been normalized to 1 in the region $1 \leq Q^2 \leq 2$, therefore the functions for opposite charged pairs and same charged pairs do not add up to the function for all charged pairs. All particles have been assumed to be pions. The first peak in the dotted line is mainly due to the decay $K_S^0 \rightarrow \pi^+\pi^-$, the second broad peak is due to $\rho^0 \rightarrow \pi^+\pi_-$.

Summary

1. A preliminary 2-dimensional intermittency analysis (in η and ϕ) shows that the data behave as expected if intermittency occures in the higher dimensional phase space: we observe almost linearity in the log-log representation and larger slopes (by a factor up to 8) compared to the 1-dimensional ones.

2. The Bose-Einstein correlations show a very steep rise for small decreasing Q^2 in the region below 0.2 $(\mathrm{GeV})^2$. This is most clearly shown for the orders $i = 3, 4$. A Gaussian shape is ruled out, but a simple exponential shape fits well the data. A power law behaviour cannot be ruled out yet.

3. The two-body correlation function, dependent on Q^2 is presented for all charged particles and opposite charged particles in the same way as for same charged particles (Bose-Einstein effect). A possible development of a pole would be due to the contribution of same charged particles only. If the latter is due to the Bose-Einstein interference effect it will have interesting consequences for the distribution of the source radii [23].

Acknowledgements

We want to thank A. Bartl and M. Markytan for initiating the investigations of higher order Bose Einstein correlations and T. Gajdosik for his work at the beginning. We are grateful to A. Bialas, W. Kittel, W. Ochs, R. Peschanski, J. Seixas and R.M. Weiner for many interesting remarks.

This work is partly supported by the Fonds zur Förderung der Wissenschaftlichen Forschung in Österreich (P8259).

References

[1] C. Albajar et al. (UA1 Collab.), *Z. Phys.* **C44** (1989 15;
M. Calvetti et al, *IEEE Trans. Nucl. Sci.* **NS-30** (1983) 71.

[2] A. Bialas and R. Peschanski, *Nucl. Phys.* **B273** (1986) 703 and **B308** (1988) 857.

[3] M.I. Adamovich et al. (EMU01 Collab.), *Phys. Rev. Lett.* 62 (1989) 2801.

[4] N. Lieske, this conference and
D. Decamp et al. (ALEPH Collab.), preprint CERN-PPE/91-121.

[5] F. Meijers (NA22 Collab.), Thesis, Univ. of Nijmegen, The Netherlands, 1987.

[6] C. Albajar et al. (UA1 Collab.), *Nucl. Phys.* **B345** (1990) 1.

[7] G.J. Alner et al. (UA5 Collab.), *Phys. Lett.* **160B** (1985) 193.

[8] D. Weselka et al. (UA1 Collab.), to be published.

[9] P. Lipa et al., Arizona Preprint AZPH/TH/91-14.

[10] P. Abreu et al. (DELPHI Collab.), *Phys. Lett.* **B247** (1990) 137;
A. De Angelis, *Mod. Phys. Lett.* **A5** (1990) 2395.

[11] I. Derado, G. Jancso, N. Schmitz and P. Stopa (EMC Collab.), *Z. Phys.* **C47** (1990) 23, and Santa Fe Workshop on "Intermittency in High Energy Collisions", eds. F. Cooper, R.C. Hwa and I. Sarcevic, World Scientific, Singapore, 1990.

[12] I.V. Ajinenko et al. (NA22 Collab.), *Phys. Lett.* **B222** (1989) 306 and **B235** (1990) 373.

[13] R. Holynski et al. (KLM Collab.), *Phys. Rev. Lett.* **62** (1989) 733 and *Phys. Rev.* **C40** (1989) R2449.

[14] See also talks of I. Derado (NA35 Collab.), E. Sternlund (EMU01 Collab.), N. Schmitz (EMC), O. Podobrin (CELLO), V. Raab (ALEPH), N. Geddes (OPAL) at this conference.

[15] W. Ochs, *Phys. Lett.***B247** (1990) 101 and *Z. Phys.***C50** (1991) 339.

[16] A. Bialas and J. Seixas, *Phys. Lett.* **B250** (1990) 161.

[17] T. Sjöstrand and M. van Zijl, *Phys. Rev.* **D36** (1987) 2019.

[18] T. Sjöstrand, Multiparticle Production, Eds. R. Hwa, G. Panchieri, Y. Srivastava, World Scientific, Singapore, 1989.

[19] N. Neumeister (UA1 Collab.), to be published.

[20] M. Biyajima et al., *Progress of Theoret. Phys.* **84** (1990) 931.

[21] N. Neumeister, XXVth Rencontre de Moriond (High Energy Hadronic Interactions), March 17-23, 1991, Les Arcs, France, and UA1-Collaboration, to be published.

[22] M. Gyulassy, contribution to the Festschrift L. Van Hove "Multiparticle Dynamics", eds. A. Giovannini and W. Kittel (World Scientific, Singapore, 1990).

[23] A. Bialas, invited talk at "Quark Matter '90", Menthon, France, May 7-11, 1990, preprint CERN-TH 5791-90.

CORRELATIONS and INTERMITTENCY in NA22

EHS/NA22 Collaboration
presented by

F. BOTTERWECK and W. KITTEL
Physics Department, University of Nijmegen, Toernooiveld 1
6525 ED Nijmegen, the Netherlands

and

F.K. RIZATDINOVA and L.N. SMIRNOVA
Nuclear Physics Inst., Moscow State University,
SU-117234 Moscow, USSR

ABSTRACT:

Density fluctuations are studied in rapidity, separately for low and intermediate transverse momentum particles in π^+p and K^+p collisions at 250 GeV/c. In our data, the effect of intermittency is increased when the analysis is restricted to low p_T particles. A multi-dimensional analysis of two-particle correlations shows interesting structure. Again, particularly strong positive short-range rapidity and azimuthal correlations are observed for low-p_T particles. This observation is in contradiction with expectations from models used for comparison (DPM, quark gluon (multi) string model). The rise of factorial moments with improving resolution is stronger when evaluated in 3 than in lower dimensions. The ratios of slopes are easier to obtain than the slopes themselves and turn out to be largely independent of the dimension and of the type of interaction. The numerical values of these ratios are better described by a Levy stable law solution with $\mu = 1.6$ than by the α-model ($\mu = 2$) or a phase transition ($\mu = 0$).

1. Intermittency at low p_T

Density fluctuations in multiparticle production are presently gaining considerable attention. To study these fluctuations in detail, Białas and Peschanski [1] have suggested to analyze scaled factorial moments of the multiplicity distribution in smaller and smaller phase-space bins, down to the experimental resolution. Including the case of a non-constant overall rapidity distribution, these moments can be defined as a double average starting "vertically" (over events):

$$\langle F_q \rangle_v = \frac{1}{M} \sum_m \frac{\frac{1}{N_{evt}} \sum_{evt} F_{q,m}}{\langle n_m \rangle^q} \quad . \tag{1}$$

Here, M is the number of phase-space bins $\delta y = \Delta y/M$ into which an original region Δy is subdivided, n_m is the multiplicity in bin m ($m = 1, ..., M$), $F_{q,m} =$

$n_m(n_m - 1).....(n_m - q + 1)$ is the factorial of order q in bin m, and $\langle n_m \rangle = \frac{1}{N_{evt}} \sum_{evt} n_m$ is the mean multiplicity in bin m.

If self-similar fluctuations of many different sizes ("intermittency") exist, then the dependence of the moment on the size of the phase-space bin follows the power law

$$\langle F_q \rangle \propto \delta y^{-f_q}, \quad (2)$$

with the intermittency strength $f_q > 0$. Otherwise, saturation of the moments at small δy is expected.

Intermittency has been studied in e^+e^- [2-6], μp [7], ν-nucleus [8], hadron-hadron [9,10], hadron-nucleus [11, 12] and nucleus-nucleus [11,13-16] collisions. Brief reviews are given in [17,18,19,20]. In general, the effect is not fully reproduced by presently used models, so that further experimental input is needed for necessary improvements.

One of the most urgent questions in this respect is whether semi-hard effects [21,22], observed to play a rôle in the transverse momentum behaviour even at our energies [23], or low-p_T effects [24,25,26] are at the origin of intermittency. A first indication for at least some contribution from low-p_T effects comes from our most prominent "spike" event [9a], where 5 out of 10 tracks involved have $p_T <$ 0.15 GeV/c. Furthermore, while currently used hadron-hadron models predict intermittency to vanish for small transverse momenta [10a], the UA1 data show a slight increase of the intermittency strengths f_q when transverse momenta are restricted to $0.15 < p_T < 0.5$ GeV/c. Because of the UA1 bias against $p_T <0.15$ GeV/c, low-p_T effects cannot be studied in the UA1 data, however.

In this section we report on the study of intermittency in different regions of low and intermediate transverse momentum in the NA22 experiment. In this CERN experiment, the European Hybrid Spectrometer (EHS) is equipped with the Rapid Cycling Bubble Chamber (RCBC) as a vertex detector and exposed to a 250 GeV/c tagged positive, meson enriched beam. In data taking, a minimum bias interaction trigger is used. The details of the spectrometer and the trigger can be found in previous publications [27,28].

Charged particle tracks are reconstructed from hits in the wire- and drift-chambers of the two lever-arm magnetic spectrometer and from measurements in the bubble chamber. The average momentum resolution $\langle \Delta p/p \rangle$ varies from a maximum of 2.5% at 30 GeV/c to around 1.5% above 100 GeV/c. In the rapidity region Δy under consideration ($-2.0 < \Delta y < 2.0$), the averaged experimental resolution in rapidity varies between 0.01 and 0.05 units.

Events are accepted for the analysis when measured and reconstructed charge multiplicity are consistent, charge balance is satisfied, no electron is detected among the secondary tracks and the number of badly reconstructed (and therefore rejected) tracks is 0. The loss of events during measurement and reconstruction is corrected for in agreement with the topological cross section data [27]. Elastic events are excluded. Furthermore, an event is called single-diffractive and excluded from the sample if the total charge multiplicity is smaller than 8 and at least one of the positive tracks has $|x| > 0.88$. After these cuts, our "cleaned" inelastic non-single-diffractive sample consists of 59 200 π^+p and K^+p events. The average in (1) is normalized to this sample, including events with no tracks in $-2.0 < y < 2.0$.

For momenta $p_{LAB} < 0.7$ GeV/c, the range in the bubble chamber and/or the change of track curvature is used for proton identification. In addition, a visual ionization scan has been used for $p_{LAB} < 1.2$ GeV/c on the full K^+p and 62% of the π^+p sample. Positive particles with $p_{LAB} > 150$ GeV/c are given the indentity of the beam particle. Other particles with momenta $p_{LAB} > 1.2$ GeV/c are not identified in the present analysis and are treated as pions.

Earlier results on the full p_T range using "horizontal" averaging are presented in refs. [9b].

Our present results for $\ln\langle F_q\rangle$ versus $-\ln\delta y$ are given in Figs.1a,b for particles with transverse momentum p_T below and above 0.15 GeV/c, in Figs.1c,d for p_T below and above 0.3 GeV/c. Particles with p_T below the cut (Figs.1a,c) show an effect much stronger than those above the cut (Figs.1b,c).

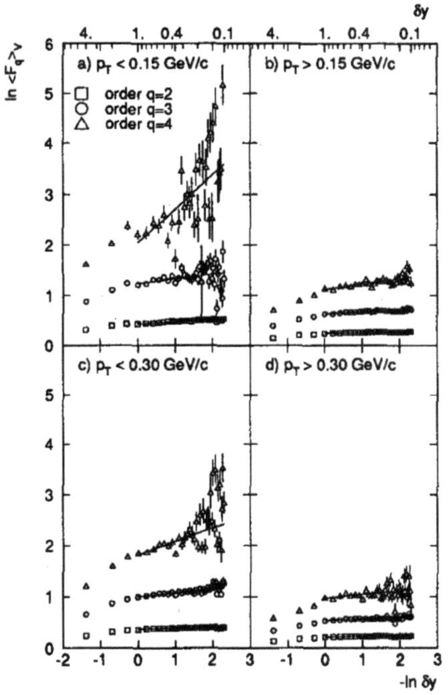

Fig.1. $\ln\langle F_q\rangle_v$ as a function of $-\ln\delta y$ for various p_T cuts as indicated.

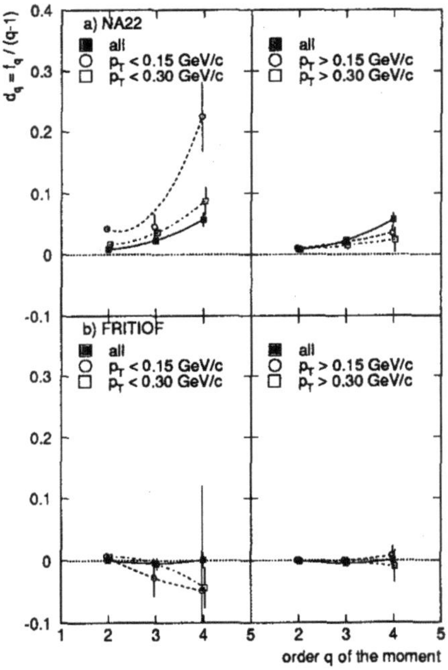

Fig.2. $d_q = f_q/(q-1)$ as a function of the order q for p_T cuts as indicated. Lines are to guide the eye.

We do not claim straight lines in Fig.1, but use fits as an indication to measure the increase of $\ln\langle F_q\rangle_v$ over the region $1 > \delta y > 0.1$. In Fig.2a, the fitted values of f_q are compared to those obtained in the full p_T range, in the reduced form $d_q = f_q/(q-1)$. Restricting the analysis to particles with $p_T < 0.15$ or 0.30 GeV/c indeed leads to an *increase* of the d_q, while a *decrease* is observed for $p_T > 0.15$ or 0.30 GeV/c.

We have verified that a similar dependence on p_T is visible in the full event sample (including events with track losses). For further analysis of possible biases see [29].

In Fig.2b, we present expectations from the FRITIOF 2.0 [30] model at similar statistics, again for all tracks and those with p_T smaller and bigger than 0.15 or 0.3 GeV/c, respectively. Besides the known fact [9] that FRITIOF gives too low slopes already for the full sample, this model does not reproduce the p_T dependence of the effect.

We conclude that the intermittency observed in our data is due to low transverse momentum particles and not due to semi-hard effects. Since, on the contrary, hard effects dominate in e^+e^- and lh collisions, a study of the p_T dependence in these could be of utmost importance in the search for the origin of intermittency.

2 Two-particle Correlations

Two-particle correlation functions have large positive values in the central region of hadron-hadron, lepton-hadron and e^+e^--collisions (see [31-34] and references given therein). Presently existing low-p_T Monte Carlo models for hadron interactions are not able to describe the short-range correlations observed [31]. To find the origin of this failure, more detailed analysis of correlation phenomena is required, including that of correlation functions in more dimensions.

The recent revival of interest in a detailed investigation of two-particle correlations is connected with the observation of an intermittency effect in hadron production, for which the importance of a multidimensional analysis has recently been established [35]. Interesting effects in intermittency, in particular on the p_T dependence are observed in the NA22 experiment (see Sect.1 above). So, a detailed investigation of two-particle correlations and intermittency can be performed in the same experiment.

The results on two- and three-particle rapidity correlations in π^+p, K^+p and pp-interactions at 250 GeV/c are published in [31]. In this section, azimuthal correlations and rapidity correlations are presented for π^+p-interactions at 250 GeV/c, both for all particles and for particles in different transverse momentum intervals. The results are compared with model predictions.

Events are accepted for this analysis under the conditions as in Sect.1, except that the number of badly reconstructed (and therefore rejected) tracks is allowed to be at most 0, 1, 1, 2 and 3 for events with charge multiplicity 2, 4, 6, 8 and >8, respectively. After these cuts, our inelastic non-single-diffractive sample consists of about 84 500 π^+p events. Corrections for rejected tracks amount to 6-7% of the tracks. They are approximately independent of the multiplicity n and are assumed to be independent of rapidity y. It has been verified that the results do not change if the analysis is restricted to the sub-sample used in Sect.1.

A comparison is performed with two models: a two string version of the dual parton model (2-string DPM) [36] and the quark-gluon string model (QGSM) [37].

In the DPM version used [38], two strings are formed in the interaction. One string is stretched between the meson quark and the proton diquark, the other between the meson antiquark and the residual single proton quark. Quark fragmentation is performed according to the symmetric LUND model [39]. The primordial transverse momentum of valence quarks corresponds to a Gaussian distribution with average $\langle k_t \rangle$= 0.42 GeV/c. Parameter values are default, except for the width of the Gaussian p_x and p_y transverse momentum distribution for primary hadrons (σ=0.44 GeV/c). The model, furthermore, includes the production of tensor mesons in the ratio PS:V:T=50:35:15. Monte Carlo events satisfying the "diffractive" criteria (see above) are excluded.

The quark-gluon string model [37], just as DPM, is based on dual topological unitarization. In addition to the two strings used in our DPM version, in QGSM strings are formed between sea quarks and antiquarks of the colliding hadrons. The transverse momentum distributions of valence and sea quarks in the hadron are of the form $P(k_t^2)$=A $\exp(-bk_t^2)$ with b= 6.25 (GeV/c)$^{-2}$ for the cylindrical diagram. The string breaking algorithm of QGSM is described in [37]. The last break-up of the string is an isotropic two-particle cluster decay. A ratio PS:V=1:1 is used, without addition of tensor mesons.

Comparison of our charge multiplicity distributions and single-particle inclusive spectra have been performed with 2-string DPM (and FRITIOF [30]) in [28] and with QGSM in [37]. In general, both models reproduce the single-particle inclusive distributions. The multiplicity distributions are reproduced best in QGSM. In particular, this is true at large multiplicity n, where multi-string formation is important.

2.1 Rapidity correlations

The rapidity correlation function can be written in the form

$$C_2(y_1, y_2) = \rho(y_1, y_2) - \rho(y_1)\rho(y_2) \tag{3}$$

where

$$\rho(y) = (1/\sigma_{in})d\sigma/dy, \qquad \rho(y_1, y_2) = (1/\sigma_{in})d\sigma/dy_1dy_2$$

are densities normalized to the particle multiplicity and particle pair multiplicity, respectively.

The second order normalized factorial cumulant is defined as

$$K_2(y_1, y_2) = \frac{C_2(y_1, y_2)}{\rho(y_1)\rho(y_2)} = \frac{\rho(y_1, y_2)}{\rho(y_1)\rho(y_2)} - 1 \ . \tag{4}$$

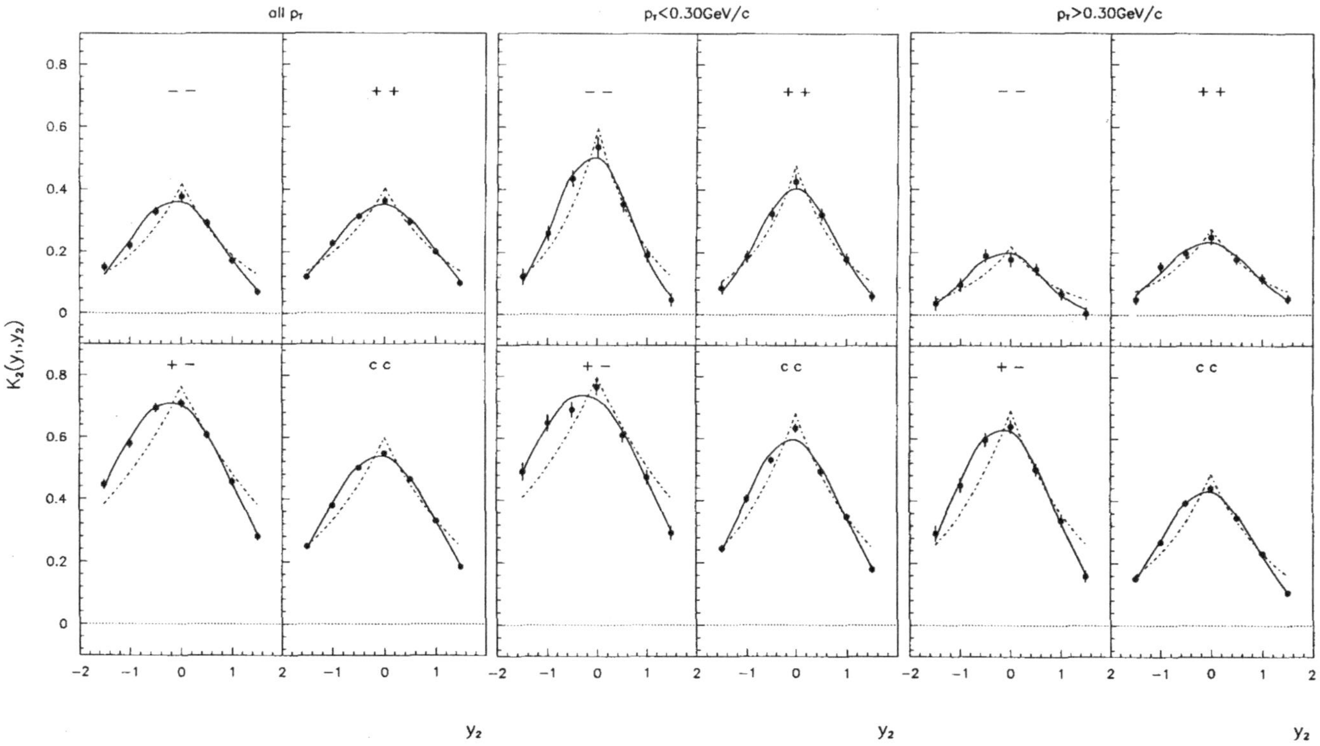

Fig.3 Normalized cumulant functions $K_2(0, y)$ for particles with all p_T, p_T <0.30 GeV/c, p_T >0.30 GeV/c as compared to functions (6) and (7), solid and dot-dashed lines, respectively.

For normalized factorial moments we have

$$\langle F_2(\delta)\rangle = 1 + \int_{y-\delta/2}^{y+\delta/2} dy_1 \int_{y-\delta/2}^{y+\delta/2} dy_2 K_2(y_1, y_2) \quad . \tag{5}$$

The cumulant $K_2(0, y_2)$ may be used for the analysis of the p_T-dependence of correlation phenomena. In Fig.3 the $K_2(0, y_2)$ function is shown for all particles, for particles with $p_T < 0.30$ GeV/c and $p_T > 0.30$ GeV/c.

To calculate intermittency analytically, it is necessary to know the analytical expression for the correlation function. In Fig.3 $K_2(0, y_2)$ is fitted by the functions

$$f_1 = c\exp[-((y - y_0)/\sigma)^2/2] \qquad \text{(full line)} \tag{6}$$

$$f_2 = a\exp(-b|y|) \qquad \text{(dashed)} \quad . \tag{7}$$

Even though for low p_T the point at $y_2 = 0$ is systematically above the Gaussian, $K_2(0, y_2)$ is well fitted by the Gaussian (6), but not by the exponential (7), when using this one-dimensional projection onto rapidity. So, a higher dimensional analysis is required.

For low-p_T particles $K_2(0, y_2)$ has larger values in the maximum and more pronounced peaks than for larger p_T, in particular for the $(--)$ combination. This latter observation is in agreement with the NA22 intermittency results obtained in Sect.1.

2.2 Azimuthal correlations

In hh-interactions no distinguished direction exists in the plane transverse to the beam and the distribution in the azimuthal angle ϕ is uniform. A two-particle correlation, however, exists also in ϕ and can be expressed as the distribution $W(\Delta\phi)$ in $\Delta\phi = |\phi_1 - \phi_2|$, the azimuthal angle between two particles, defined between 0 and π. The azimuthal correlation may depend on the charge of the particles in the pair on the rapidity distance $\Delta y = |y_1 - y_2|$ between these particles and on transverse momentum.

In Fig.4, the distribution $W(\Delta\phi, \Delta y)$ normalized to unity is shown as a function of $\Delta\phi$, for all charge combinations, in the Δy intervals for $\Delta y < 1$, $1 < \Delta y < 2$ and $2 < \Delta y < 3$. The horizontal line shows the average value $1/\pi$ corresponding to a flat distribution in $\Delta\phi$. The distribution is influenced by conservation of transverse momentum, by decay of resonances (for unlike sign particles) and by Bose-Einstein correlations (for identical particles). In all cases, W is larger than $1/\pi$ for $\Delta\phi > \pi/2$ and has a maximum at $\Delta\phi = \pi$. Except for $(--)$ pairs at $\Delta y < 1$, the W function is smaller than $1/\pi$ for $\Delta\phi < \pi/2$. The source of such a global anti-correlation is transverse momentum conservation.

Model predictions are shown in Fig.4 for two string DPM (dashed) and multi-string QGSM (dot-dashed). The comparison with the data shows that it is much easier to predict azimuthal correlations at large than at small Δy. At small Δy

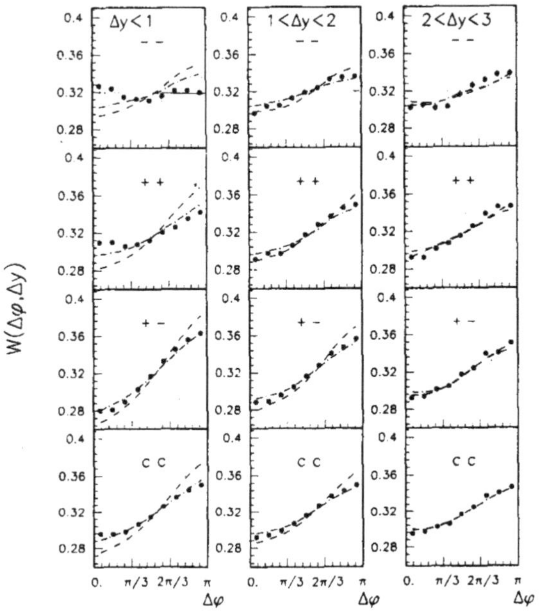

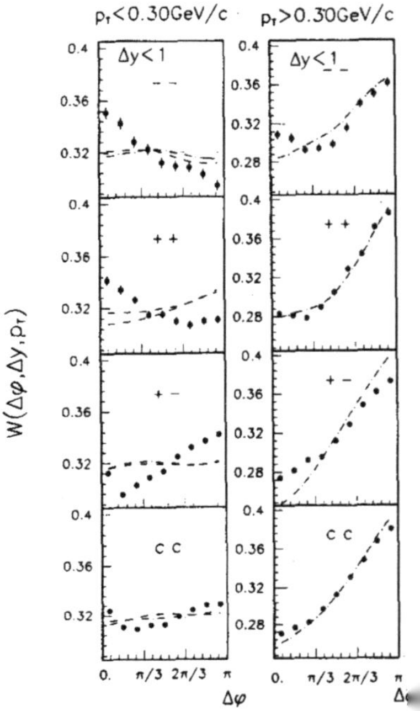

Fig.4. $W(\Delta\phi, \Delta y, p_T)$ for inclusive non-single-diffractive π^+p interactions at 250 GeV/c as compared to DPM (dashed) and QGSM (dot-dashed).

Fig.5. $W(\Delta\phi, \Delta y, p_T)$ as compared with calculations in DPM (dashed) and QGSM (dot-dashed) for $\Delta y < 1$ and p_T cut as indicated.

the models differ from each other and from the experimental data. The QGSM shows somewhat better agreement with the experimental distributions than the other models. This is a consequence of the multi-string structure of QGSM, where strong azimuthal correlations existing within one string are destroyed and the $\Delta\phi$-dependence is weaker than in two-string DPM.

Differences between experiment and both models exist at small $\Delta\phi$ and $\Delta y < 1$, for all charge combinations and in particular for $(--)$ pairs. Bose-Einstein correlations are not included in the models may well be the most important source of this disagreement.The influence of Bose-Einstein correlation can also be observed for the $(++)$ combination, but is reduced there due to the influence from the (positive) beam particles.

In Fig.5(a-b) the azimuthal distributions are shown for particles with $\Delta y < 1$ $p_T < 0.30$ GeV/c and $p_T > 0.30$ GeV/c, in comparison with the model calculations. A comparison of these figures shows that the azimuthal distributions have a strong p_T-dependence. Large positive azimuthal correlations exist at $\Delta y < 1$ for particles with small p_T. As the transverse momentum of particles increases, the peak at $\Delta\phi = \pi$ becomes more pronounced. This is reproduced by the models and reflects momentum compensation.

In conclusion:

- The slope of the correlation functions $K_2(0, y_2)$ is better described by a Gaussian than by a exponential in y_2.
- The correlation functions depend strongly on transverse momentum and are largest for small-p_T particles.
- Positive correlations are observed at large values of the azimuthal angle $\Delta\phi$, as expected from transverse momentum conservation.
- Significant deviations from DPM and QGSM model predictions are observed at small values of $\Delta\phi$ and Δy for like-charge particle pairs, where Bose-Einstein effects should contribute. These effects are enhanced for particles with small transverse momentum.

3 Intermittency Analysis in More Dimensions

One cannot do a three-dimensional analysis with the usual phase-space variables rapidity y, azimuthal angle ϕ, and transverse momentum p_T, since especially the latter one is far from a flat distribution. Therefore, Ochs [35a] as well as Białas and Gazdzicki [35b] have proposed to look at normalized cumulative variables

$$X(y) = \frac{\int_{y_{min}}^{y} \rho(y')dy'}{\int_{y_{min}}^{y_{max}} \rho(y')dy'}. \tag{8}$$

During our first investigations it turned out that there was a strong correlation between the magnitude of the factorial moments and the number of bins chosen to define the grid of the cumulative distribution. A stable situation was reached when 100000 bins were used in $X(y)$.

To go to higher dimensions, Ochs assumed that the density function in 3 dimensions factorizes

$$\rho(y, \phi, p_T) = \rho(y)\rho(\phi)\rho(p_T). \tag{9}$$

Using this rather strong assumption it is sufficient to calculate the cumulative distribution for all 3 variables, independently. On the other hand, this assumption is not necessary in [35b]. Since in our data both methods lead to rather similar results [42], we here restrict ourselves to the results obtained with [35a].

In Fig.6 a compilation of the factorial moments is shown. For each order the moments have the same value for the 1, 2 and 3 dimensional distributions in the first point. This is due to the fact that for all the distributions the same cuts on the initial interval has been used ($-2 < y < 2$, $0 \le \phi \le 2\pi$, $-18 < \log(p_T^2) < 6$).

Each value on the x-axis corresponds to the number of divisions of one variable. Thus, for $M = 10$ the number of intervals in 1 dimension is 10, in 2 dimensions 100, and in 3 dimensions 10^3.

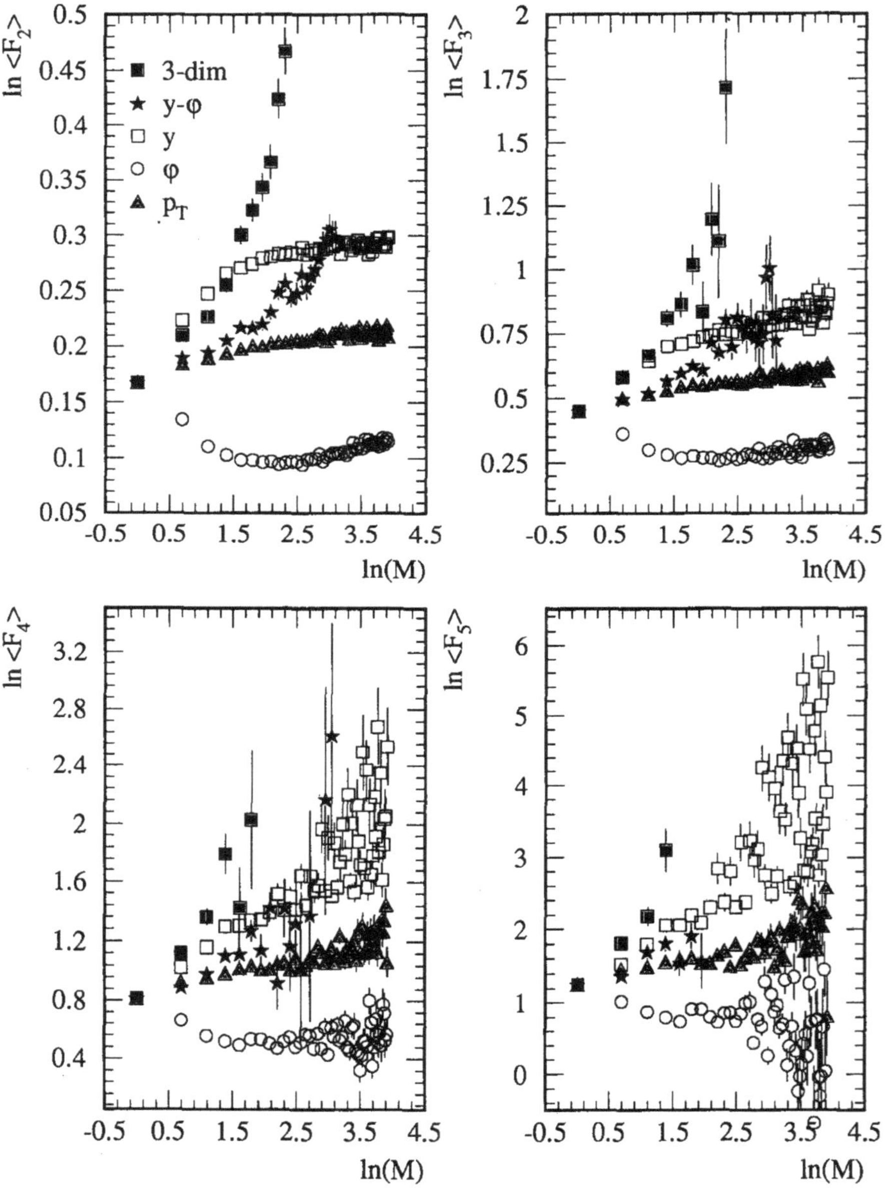

Fig.6 Comparison of $\langle F_q \rangle$ for different dimensions. The clean sample of section 1 has been used, with event weight.

The rise of the moments is clearly strongest in the 3 dimensional case. Since it is faster than a power law, no slopes have been fitted. In order to be able to compare our results with other experiments and to study a modified power law, the Ochs plots are presented in Figs.7 and 8. While in [35a] data at small scales, where F_2 does not vary significantly, are rejected, in Fig.7 all points are presented, leading to a somewhat messy, but honest plot. Numerically fitting a straight line does not make sense, but an indication of the slope $r_q = f_q/f_2$ can be obtained from an 'eye-ball' fit as shown in Fig.8. In table 1 the results as published in [35a] are shown, together with the new results, obtained from Fig.8. The behavior of the ratio of the anomalous dimensions $D_q(= f_q/(q-1))$ and D_2 are shown in Fig.9.

		r_3	r_4	r_5
e^+e^-	TASSO, DELPHI	2.83±0.	5.32±0.20	8.33±0.33
μp	EMC	2.60±0.10	4.76±0.22	7.9 ±0.3
hh	UA1, UA5, NA22	2.43±0.10	4.3 ±0.2	6.9 ±0.3
hh	NA22 new variable	2.8 ±0.2	5.0 ±0.2	8.0 ±1.0
pA, AA	KLM, EMU01	2.86±0.07	5.22±0.22	8.2 ±0.3

Table 1: Values for $r_q = f_q/f_2$.

Exclusion of possibly undetected Dalitz decay and γ conversions reduces the value of $\langle F_q \rangle$ in particular at large $\ln M$, but the rise remains faster than a power law and also the other conclusions remain the same. The behaviour of the slopes as a function of the order can give insight into the process leading to intermittent behaviour. If it is a cascade mechanism that leads to the power law behaviour, then in the Gaussian approximation, the moments are assumed to fulfil the relation:

$$\frac{D_q}{D_2} = \frac{f_q}{f_2}\frac{1}{q-1} = \frac{q}{2}. \tag{10}$$

It is shown in [43] that the central limit theorem is only a very first approximation to a multiplicative process, as for example the α-model. Brax and Peschanski [44] argued that a better approximation might be obtained from Levy stable law, yielding a generalized behavior of the form

$$\frac{D_q}{D_2} = \frac{1}{2^\mu - 2}\frac{q^\mu - q}{q-1}, \tag{11}$$

of which (10) is the special case of $\mu = 2$.

If, on the other hand, the underlying mechanism is a second order phase transition, then the multifractal behavior (10) and (11) is replaced by a monofractal

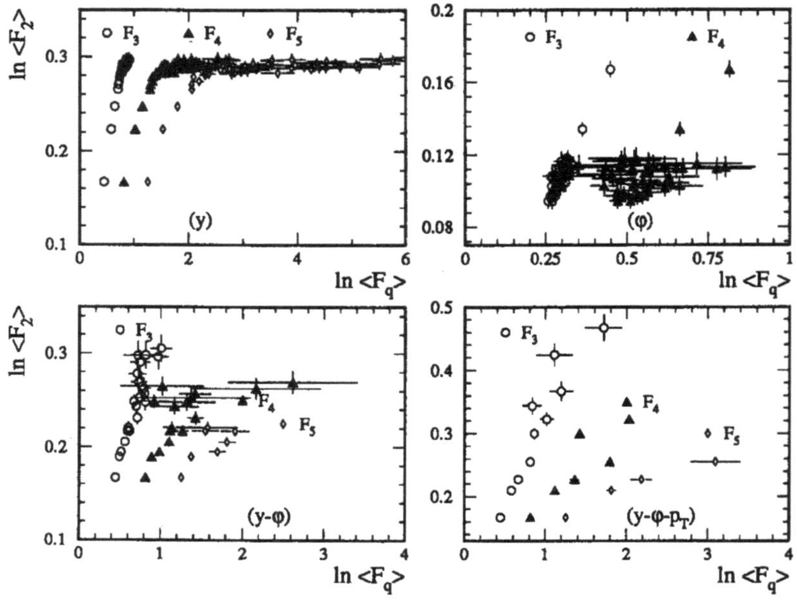

Fig.7 $\ln\langle F_2\rangle$ versus $\ln\langle F_q\rangle$ dependence for bins in $y, \phi, y-\phi$ and $y-\phi-p_T$.

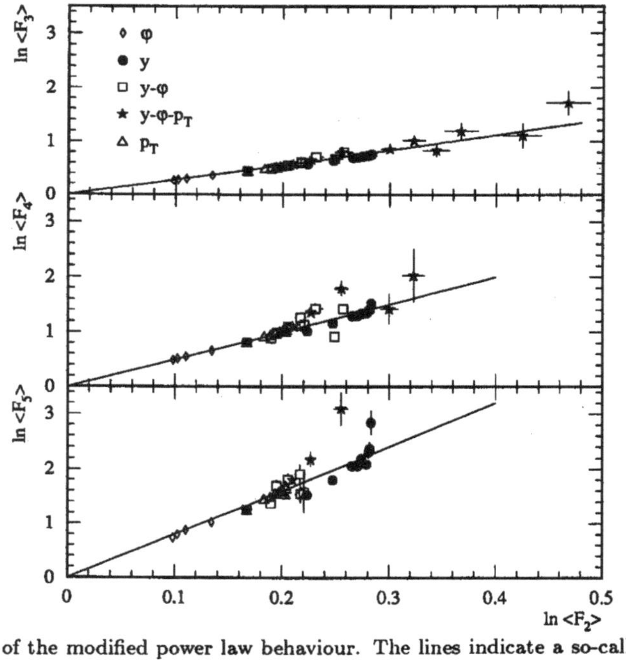

Fig.8 Illustration of the modified power law behaviour. The lines indicate a so-called 'eye-ball' fit to the data. Only the data in bins where $\langle F_2\rangle$ varies strongly are used.

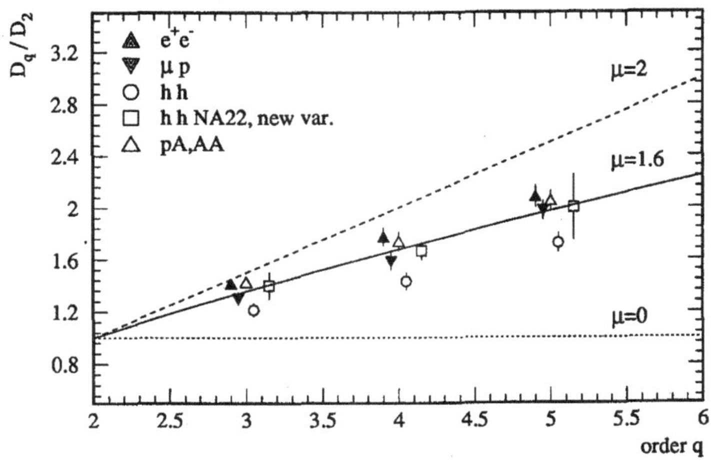

Fig.9 Ratio of anomalous dimensions as a function of order q.

behavior

$$\frac{D_q}{D_2} = 1 \tag{12}$$

corresponding to $\mu = 0$ and yielding an order independent anomalous dimension. Such a behavior could indicate the existence of a quark gluon plasma phase transition.

The data are best fitted with the Levy stable law solution with 1.6, lower than expected from the Gaussian approximation, but definitely higher than from a phase transition.

4 Conclusions

1. In NA22 data, the effect of intermittency in rapidity is increased when the analysis is restricted to low p_T particles.
2. A corresponding increase is observed in the two-particle rapidity and azimuthal correlation for small p_T. This effect is not reproduced by currently used models for multiparticle production in hadronic collisions.
3. The rise of the factorial moments with improving resolution is stronger when evaluated in 3 than in lower dimensions. The ratios of slopes are easier to obtain than the slopes themselves and turn out to be largely independent of the dimension and of the type of interaction. The numerical values of these ratios are better described by a Levy stable law solution with $\mu = 1.6$ than by the α-model ($\mu = 2$) or by a phase transition ($\mu = 0$).

References

1. A. Białas and R.Peschanski, *Nucl. Phys.* **B273** (1986) 703; **B308** (1988) 857.
2. B. Buschbeck, P.Lipa, R. Peschanski, *Phys. Lett.* **B215** (1988) 788.
3. W. Braunschweig et al. (TASSO), *Phys. Lett.* **B231** (1989) 548, **B231** (1989) 548.
4. S. Abachi et al. (HRS), "Study of Intermittency in e^+e^- Annihilations at 29 GeV", preprint ANL-HEP-CP-90-50 (1990).
5. H.-J. Behrend et al. (CELLO), *Phys. Lett.* **B256** (1991) 97.
6. P. Abreu et al. (DELPHI), *Phys. Lett.* **B247** (1990) 137;
   A. De Angelis (DELPHI), *Mod. Phys. Lett.* **A5** (1990) 2395;
   M.Z. Akrawy et al. (OPAL), *Phys. Lett.* **B262** (1991) 351;
   D. Decamp et al. (ALEPH), "An Investigation into Intermittency", CERN-PPE/91-121.
7. I. Derado, G. Jancso, N. Schmitz and P. Stopa, *Z. Phys.* **C47** (1990) 23;
   I. Derado et al., this workshop.
8. L. Verluyten et al. (WA59 and E180), "Study of Factorial Moments in Neutrino Neon and Deuterium Charged Current Interactions", paper submitted to the XXVth Int. Conf. on High Energy Physics, Singapore 1990.
9. a) M. Adamus et al. (NA22), *Phys. Lett.* **B185** (1987) 200;
   b) I.V. Ajinenko et al. (NA22), *Phys. Lett.* **B222** (1989) 306, **B235** (1990) 373 and **B261** (1991) 165;
   c) V. Arena et al (IHSC), "Study of Intermittency in Hadron-Hadron Collisions at 16.7 GeV".
10 a) B. Buschbeck (UA1),"Intermittency Measured in $p\bar{p}$ Collisions at 630 GeV", Festschrift Léon Van Hove, eds. A. Giovannini and W. Kittel (World Scientific 1990), p.211 and private communication;
   b) C. Albajar et al. (UA1), *Nucl. Phys.* **B345** (1990) 1.
11. R. Hołynski et al. (KLM), *Phys. Rev. Lett.* **62** (1989) 733; *Phys. Rev.* **C40** (1989) R2449.
12. F. Botterweck et al. (NA22), *Z. Phys.* **C51** (1991) 37.
13. I. Derado (NA35), "Investigations of Intermittency in the Reaction $^{32}S+^{32}S$ at 200 GeV per Nucleon", Festschift Léon Van Hove, eds. A. Giovannini and W. Kittel (World Scientific 1990) p.257.
14. M.I. Adamovich et al. (EMU01), *Phys. Rev. Lett.* **65** (199) 412.
15. K. Sengupta et al. (EMU08), *Phys. Lett.* **B236** (1990) 219;
   P.L. Jain and G. Singh, "One- and Two- Dimensional Analysis of Intermittency in Ultrarelativistec Nucleus-Nucleus Interactions", to be publ.in Phys. Rev. C.
16. T. Åkesson et al. (HELIOS), *Phys. Lett.* **B252** (1990) 303.
17. W. Kittel and R. Peschanski, *Nucl. Phys. B* **(Proc. Suppl.) 16** (1990) 445;
   W. Kittel, "Intermittency (A Review of Experimental Results)", Proc. XXth Int. Symp. on Multiparticle Dynamics, eds. R. Baier and D. Wegener (World Scientific) to be publ., Nijmegen preprint HEN-335/90.

18. A. Białas, "Intermittency '90", preprint CERN-TH-5791/90 (1990).
19. R. Peschanski, "Intermittency in Particle Collisions", CERN-TH-5891/90, to be publ. in *Int. J. Mod. Phys. A*.
20. B. Buschbeck, "Intermittency, a Short Experimental Review", Proc. XXV Rencontres de Moriond, ed. J. Tran Thanh Van, to be publ. (Vienna preprint HEPHY-PUB 554/91).
21. W. Ochs and J. Wosiek, *Phys. Lett.* **B214** (1988) 617; B 232 (1989) 271.
22. T. Sjöstrand, "Multiple Interactions, Intermittency, and other Studies", Proc. Perugia Workshop on Multiparticle Production, eds. R. Hwa, G. Pancheri and Y. Srivastava (World Scientific 1989) p.327.
23. I.V. Ajinenko et al. (NA22), *Phys. Lett.* **B197** (1987) 457.
24. L. Van Hove, *Ann. Phys.* **192** (1989) 66.
25. A. Białas et al., *Phys. Lett.* **B229** (1989) 398.
26. F. Botterweck et al. (NA22), *Z. Phys.* **C51** (1991) 541.
27. M. Adamus et al. (NA22), *Z. Phys.* **C32** (1986) 475.
28. M. Adamus et al. (NA22), *Z. Phys.* **C39** (1988) 311.
29. F. Botterweck (NA22), "Intermittency in NA22", Proc. XXV Rencontre de Moriond, Les Arcs (France), ed. J. Tran Thanh Van (Editions Frontières 1990) p.455; W. Kittel (NA22), "Low p_T Intermittency in π^+p and K^+p Collisions at 250 GeV/c", Proc. Santa Fe Workshop on Intermittency in High-Energy Collisions, eds. F. Cooper et al. (World Scientific 1991) p.83.
30. B. Andersson, G. Gustafson and B. Nilsson-Almqvist, *Nucl. Phys.* **B281** (1987) 289; "A High Energy String Dynamics Model for Hadronic Interactions", Lund preprint LUTP 87-6.
31. V.V. Aivazyan et al. (NA22): *Z. Phys.* **C51** (1991) 167.
32. M. Althoff et al.: *Z. Phys.* **C29** (1985) 347.
33. R.E. Ansorge et al. (UA5): *Z. Phys.* **C37** (1988) 191.
34. J. Figiel: Krakov preprint IFJ 1398/Ph (1988)
35. a) W. Ochs: *Phys. Lett.* **B247** (1990) 101; *Z. Phys.* **C50** (1991) 339;
b) A. Białas and M. Gazdzicki: *Phys. Lett.* **B252** (1990) 483.
36. A. Capella: Europ. Study Conf. on Proton and Soft Hadronic Int. (Erice), R.T. Van de Walle (ed.), (World Scientific, Singapure, 1982) p.199.
37. N.S. Amelin, L.V. Bravina, L.N. Smirnova: *Sov. J. Nucl. Phys.* **52** (1990) 362.
38. E.A. De Wolf: Proc. XV. Int. Symp. on Multiparticle Dynamics (Lund), G. Gustafson (ed.), (World Scientific, Singapore,1984) p.1.
39. T. Sjöstrand: *Comp. Phys. Comm.* **27** (1982) 243.
40. E.A. De Wolf: *Acta Phys. Pol. B21* (1990) 611.
41. J. Bailly et al. (NA23): *Z. Phys.* **C40** (1988) 13.
42. F. Botterweck, Ph.D. thesis in preparation.
43. J. Alberty and A. Białas: *Z. Phys.* **C50** (1991) 315.
44. Ph. Brax and R. Peschanski: *Phys. Lett.* **B253** (1991) 225.

FACTORIAL CORRELATORS AND FACTORIAL CUMULANTS IN π^+p AND K^+p INTERACTIONS AT 250 GeV/c

EHS–NA22 Collaboration

Presented by

M. CHARLET
E.C. Guest Scientist, Physics Department, University of Nijmegen,
Toernooiveld 1, NL-6525 ED Nijmegen, The Netherlands

Abstract

A selected sample of 59232 π^+p and K^+p non-single-diffractive interactions with $\sqrt{s} = 22\,GeV$ is used to investigate one and two dimensional factorial correlators in rapidity and azimuthal angle. In the study of the one dimensional factorial correlators a shifting-technique is used. Furthermore, the factorial cumulants are investigated, the multi-particle contributions to the factorial moments are calculated, and the linked pair ansatz is tested and is found not to be valid for our interactions. An expansion of the multiplicity distributions in factorial cumulants allows us to test the significance of the factorial moments.

1 Introduction

While factorial moments measure local density fluctuations in phase space ([1], [2], [3] and a review can be found in [4]), the correlation between these fluctuations within a given event is measured by the factorial correlators ([2], [5] and [11]). Additional information can be obtained from the factorial cumulants ([11], [13], [14] and [9]), measuring the genuine multi-particle correlations. Here, the combinations of lower order correlations, present in the factorial moments, are removed.

This paper is organized as follows. In Section 2 the factorial correlators are analysed: definitions and notations are explained in §2.1; in the following paragraphs, results of the analysis in rapidity (§2.2), in azimuthal angle (§2.3) and the two-dimensional analysis (§2.4), are shown. In Section 3 we study the factorial cumulants: an introduction is given in §3.1, the multi-particle contributions are calculated in §3.2, a test of the linked pair approximation ([11], [13] and [14]) is shown in §3.3 and in §3.4 the factorial cumulants are used in an expansion of the multiplicity distributions ([15] and [16]), in order to check the significance of the cumulants calculated from our data. A Summary is given in Section 4.

The selection of the event sample is described in [8].

2 Factorial Correlators

2.1 Definitions

As for the calculation of the factorial moments, an initial interval Δy is divided into M bins, numbered $m = 1, 2, \ldots, M$; with bin size δy (see fig. 1a). Then the factorial correlator

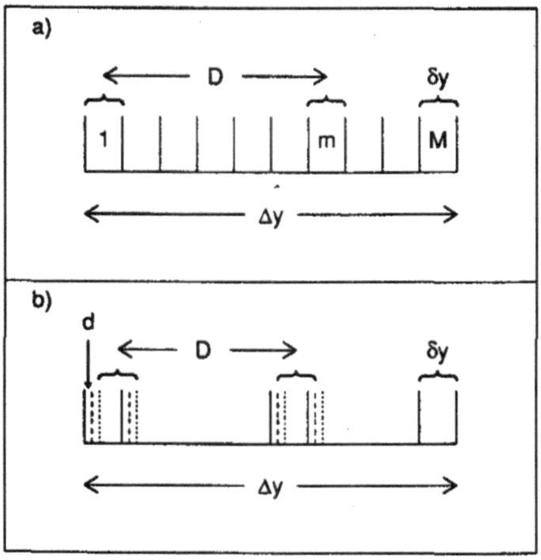

Figure 1: a) The initial interval Δy is divided in $M = \Delta y / \delta y$ bins with size δy; the distance D is measured from bin center to bin center and is a multiple of the bin size δy. b) Illustrates the bin shifting method; now there are $M = \Delta y / d$ overlapping bins of size $\delta y > d$, d being the size of the shift; the distance D is a multiple of d.

of order ij in bins m and m' is defined as

$$\langle F_{ij}^{(mm')} \rangle = \frac{\langle f_i^{(m)} f_j^{(m')} \rangle}{\langle f_i^{(m)} \rangle \langle f_j^{(m')} \rangle} \quad , \tag{1}$$

with

$$f_i^{(m)} = n_m(n_m - 1) \ldots (n_m - i + 1) \tag{2}$$

the unnormalized factorial of order i in bin m, n_m the multiplicity in bin m and $n_{m'}$ that in bin m'. The brackets $\langle\rangle$ stand for averaging over the events. The correlators are calculated at fixed δy for every combination mm' and then averaged over all combinations with the same distance $D = k\,\delta y$ between the bins (see fig. 1a).

$$\langle F_{ij}(D) \rangle = \frac{1}{M-k} \sum_{m=1}^{M-k} \langle F_{ij}^{(m,m+k)} \rangle \quad . \tag{3}$$

The maximum and minimum possible distance between the bins is determined by the bin size δy: $D_{max} = \Delta y - \delta y$ and $D_{min} = \delta y$. Since the π^+p and K^+p systems are not symmetric with respect to $y = 0.$, the factorial correlators are not symmetric w.r.t. the order ij; therefore we have to symmetrize them:

$$\langle F_{ij}^S \rangle = \frac{\langle F_{ij} \rangle + \langle F_{ji} \rangle}{2} \quad . \tag{4}$$

Henceforth, we omit the superscript S as we always use the symmetrized factorial correlators, except when explicitly stated otherwise.

In the calculation of the statistical errors on the factorial correlators, we first estimate the errors on the probabilities P_{n_m} (the probability to find n_m particles in bin m) and $P_{n_m,n'_{m'}}$ (the probability to find n_m particles in bin m and $n'_{m'}$ particles in bin m'):

$$\Delta P = \sqrt{\frac{1}{N_{ev}} P(1-P)} \quad , \tag{5}$$

then we apply the error propagation as if the nominator and the denominator were independent.

We investigate the one-dimensional correlators in rapidity and in azimuthal angle, using a technique of "shifting". The bin size δy is limited by statistics, but we can take the bin size quite large, because, as earlier shown in [5] and [2], the factorial correlators are independent of the bin size δy. Unfortunately, using large bin sizes has disadvantages too: when keeping the bins fixed as in fig. 1a, only a few different distances $D = k\,\delta y$ (with k integer) are possible. Furthermore, there are only few bin combinations mm' with the same D-value, especially for the largest distances.

A possible alternative is shown in fig. 1b. We now have $M = \Delta y/d$ overlapping bins of size $\delta y > d$ (numbered $m = 1, \ldots, M$). For each distance $D = ld$ (with l integer), the initial interval is scanned with the bin combination $m, m + l$, with shifts over a distance d, independently in bins m and $m + l$. Now, the distance D is not a multiple of the bin size δy anymore, but of a smaller size d, giving more values of D; the shifting also gives more combinations m, m' thus allowing us to look at larger bin distances.

The shifting technique is not applied for the calculation of the two-dimensional factorial correlators; those are investigated in rapidity and azimuthal angle for a few different definitions of distance.

2.2 Factorial Correlators in Rapidity

In the initial rapidity interval $\Delta y = [-2, 2]$ we use bins of size $\delta y = 0.4$, so $D_{max} = 3.6$ and $D_{min} = 0.4$. We use as shift parameter $d = 0.08$, i.e. 1/5 of the bin size. In fig. 2a the $\ln \langle F_{ij} \rangle$ are shown as a function of $-\ln D$. Just as in case of fixed bins the factorial correlators ([5] and [2]), $\langle F_{ij} \rangle$ increase with decreasing bin distance D.

At large bin distances D the factorial correlators become unstable. The fluctuations are stronger and already visible at smaller D for the higher orders.

Instead of the rapidity y, whose single particle distribution is not completely flat, one can use an integral variable $X(y)$ proposed by Białas and Gazdzicki [6], and Ochs [7].

$$X(y) = \frac{\int_a^y \rho(y')\,dy'}{\int_a^b \rho(y')\,dy'} \quad , \tag{6}$$

where $\rho(y')$ is the single particle distribution in rapidity, and $[a, b]$ is the initial rapidity interval. The single particle distribution in X is uniform in the interval $[0, 1]$. In fig. 2b the factorial correlators are shown in this new variable, they are still unstable at large distances, but the largest fluctuations have disappeared.

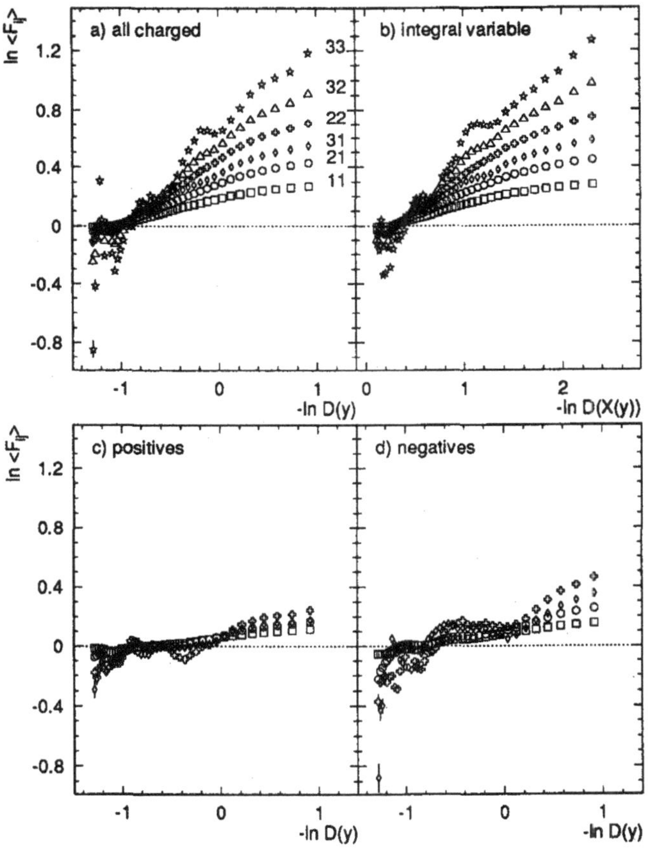

Figure 2: $\ln \langle F_{ij} \rangle$ as a function of $-\ln D$ for: (a) all charged particles, (b) positive charged particles, (c) negative charged particles, (d) all charged particles, using the integral variable (6).

For positive (fig. 2c) and negative (fig. 2d) separately, the factorial correlators behave the same, but become instable at lower orders already.

As said earlier, the non-symmetrized $\langle F_{ij} \rangle$ are not symmetric w.r.t. the order ij. Therefore we show in fig. 3a,b,c the quantity $(\langle F_{ij} \rangle - \langle F_{ji} \rangle)$ / $(\langle F_{ij} \rangle + \langle F_{ji} \rangle)$ as a function of $-\ln D(y)$, were the F_{ij} is not symmetrized. The asymmetry increases with increasing bin distance D and with increasing order ij, but is always smaller than 10%. In fig. 3d,e,f, we show the same quantity, but now the F_{ij} are calculated for the integral variable $X(y)$ (6), one can see that the asymmetry remains when using a variable which has a flat single particle distribution.

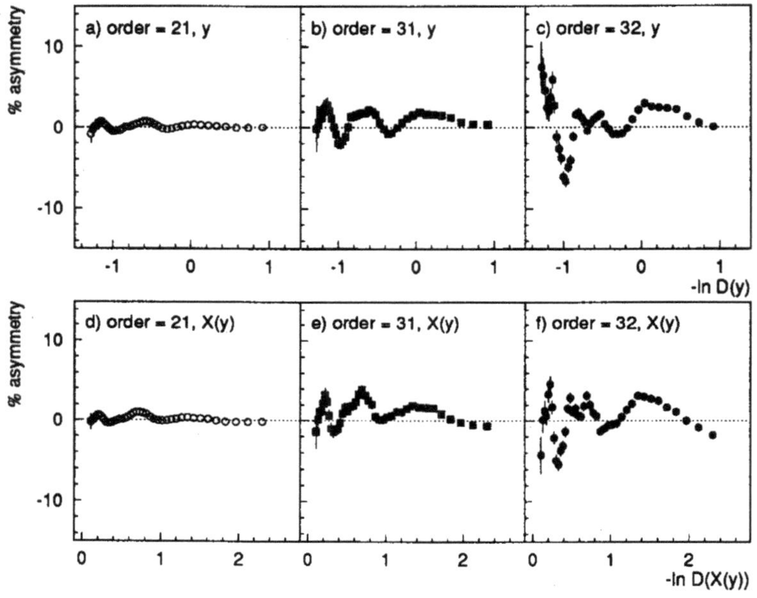

Figure 3: The amount of asymmetry, defined as $(\langle F_{ij} \rangle - \langle F_{ji} \rangle)$ / $(\langle F_{ij} \rangle + \langle F_{ji} \rangle)$, were F_{ij} is not symmetrized, as a function of $-\ln D$; for rapidity ((a) order 21, (b) order 31, (c) order 32) and for the integral variable (6) ((d) order 21, (e) order 31, (f) order 32).

2.3 Factorial Correlators in Azimuthal Angle

When one wants to investigate the factorial correlators in azimuthal angle φ, one has to be aware of the fact that φ is a cyclic variable, the maximum bin distance D being π (see fig. 4).

The initial $\Delta\varphi$ interval equals 2π. We take the bin size $\delta\varphi = \pi/5$ and the shift parameter $d = \pi/25$, again 1/5 of the bin size. The particles selected for the investigation in azimuthal angel have rapidity in the interval $\Delta y = [-2, 2]$, the initial interval of the study in rapidity. In fig. 5 we show $\ln \langle F_{ij} \rangle$ as a function of $-\ln D(\varphi)$, contrary to the behaviour in rapidity y, the factorial correlators in φ decrease with decreasing bin distance D. Even when we use a bin size $\delta\varphi = \pi/15$ in order to look at smaller bin distances, the factorial correlators do not rise with decreasing bin distance (not shown). Note that also the factorial moments in φ first decrease with decreasing bin size $\delta\varphi$, till $\delta\varphi$ reaches the value $2\pi/5$ and then increase for smaller $\delta\varphi$ ([8],[10]). It is believed that this behaviour is caused by local momentum conservation and resonances, producing particles far apart from each other in azimuth.

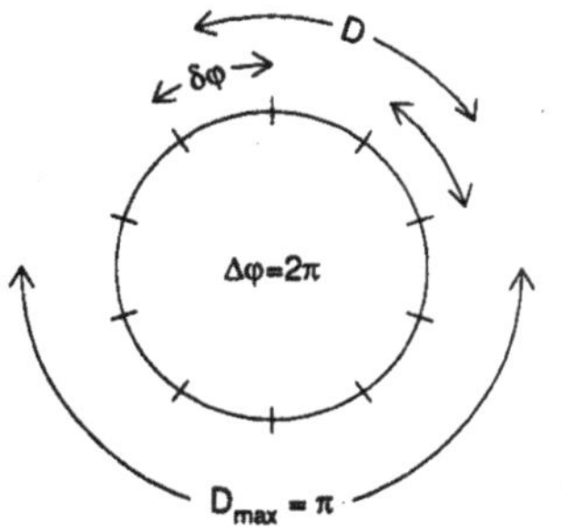

Figure 4: Illustration of the variables used in the calculation of the factorial correlators in φ: the initial interval $\Delta\varphi = 2\pi$, the bin size $\delta\varphi = \pi/5$ and the bin distance D, the maximum bin distance equals π. The shiftparameter $d = \delta\varphi/5 = \pi/25$ is not shown.

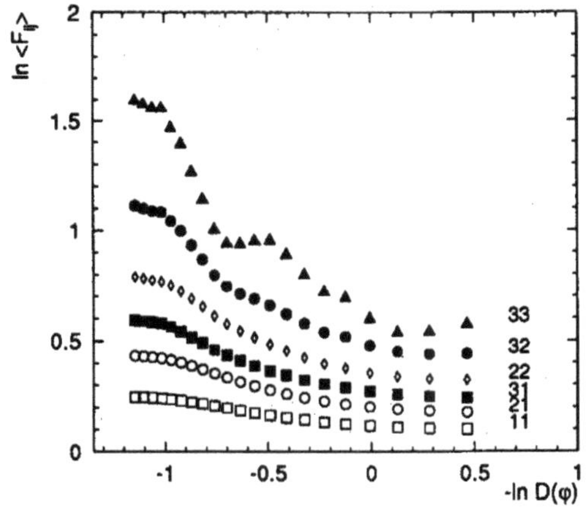

Figure 5: $\ln\langle F_{ij}\rangle$ as a function of $-\ln D(\varphi)$. Contrary to the factorial correlators in y, those in φ decrease with decreasing bin distance $D(\varphi)$.

2.4 Two-Dimensional Factorial Correlators

A few problems arise in the study of the two-dimensional factorial correlators. Firstly, there is no unambiguous definition of distance anymore. Secondly, the opposite behaviour in y and in φ of the one-dimensional factorial correlators and the cyclic character of the variable φ, give rise to discontinuous behaviour of the factorial correlators, when one takes equal numbers of bins in both variables. This problem is tackled by choosing a binning that yields an equal number of possible bin distances. We choose:

$$D_y = k_y \, \delta y = k_y \, 0.4 \quad , \tag{7}$$

$$D_\varphi = k_\varphi \, \delta\varphi = k_\varphi \frac{\pi}{9} \quad , \tag{8}$$

this means 10 rapidity bins, 18 bins in azimuthal angle and 9 possible distances in both variables. We consider the following three definitions of distance:

$$D = \sqrt{{D_y}^2 + {D_\varphi}^2} \quad , \tag{9}$$

$$D = \sqrt{{k_y}^2 + {k_\varphi}^2} \quad , \tag{10}$$

and a kind of "nearest neighbours" (see fig. 6). (If we use the second definition, we can see where the discontinuous behaviour, in the case of taking an unequal number of distances, comes from: if we divide both intervals in 10 bins, then, for $k_y = 1$ and $k_\varphi = 5$, the factorial correlator is large. The counterpart, that gives the same distance, is the combination $k_y = 5$ and $k_\varphi = 1$ and now the factorial correlator is small. After averaging over constant D we have an intermediate value. But the combination $k_y = 9$ and $k_\varphi = 1$, leading to a small value, has no counterpart $k_y = 1$ and $k_\varphi = 9$ with a large value of the factorial correlator. Thus after averaging the small value remains.)

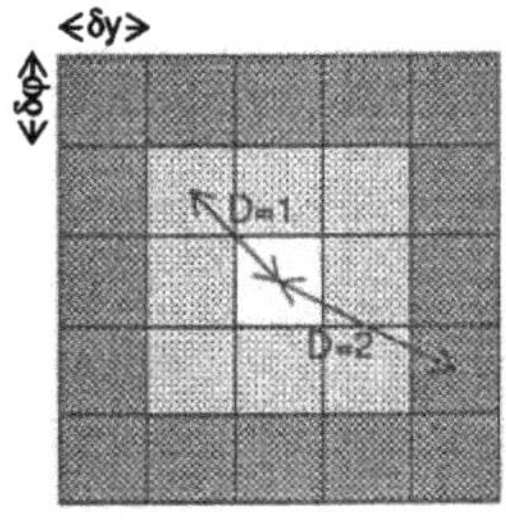

Figure 6: Definition of distance in the case of nearest neighbours. All squares with the same collour, have the "same" distance from the central blanc square.

In fig. 7 we show the $\ln \langle F_{ij} \rangle$ as a function of $-\ln D$. One can see that the definition of distance is important. Definition (9), shown in fig. 7a, gives the factorial correlators a rather random appearance, with the use of definition (10) or the "nearest neighbour"

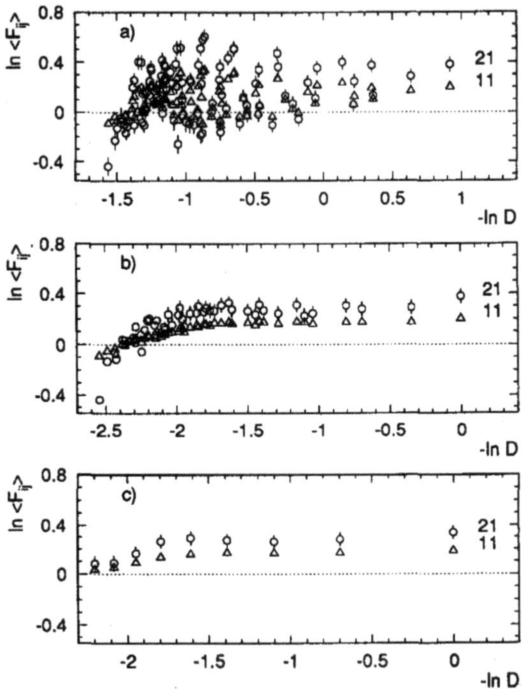

Figure 7: The ln of the two-dimensional factorial correlators as a function of $-\ln D$, with a) $D^2 = {D_y}^2 + {D_\varphi}^2$, b) $D^2 = {k_y}^2 + {k_\varphi}^2$, c) D defined for the "nearest neighbours" (fig. 6).

definition, the factorial correlators first increase a little with decreasing distance, but soon saturate.

Maybe, one could use another definition of distance, but the different behaviour of the one-dimensional correlators in y and in φ will always be hard to deal with.

3 Factorial Cumulants

3.1 Definitions

We recall that the unnormalized factorial moment of order i in bin m is defined as

$$\langle f_i^{(m)} \rangle = \langle n_m(n_m - 1)\ldots(n_m - i + 1) \rangle \quad , \tag{11}$$

and the vertically averaged normalized moment as

$$\langle F_i \rangle_V^{(M)} = \frac{1}{M} \sum_{m=1}^{M} \frac{\langle f_i^{(m)} \rangle}{\langle n_m \rangle^i} = \left[\frac{\langle f_i^{(m)} \rangle}{\langle n_m \rangle^i} \right]_M \quad , \tag{12}$$

where the $\langle\,\rangle$ brackets denote event averaging and the $[\,]$ brackets the bin averaging.

A general relation between the factorial moments $\langle f_j\rangle$ and the factorial cumulants $\langle k_i\rangle$ can be found in [12].

$$\langle k_i\rangle = i! \sum_{n_j} \left\{ (-1)^{(n_1-1)}(n_1-1)!\,\delta(i-\sum_{j=1}^{i} n_j) \prod_{j=1}^{m} \frac{1}{(n_j-n_{j+1})!}\left(\frac{\langle f_j\rangle}{j!}\right)^{(n_j-n_{j+1})} \right\} \quad , \quad (13)$$

where the first sum runs over all $n_j \geq n_{j+1}$, $n_1 \geq 1$ and $j = 1,\ldots,i$. For the orders $i = 1,\ldots,5$ one finds:

$$\langle k_1\rangle = \langle f_1\rangle \quad (14)$$

$$\langle k_2\rangle = \langle f_2\rangle - \langle f_1\rangle^2 \quad (15)$$

$$\langle k_3\rangle = \langle f_3\rangle - 3\langle f_2\rangle\langle f_1\rangle + 2\langle f_1\rangle^3 \quad (16)$$

$$\langle k_4\rangle = \langle f_4\rangle - 4\langle f_3\rangle\langle f_1\rangle - 3\langle f_2\rangle^2 + 12\langle f_2\rangle\langle f_1\rangle^2 - 6\langle f_1\rangle^4 \quad (17)$$

$$\begin{aligned}\langle k_5\rangle &= \langle f_5\rangle - 5\langle f_4\rangle\langle f_1\rangle - 10\langle f_3\rangle\langle f_2\rangle + 20\langle f_3\rangle\langle f_1\rangle^2 \\ &\quad +30\langle f_2\rangle^2\langle f_1\rangle - 60\langle f_2\rangle\langle f_1\rangle^3 + 24\langle f_1\rangle^5 \quad . \end{aligned} \quad (18)$$

Then the vertically averaged normalized factorial cumulants are defined as

$$\langle K_i\rangle_V^{(M)} = \frac{1}{M}\sum_{m=1}^{M}\frac{\langle k_i^{(m)}\rangle}{\langle n_m\rangle^i} = \left[\frac{\langle k_i^{(m)}\rangle}{\langle n_m\rangle^i}\right]_M \quad . \quad (19)$$

The cumulant of order i is a measure of the genuine i-particle correlation, combinations of lower order correlations are removed.

3.2 Contributions of Multi-particle Correlations

Formula (13) can be inverted ([12])

$$\langle f_i\rangle = i! \sum_{n_j} \left\{ \delta(i-\sum_{j=1}^{i} n_j) \prod_{j=1}^{m} \frac{1}{(n_j-n_{j+1})!}\left(\frac{\langle k_j\rangle}{j!}\right)^{(n_j-n_{j+1})} \right\} \quad , \quad (20)$$

where the first sum runs over all $n_j \geq n_{j+1}$ and $j = 1,\ldots,i$. Or for the orders $i = 2,\ldots,5$:

$$\langle f_2\rangle = \langle k_1\rangle^2 + \langle k_2\rangle \quad (21)$$

$$\langle f_3\rangle = \langle k_1\rangle^3 + 3\langle k_2\rangle\langle k_1\rangle + \langle k_3\rangle \quad (22)$$

$$\langle f_4\rangle = \langle k_1\rangle^4 + 6\langle k_2\rangle\langle k_1\rangle^2 + 3\langle k_2\rangle^2 + 4\langle k_3\rangle\langle k_1\rangle + \langle k_4\rangle \quad (23)$$

$$\begin{aligned}\langle f_5\rangle &= \langle k_1\rangle^5 + 10\langle k_2\rangle\langle k_1\rangle^3 + 15\langle k_2\rangle^2\langle k_1\rangle + 10\langle k_3\rangle\langle k_1\rangle^2 \\ &\quad +10\langle k_3\rangle\langle k_2\rangle + 5\langle k_4\rangle\langle k_1\rangle + \langle k_5\rangle \quad . \end{aligned} \quad (24)$$

One can use (20) to study the contribution to the factorial moments of the multi-particle correlations up to some order. If in (20) one sets $\langle k_j\rangle = 0$ for $j > 2$, the two-particle correlation contribution to the factorial moments is found; setting $\langle k_j\rangle = 0$ for $j > 3$ gives the sum of two and three particle correlation contributions and so on (see fig. 8a, b, c). The difference between the factorial moment $\langle F_i\rangle$ of order i and the contribution of the particle correlations of order $\leq i-1$ is the cumulant $\langle K_i\rangle$ of order i. The cumulants of orders 3, 4 and 5 are shown in fig. 8d, e, f; we do not show $\langle K_2\rangle = \langle F_2\rangle - 1$.

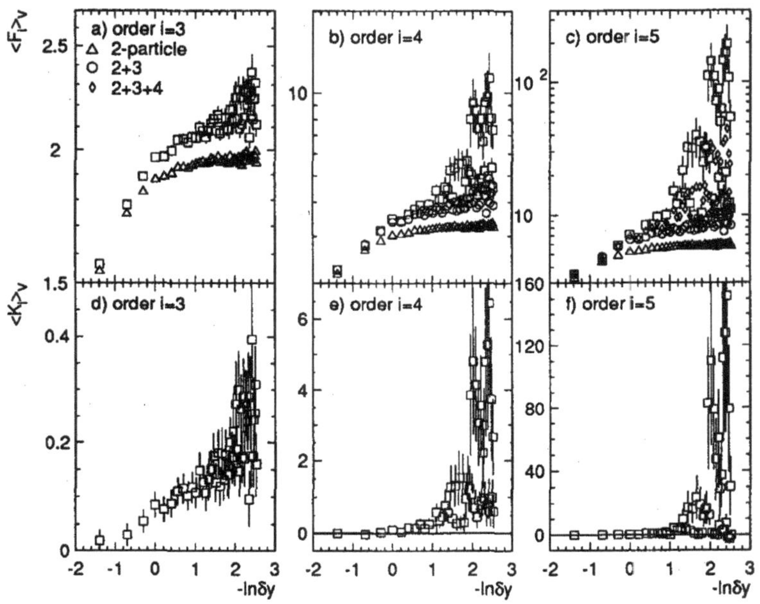

Figure 8: The factorial moments $\langle F_i \rangle_V$ (□) together with the contributions of : the 2-particle correlations (△), the sum of 2- and 3-particle correlations (○) and the sum of 2,3 and 4-particle correlations (◇); for order $i = 3$ (a), $i = 4$ (b), $i = 5$ (c).

Due to low statistics, the errors on the contributions are quite large and are omitted for clearness; furthermore, for small δy, there are large fluctuations, so only a general trend can be seen. The contributions of the multi-particle correlations of higher orders increase with decreasing bin size δy; for large δy the factorial moments are almost completely determined by the lowest order correlations. For small bin sizes (i.e. large $-\ln \delta y$) there is a sizeable amount of genuine higher order correlations.

3.3 Linked Pair Approximation

In the linked pair approximation one assumes that the higher order normalized correlation functions are combinations of the second order normalized correlation function. In combination with the strip approximation one finds [13] that the following ratio should be independent of the bin size δy:

$$A_i = \frac{\left[\frac{\langle k_i^{(m)} \rangle}{\langle n_m \rangle^i}\right]_M}{\left[\left(\frac{\langle k_2^{(m)} \rangle}{\langle n_m \rangle^2}\right)^{(i-1)}\right]_M} \quad . \tag{25}$$

In fig. 9 these ratios are shown for (a) $i = 3$, (b) $i = 4$, (c) $i = 5$. We see that the ratios increase with decreasing δy (only shown for $\delta y \geq 0.167$) and are not at all constant.

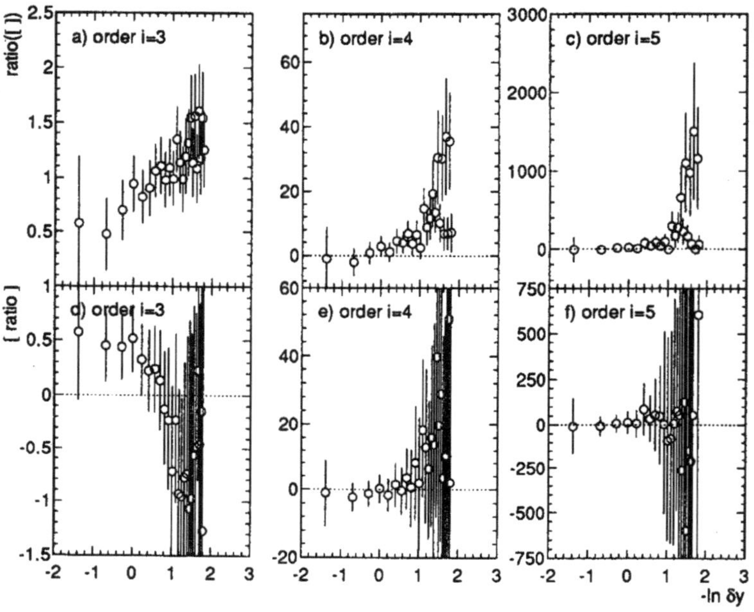

Figure 9: The ratios (25) for $i = 3$ (a), $i = 4$ (b), $i = 5$, (c) and the ratios (26) for $i = 3$ (d), $i = 4$ (e), $i = 5$ (f).

Instead of taking the ratio of the bin averages, we also tried the inverse, taking the ratios first and averaging over the bins; so we considered

$$\left[\frac{\left(\frac{\langle k_i^{(m)} \rangle}{\langle n_m \rangle^i} \right)}{\left(\frac{\langle k_2^{(m)} \rangle}{\langle n_m \rangle^2} \right)^{(i-1)}} \right]_M , \tag{26}$$

as function of δy. In fig. 9d, e, f, we can see that the picture has changed completely, the errors have become much larger and the values fluctuate a lot, making it hard to draw any conclusion.

3.4 Expansion of the Multiplicity Distribution in Factorial Cumulants

In [15], Buras derives an expansion of the multiplicity distribution in terms of the cumulants $\langle k_i \rangle$. If one takes the cumulants of the first m orders in the expansion, this is called the multiplicity distribution of a generalized Feynman gas of order m. Such an expansion contains a sum of products of Poisson distributions; the probability $P_n^{(m)}$ to find n particles for an expansion of order m is given by:

$$P_n^{(m)} = \sum_{\{m_i\}} \left[\bar{\theta}(t)\mathcal{P}_t\left(\gamma_1^{(m)}\right) \prod_{i=2}^{m} \mathcal{P}_{m_i}\left(\gamma_i^{(m)}\right) \right] \quad , \tag{27}$$

where the summation runs over all posible integers $m_i \geq 0$ and $i = 2, \ldots, m$; and

$$t = n - \sum_{i=2}^{m} i m_i \quad , \tag{28}$$

$$\begin{aligned} \bar{\theta}(t) &= 1 \quad t \geq 0 \\ &= 0 \quad t < 0 \quad , \end{aligned} \tag{29}$$

$$\mathcal{P}_n(x) = \frac{x^n}{n!} e^{-x} \quad , \tag{30}$$

the Poisson distribution and

$$\gamma_l^{(m)} = \frac{1}{l!} \sum_{i=l}^{m} (-1)^{(i-l)} \frac{\langle k_i \rangle}{(i-l)!} \quad , \tag{31}$$

where the dependence of the cumulants $\langle k_i \rangle$ shows up.

If $m = 1$, the expansion reduces to a simple Poisson distribution, which is known to be too narrow to fit multiplicity distributions (see fig. 10(a) for the initial rapidity interval $[-2,2]$ and for $M = 41$, (c), (h) and (m) for bin 19, bin 20 and the bin averaged multiplicity distribution). Therefore we have to use higher order expansions to get a good description of the experimental multiplicity dstributions.

Formula (27) can be used to judge the significance of the factorial cumulants calculated from the data ([16]). The idea is that the description of the multiplicity distribution will improve when increasing the order m, this implies the use of factorial cumulants up to order m. When the higher order factorial cumulants are unreliable, no further improvement will be found. Unphysical values of $P_n^{(m)}$, such as $P_n^{(m)} < 0$ or $P_n^{(m)} > 1$, usually indicate that one must go further in the approximation (use of higher order factorial cumulants) or just the opposite, one has gone too far and factorial cumulants with too large errors have been used.

The multiplicity distribution depends strongly on the size and the center of the considered rapidity interval. This is expressed by many different values of $\langle k_i \rangle$, calculated from the experimental multiplicity distributions, and the precision with which they are determined. So also the $P_n^{(m)}$ behave differently for the various intervals and general conclusions are difficult to draw, though following observations can be made.

- $P_n^{(m)}$ oscillates around the experimental values. The oscillations are small when the approximation is good (e.g. fig. 10b: P_n and $P_n^{(3)}$ for $-2 \leq y \leq 2$) and become larger for bad approximations (fig. 10e: P_n and $P_n^{(3)}$ for the 19th bin, $M = 41$).

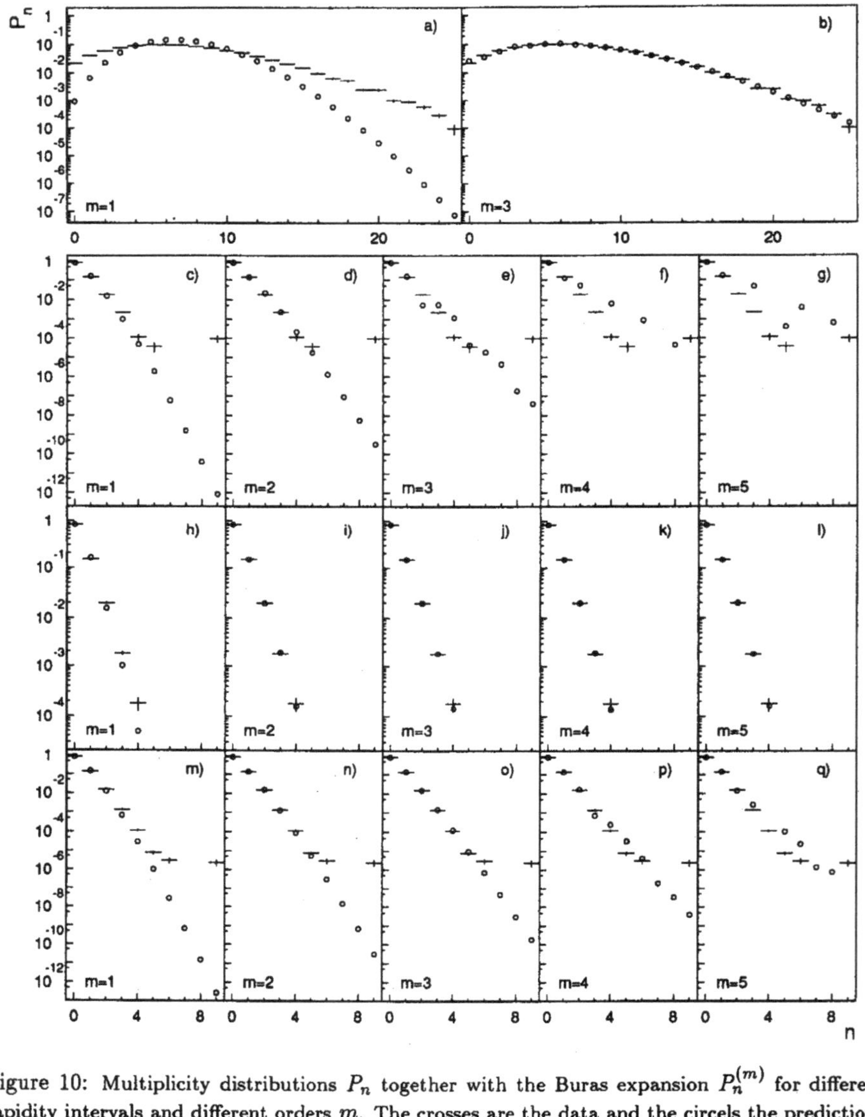

Figure 10: Multiplicity distributions P_n together with the Buras expansion $P_n^{(m)}$ for different rapidity intervals and different orders m. The crosses are the data and the circels the predictions $P_n^{(m)}$. (a) and (b) show the distributions for the rapidity interval $[-2, 2]$, then the interval is divided into 41 bins and in (c-g) bin 19 is shown and in (h-l) bin 20. The bin averaged multiplicity distribution is shown in (m-q). When no value of $P_n^{(m)}$ is shown, the value is found to be negative.

- For most of the multiplicity distributions, the expansion to the third order gives a good approximation. (see fig. 10)

- When a "spike" with large multiplicity is present in the bin under consideration, one expects the higher order cumulants to contribute a lot to the probability expansion. However, the tail of the distribution is never reproduced, on the contrary, for higher orders one gets unphysical results (negative probabilities, see fig. 10f,g).

- In the absence of spikes, the multiplicity distribution is smooth and can easily be reconstructed by the expansion (27). Using higher order factorial cumulants in the approximation usually does not lead to a strange behaviour of $P_n^{(m)}$. There is even a slight improvement for higher order $P_n^{(m)}$ (e.g. fig. 10h-l).

- After averaging the multiplicity distribution over the bins, the tail caused by the spike has not completely disappeared. Again the Buras expansion is not able to reproduce the distribution and negative values of higher order $P_n^{(m)}$ can be found (see fig. 10m-q).

4 Summary

- The technique of bin-shifting allows to look at bin distances not equal to a multiple of the bin size and at larger bin distances. The factorial correlators in rapidity still increase with decreasing bin distance.

- At large bin distances the factorial correlators are not stable, they fluctuate. The fluctuations are largest for the highest orders and they become stronger when positive and negative tracks are taken separately. When we transform our rapidity distribution to a flat distribution, only the largest fluctuations disappear.

- In our π^+ and K^+ data the factorial correlators are not symmetric w.r.t. the order ij. The asymetry increases with increasing bin distance and with increasing order and remains after transformation to a flat distribution.

- Contrary to the factorial correlators in y the factorial correlators in azimuthal angle φ decrease with decreasing bin distance.

- Due to the different behaviour in y and φ of the factorial correlators, it is not obvious how to study the two-dimensional factorial correlators. A good definition of distance is needed.

- The contributions of genuine higher order particle correlations to the factorial moments increase with decreasing bin size.

- The linked pair approximation predicts the ratio A_i to be constant, while for our experiment this ratio increases with decreasing bin size. If we take the ratio before bin averaging, the picture becomes completely different: large fluctuations and very large errors.

- The Buras expansion of the multiplicity distribution in factorial cumulants provides a tool to check the significance of the factorial cumulants. The big variety in multiplicity distributions makes it difficult to draw general conclusions. Usually the third order expansion follows the experimental distribution quite well. If the tail of the distribution is "spiky" the expansion is not able to reconstruct it.

References

[1] A. Białas and R. Peschanski, *Nucl. Phys.* **B273** (1986) 703.

[2] A. Białas and R. Peschanski, *Nucl. Phys.* **B308** (1988) 857.

[3] A. Białas and R. Peschanski, *Phys. Lett.* **B207** (1988) 59.

[4] W. Kittel and R. Peschanski, *Nucl. Phys.* B(Proc. Suppl.)**16**(1988) 445.

[5] V.V. Aivazyan et al., *Phys. Lett.* **B258** (1991) 487.

[6] A. Białas and M. Gazdzicki, *Phys. Lett.* **B252** (1990) 483.

[7] W. Ochs, *Phys. Lett.* **247** (1990) 101.

[8] F. Botterweck, W. Kittel, F.K. Rizatdinova and L. Smirnova, *contribution to these proceedings.*

[9] P. Lipa, *contribution to these proceedings.*

[10] F. Botterweck, *PhD. Thesis in progress.*

[11] H.C. Eggers, *PhD. Thesis.*

[12] A.H. Mueller, *Phys. Rev.* **D4** (1971) 150.

[13] P. Carruthers, H.C. Eggers and Ina Sarcevic, *Phys. Lett.* **B254** (1991) 258.

[14] P. Carruthers and Ina Sarcevic *Phys. Rev. Lett.* **63** (1989) 1562.

[15] A.J. Buras, *Nucl. Phys.* **B56** (1973) 275.

[16] S. Hegyi and S. Krasznovszky, *Phys. Lett.* **B251** (1990) 197.

Fluctuations and correlations in deep-inelastic muon-nucleon scattering at 280 GeV

I. Derado, J. Figiel, G. Jancso, N. Schmitz,
A. Söldner-Rembold, P. Stopa

Max-Planck-Institut für Physik
München, Germany

presented by N. Schmitz

Abstract

Using data of the European Muon Collaboration on muon-proton and muon-deuteron scattering at 280 GeV, an intermittency analysis in one, two and three phase space dimensions is carried out. In addition, a factorial correlator analysis and a multifractal analysis, both in one dimension (rapidity), are presented.

1. Introduction

This paper continues our previous investigation [1, 2] of fluctuations (intermittency) in hadron production by muon–nucleon interactions at 280 GeV. The data come from the NA9 experiment of the European Muon Collaboration (EMC), using the EMC spectrometer [3] in the 280 GeV muon beam at the CERN SPS. Data were taken both with a hydrogen and a deuterium target.

The same selection criteria were applied to the event samples as in refs. [1, 2] , namely

$$\begin{array}{ll} 4 < Q^2 < 200 \text{ GeV}^2 & E'_\mu > 20 \text{ GeV} \\ 4 < W < 20 \text{ GeV} & \nu/E_\mu < 0.9 \\ 20 < \nu < 260 \text{ GeV} & \Theta_\mu > 0.75^\circ \end{array}$$

After these cuts the two event samples consist of $\sim 25500\, \mu p$ events and $\sim 20600\, \mu d$ events, with an average hadronic mass of $\langle W \rangle \approx 13.4$ GeV and an average charged multiplicity of $\langle n \rangle \approx 6.3$. About one third of the events have a $W > 16$ GeV. For the subsequent analyses the two event samples were combined, since no significant differences between the μp and μd results were found.

As in refs. [1, 2] the corrections for various effects were extensively studied by Monte Carlo simulations. It was found that the corrections did not affect the results of these analyses. They were therefore not applied to the data, but the measured data were rather used directly.

In this paper we present an intermittency analysis in 1, 2 and 3 dimensions, a factorial correlator analysis and a multifractal analysis of our data.

2. Intermittency

In the field of intermittency progress has been made recently in the following three aspects:

i.) Ochs [4–6] has pointed out the importance of carrying out intermittency analyses not only in one phase space dimension (usually rapidity y), but also in two and three dimensions (azimuthal angle φ and transverse momentum p_T in addition to y). This is because particle density fluctuations are expected to occur in the full phase space such that the intermittency signal should be stronger in a higher phase space dimension and should be washed out when projecting onto a lower dimension. We therefore present an intermittency analysis also in two and three dimensions (y, φ, lnp_T) where the following intervals for the kinematic variables are selected:

$$-3 < y < 3$$
$$-\pi < \varphi < \pi$$
$$-5 < lnp_T < 0.7 \quad (0.007 < p_T < 2.0 \text{ GeV}) .$$

In the following we denote by M the number of linear bins in one variable (we choose $M = M_y = M_\varphi = M_{lnp_T}$) such that $M_d = M^d$ is the number of boxes in d dimensions (d = 1, 2, or 3).

ii.) In general the single particle density distributions $\rho(x)$ in the original phase space variables $x = y, \varphi, lnp_T$ are not flat, which leads to trivial density fluctuations. In order to eliminate such fluctuations Ochs [6] has proposed to use transformed variables $\tilde{x} = \tilde{y}, \tilde{\varphi}, \widetilde{lnp_T}$ with flat density distributions. The transformed variables are defined as

$$\tilde{x}(x) = \int_a^x \rho(x')dx' \Big/ \int_a^b \rho(x')dx' \tag{1}$$

with the interval $a < x < b$. A more sophisticated method of transformation, which takes into account the correlation between the three variables (i.e. $\rho(y, \varphi, lnp_T) \neq \rho(y) \cdot \rho'(\varphi) \cdot \rho''(lnp_T)$), has been introduced in ref. [7].

Throughout our intermittency analysis presented here we use transformed variables according to eq. (1), thus neglecting the correlations.

iii.) In a log–log plot of a factorial moment F_i vs. the number of boxes M_d the data points are strongly correlated since they originate from the same distribution of the particles in phase space. These correlations should be taken into account, as pointed out by B. Wosiek [8], by computing the experimental covariance matrix, when estimating realistically the errors on the data points and when determining intermittency indices (slopes).
As it turns out, a log–log plot of F_i vs. M_d is in general not a straight line over a wider range of subdivisions as expected for ideal intermittency. In the figures below we show therefore the local slopes $\varphi_i(M_d)$, defined as

$$\varphi_i = \frac{\Delta ln F_i}{\Delta ln M^d}, \tag{2}$$

as functions of M_d, taking into account the correlations between the data points. Since the φ_i values fluctuate rather strongly, in particular at higher M_d, they were smoothened by fitting to several adjacent points in the F_i vs. M_d plot.

We now present some new results of our intermittency analysis where the normalized factorial moments are defined with the usual notation as ("horizontal averaging")

$$F_i = \frac{1}{M^d} \sum_{k=1}^{M^d} \frac{\langle n_k(n_k-1)\ldots(n_k-i+1)\rangle}{(\langle N\rangle/M^d)^i} \tag{3}$$

(N = total multiplicity, n_k = multiplicity in k-th box). Fig. 1a shows F_2 and F_3 in 3 dimensions for all charged particles, together with the corresponding local slopes (Fig. 1b). The second moment shows a concave rise which becomes nearly linear at larger M^3, leading to a slope approaching $\sim$ 0.27. It is interesting to note that $ln(F_2 - 1)$ vs. $ln M^3$ (Fig. 1c, $F_2 - 1 = K_2$ is the second normalized factorial cumulant [9]) is very close to a straight line over the whole M^3 range; this observation is further discussed by Fiałkowski [10] .

Fig. 2 shows the moments F_2 and F_3 in 3 dimensions for positive and negative hadrons separately, together with the local slopes for F_2. The slopes for positives and negatives are seen to be rather similar; they are close to the slope for all charged hadrons (Fig. 1).

In Fig. 3 we compare the moments F_2 in 3 dimensions for three different rapidity regions, namely $|y| < 1.0$, $1.0 < |y| < 1.75$ and $1.75 < |y| < 3.0$. These three regions have rather different average particle densities $\langle\rho\rangle = \langle\frac{1}{N_{ev}}\frac{dN}{dy}\rangle$, namely $\langle\rho\rangle$ = 1.62, 1.27 and 0.45 respectively. Although the F_2 values themselves are different in the three y regions, the local slopes φ_2 are rather similar. Thus, φ_2 seems to be approximately independent of the rapidity region (central or peripheral), i.e. of the particle density.

Fig. 4 shows log–log plots of F_3, F_4 and F_5 vs. F_2 ("Ochs plots" [4–6]) in 1, 2 and 3 dimensions. The points are seen to fall approximately on straight lines without intercept, as expected for the case that the moments obey a generalized power law (The slopes in an "Ochs plot" are given by the ratios φ_i/φ_2 of intermittency indices) [4–6]. Furthermore the points of a given higher moment in different dimensions seem to lie on the same straight line.

Finally in Fig. 5 we present the factorial moments in 2 dimensions, namely $(\tilde{y}, \tilde{\varphi})$ and $(\tilde{y}, \widetilde{lnp_T})$, together with their local slopes. The moments show a rather linear lnM^2 dependence with slopes which are in general smaller than in 3 dimensions.

3. Factorial Correlators

Factorial correlators give information about the correlations between local density fluctuations in phase space. They have been introduced by Bialas and Peschanski [11] and are defined (e.g. in rapidity) as follows:

The total rapidity interval Δ (here: $-2 < y < 2$, i.e. $\Delta = 4$) is subdivided into M bins of width $\delta y = \Delta/M$. Let D be the rapidity distance between the k-th and l-th bin, i.e. $D = d \cdot \delta y$ with $d = |k - l|$ being the correlation length. The factorial correlator F_{ij} of order i, j between the k-th and l-th bin is then defined as

$$F_{ij}(k,l,\delta y) = \frac{\langle n_k(n_k-1)\ldots(n_k-i+1)n_l(n_l-1)\ldots(n_l-j+1)\rangle}{\langle n_k(n_k-1)\ldots(n_k-i+1)\rangle\langle n_l(n_l-1)\ldots(n_l-j+1)\rangle} \tag{4}$$

where n_k and n_l are the numbers of particles in the k-th and l-th bin respectively. $\langle\rangle$ denotes averaging over all events.

Usually, in order to increase statistics, the average is taken over all bin combinations with a given distance D, i.e.

$$C_{ij}(D,\delta y) = \frac{1}{M-d}\frac{1}{2}\sum_{k=1}^{M-d}\left[F_{ij}(k,k+d,\delta y) + F_{ji}(k,k+d,\delta y)\right]. \tag{5}$$

Then the correlators are studied as functions of distance D and bin width δy.

A simple intermittency model (α –model [11]) makes the following predictions: At fixed D, the correlators C_{ij} should be independent of δy; they should show a power–law dependence on D,

$$C_{ij} \propto D^{-\varphi_{ij}}, \tag{6}$$

and the exponents (slopes) φ_{ij} should be connected with the normal intermittency indices φ_i by

$$\varphi_{ij} = \varphi_{i+j} - \varphi_i - \varphi_j = ij\varphi_2 \tag{7}$$

where the first equality holds for the α –model and the second one in the log–normal approximation [11].

The following two figures present our results for charged hadrons. Fig. 6 shows log–log plots of the factorial correlators C_{ij} for various pairs (i,j) as functions of $1/D$ for four different bin widths $\delta y = 0.4, 0.2, 0.13, 0.1$. The increase of lnC_{ij} with

$-lnD$ is not strictly linear as predicted, but the slopes φ_{ij} rather depend on D. The rise of lnC_{ij} towards small values of $-lnD$ (i.e. large D) is due to long-range correlations (probably energy-momentum conservation).

In Fig. 7 several C_{ij} are plotted vs. $1/\delta y$ for two values of $D = 0.4$ and 0.8. The correlators are seen to be rather independent of δy, as predicted. A similar behaviour has been observed by the NA22 collaboration [12] in $\pi^+ p$ and $K^+ p$ scattering.

4. Multifractality

In a series of papers [13–18] Hwa and others have proposed to investigate the multifractal properties of particle density distributions in phase space by determining the so-called G-moments. In one dimension (rapidity), let the full rapidity interval Δ (here: $-3 < y < 3$, i.e. $\Delta = 6$) be subdivided into M bins of width $\delta y = \Delta/M$. Then for a given event, the G-moments G_q of order q (with q being any integer or non-integer positive or negative number) are computed as

$$G_q(M) = \sum_k {}' p_k^q \tag{8}$$

with $p_k = n_k/N$ where n_k is the multiplicity in the k-th bin and N the total multiplicity in Δ ($\sum_k p_k = 1$). The sum in (8) is over all non-empty bins (i.e. with $n_k > 0$). For very fine subdivisions such that $n_k = 0$ or 1 one obtains saturation, i.e. independence of M:

$$G_q(M) \to N\left(\frac{1}{N}\right)^q = N^{1-q} \text{ for } M \to \infty . \tag{9}$$

Furthermore, for $M = 1 : G_q(1) = 1$ for all q; and for $q = 1 : G_1(M) = 1$ for all M. Usually, in order to increase the statistics, the average is taken over all events, yielding $\langle G_q(M)\rangle$ or $\langle lnG_q(M)\rangle$.

Once the G-moments are computed the multifractal analysis proceeds in the following manner: From $\langle lnG_q(M)\rangle$ vs. lnM one determines the slopes

$$\tau(q, M) = -\frac{\partial\langle lnG_q(M)\rangle}{\partial lnM} \tag{10}$$

the derivatives of which (at fixed M) are the Hölder exponents $\alpha(q, M)$:

$$\alpha(q, M) = \frac{\partial\tau(q, M)}{\partial q} . \tag{11}$$

The fractal dimensions $f(q, M)$ are given by

$$f(q, M) = q \cdot \alpha(q, M) - \tau(q, M) . \tag{12}$$

Eliminating q yields the spectral function $f(\alpha, M)$ vs. α with

$$\left.\frac{\partial f(\alpha, M)}{\partial\alpha}\right|_{\alpha(q,M)} = q , \tag{13}$$

i.e. $\alpha(q, M)$ is the α–value at which the derivative of the spectral function is equal to q. The generalized dimensions (Renyi–dimensions) are given by

$$D_q(M) = \frac{1}{q-1}\tau(q, M). \tag{14}$$

$D_0(M) = f(\alpha(0, M)) = f_{max}(M)$ is the Hausdorff dimension, $D_1(M) = f(\alpha(1, M)) = \alpha(1, M)$ the information dimension (at $\alpha(1, M)$ the slope of the spectral function is unity according to eq. (13)), and $D_2(M) = 2\alpha(2, M) - f(\alpha(2, M))$ is the correlation dimension.

We now present the experimental results. Fig. 8 shows the G–moments $\langle lnG_q(M)\rangle$ vs. lnM for integer q–values in the interval $-5 \leq q \leq 5$. The saturation at $(1-q)\langle lnN\rangle$ according to (9) is clearly seen. In Fig. 9 the functions $\tau(q)$ and $\alpha(q)$ vs. q are plotted as determined for the M–interval $(M_1, M_2) = (1, 2)$. Fig. 10 shows the spectral functions for the two M–intervals $(M_1, M_2) = (1, 2)$ and (2, 3). For the first M–interval one obtains: $D_0 \approx 0.86$ and $D_1 \approx 0.75$. For $M \to \infty$: $\tau(q) \to 0$, $\alpha(q) \to 0$ and $f(q) \to 0$ according to eqs. (9), (10), (11) and (12). The fact that the spectral functions in Fig. 10 do not degenerate to a single point implies multifractality in hadron production by μN interactions.

In ref. [15] Florkowski and Hwa have derived a universality relation for the G–moments which we now test using our data. In general the G–moments $G_q(\mu, \nu)$ depend on the number of rapidity bins $M = 2^\mu$ and the multiplicity $N = 2^\nu$ in Δ, expressed by the exponents μ and ν respectively (μ and ν need not be integer). Fig. 11 shows this dependence for $q = 0$ and three multiplicities ($\nu = 2, 3, 4$). At $\mu = 0$ $(M = 1), \langle lnG_q(0, \nu)\rangle = 0$ for all q and ν. For large μ, $\langle lnG_q(\mu, \nu)\rangle \to (1-q)lnN = (1-q)\nu ln2$ according to eq. (9). This saturation is clearly seen in Fig. 11.

In ref. [15] it is conjectured that the G–moments (at fixed q) show universality at the same $\xi = \mu - \nu$, i.e. at the same average multiplicity of particles per bin $N/M = 2^{\nu-\mu} = 2^{-\xi}$. In particular the universality relation

$$\Gamma_q(\xi) = lnG_q(\mu, \nu) - lnG_q(\nu, \nu) \tag{15}$$

is derived, i.e. the right hand side of eq. (15) should depend only on the difference ξ.

Relation (15) is tested in Fig. 12, using charged hadrons in the range $-1.5 < y < 1.5$. The figure shows for $q = -5$, 0 and 5 the right hand side of relation (15) as a function of ξ for various values of ν (open symbols for low multiplicities $N = 4, \ldots 8$; full symbols for high multiplicities $N = 9, \ldots 15$). Approximate universality is observed over a wide range of N, in particular at low $|\xi|$. This is also seen in Fig. 13, where for $q = 0$ and 5 and for four different ξ values ($\xi = -2, -1, 1, 2$) the right hand side of eq. (15) is plotted as a function of the multiplicity N. Universality (independence on N at fixed ξ) is seen to be reasonably well fulfilled for $|\xi| \lesssim 1$, but not so good for $|\xi| \gtrsim 1$.

In order to investigate whether there is universality of the spectral function at fixed ξ, a multifractal analysis has been carried out in fixed ξ–intervals (ξ_1, ξ_2) for different total multiplicities $N = 4, 8, 16\,(\nu = 2, 3, 4)$. Fig. 14 shows the spectral function for the three N values in each of three different ξ–intervals. For $(\xi_1, \xi_2) = (-2, -1)$ (i.e. $(\mu_1, \mu_2) = (0, 1), (1, 2), (2, 3)$ for $\nu = 2, 3, 4$ respectively) universality is observed only for the rising branch of the spectral function ($q \geq 0$), but not for the falling branch. For the two other ξ–intervals the spectral functions do not show the normal multifractal behaviour.

5. Summary

The EMC data on hadron production in μN scattering show the following features:

— In an intermittency analysis in 3 phase space dimensions the intermittency indices (slopes) φ_2 of the second factorial moment turn out to be roughly equal for positive and negative hadrons and both similar to the slope for all charged hadrons. Furthermore the slope φ_2 is roughly independent of the rapidity region selected, i.e. of the particle density. Finally, in an Ochs plot of lnF_i vs. $lnF_2\,(i = 3, 4, 5)$ the points for different dimensions fall on one straight line for a given i.

— The factorial correlators do not obey strictly a power law as functions of the rapidity distance D. At fixed D they are rather independent of the bin width δy.

— In a multifractal analysis of the G–moments of the rapidity distribution multifractality is observed. The G–moments show approximate universality, the universality function $\Gamma_q(\mu, \nu) = \Gamma_q(\mu - \nu)$ being rather independent of ν at fixed $\mu - \nu$ with $|\mu - \nu| \lesssim 1$. The spectral function shows universality only on its left branch for $\mu - \nu \lesssim -1$, whereas for $\mu - \nu \gtrsim 0$ multifractality breaks down.

REFERENCES

[1] I. Derado, G. Jancso, N. Schmitz, P. Stopa: *Z. Phys.* **C47** , 23 (1990)

[2] I. Derado, G. Jancso, N. Schmitz, P. Stopa: *Proc. Santa Fe Workshop on Intermittency in High Energy Collisions* , Santa Fe, USA, 1990, eds. F. Cooper, R.C. Hwa. I. Sarcevic (World Scientific 1991), p. 67

[3] O.C. Allkofer et al. (EMC): *Nucl. Instr. Meth.* **179**, 445 (1981); I.P. Albanese et al. (EMC): *Nucl. Instr. Meth.* **212**, 111 (1983)

[4] W. Ochs: *Phys. Lett.* **B247**, 101 (1990)

[5] W. Ochs: *Proc. Santa Fe Workshop on Intermittency in High Energy Collisions*, Santa Fe, USA, 1990, eds. F. Cooper, R.C. Hwa, I. Sarcevic (World Scientific 1991), p. 166

[6] W. Ochs: *Z. Phys.* **C50**, 339 (1991)

[7] A. Bialas, M. Gazdzicki: *Phys. Lett.* **B252**, 483 (1990)

[8] B. Wosiek: *Acta Phys. Polon.* **B21**, 1021 (1990)

[9] P. Carruthers, H.C. Eggers, I. Sarcevic: *Phys. Lett.* **B254**, 258 (1991)

[10] K. Fiałkowski: Contribution to these proceedings

[11] A. Bialas, R. Peschanski: *Nucl. Phys.* **B308**, 857 (1988)

[12] V.V. Aivazyan et al. (NA22): *Phys. Lett.* **B258**, 487 (1991)

[13] R.C. Hwa: *Phys. Rev.* **D41**, 1456 (1990)

[14] C.B. Chiu, R.C. Hwa: *Phys. Rev.* **D43**, 100 (1991)

[15] W. Florkowski, R.C. Hwa: *Phys. Rev.* **D43**, 1548 (1991)

[16] C.B. Chiu, K. Fiałkowski, R.C. Hwa: Mod. Phys. Lett. **A5**, 2651 (1990)

[17] R.C. Hwa: *Phys. Rev.* **D44**, 915 (1991)

[18] C.B. Chiu, R.C. Hwa: preprint OITS–466 (May 1991)

Figure captions

1.) a) Log–log plot of the factorial moments F_2 and F_3 vs. M^3 for charged hadrons in 3 dimensions, b) corresponding (smoothened) local slopes φ_2 and φ_3, c) log–log plot of $F_2 - 1 = K_2$ in 3 dimensions

2.) Log–log plots of the factorial moments F_2 and F_3 vs. M^3 for a) positive and c) negative hadrons, and corresponding (smoothened) local slope φ_2 for b) positives and d) negatives

3.) Log–log plot of the factorial moment F_2 vs. M^3 for charged hadrons in three rapidity regions: $|y| < 1.0$, $1.0 < |y| < 1.75$ and $1.75 < |y| < 3.0$

4.) Log–log plot of $F_i\,(i = 3,4,5)$ vs. F_2 for charged hadrons ("Ochs plots") in 1, 2 and 3 phase space dimensions

5.) Log–log plots for the factorial moments $F_i\,(i = 2,3,4)$ vs. M^2 for charged hadrons in 2 dimensions, namely a) $(\tilde{y},\tilde{\varphi})$ and c) $(\tilde{y},\widetilde{lnp_T})$ together with their (smoothened) local slopes, b) and d)

6.) Log–log plot of the factorial correlators C_{ij} for charged hadrons vs. the inverse rapidity distance $1/D$ for charged hadrons, for four different rapidity bin widths $\delta y = 0.4, 0.2, 0.13$ and 0.1

7.) Factorial correlators C_{ij} for charged hadrons vs. the inverse rapidity bin width $1/\delta y$ for two rapidity distances a) $D = 0.4$ and b) $D = 0.8$

8.) G–moments $\langle lnG_q(M)\rangle$ in rapidity vs. lnM for charged hadrons for integer q–values in the interval $-5 \le q \le 5$

9.) The function $\tau(q)$ and $\alpha(q)$ vs. q for charged hadrons as determined from $\langle lnG_q(M)\rangle$ at $M = 1$ and $M = 2$

10.) Spectral function $f(\alpha)$ vs. α for charged hadrons for $(M_1, M_2) = (1,2)$ and (2,3). The straight line $f(\alpha) = \alpha$ is the tangent to the spectral function at $\alpha(1)$

11.) G–moment $\langle lnG_0(\mu,\nu)\rangle$ vs. μ $(M = 2^\mu)$ for charged hadrons, for three total multiplicities $N = 2^\nu = 4, 8, 16$. The asymptotic values $ln2^\nu$ are indicated by arrows

12.) Universality function $\Gamma_q(\xi)\,(\xi = \mu - \nu)$ vs. ξ for $q = -5$, 0 and 5 and for various total multiplicities between 4 and 15. Charged hadrons are taken in the interval $-1.5 < y < 1.5$

13.) Universality function $\Gamma_q(\xi)$ vs. total multiplicity $N = 2^\nu$ for $q = 0$ and 5 and for four different values of $\xi = -2, -1, 1, 2$. The data are the same as for Fig. 12

14.) Spectral function $f(\alpha)$ vs. α for three total multiplicities $N = 4, 8, 16$ in three different ξ–intervals: $(\xi_1, \xi_2) = (-2, -1)$, (-1, 0) and (0, 1)

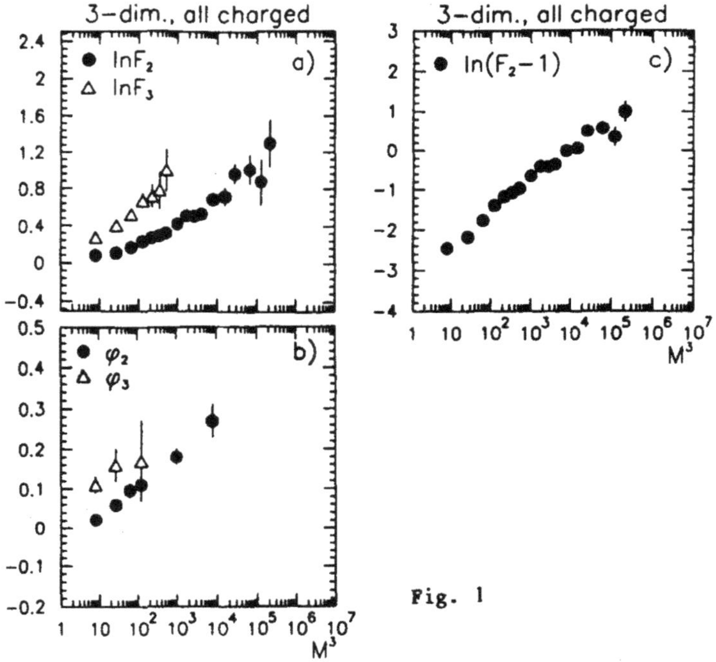
3-dim., all charged
lnF2
lnF3
a)
φ2
φ3
b)
3-dim., all charged
ln(F2-1)
c)
M3

Fig. 1

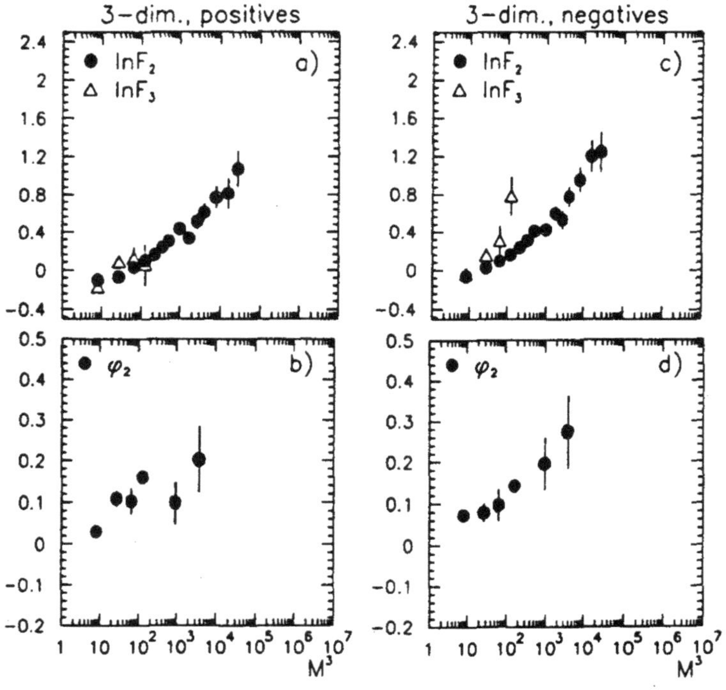
3-dim., positives
lnF2
lnF3
a)
φ2
b)
3-dim., negatives
lnF2
lnF3
c)
φ2
d)
M3

Fig. 2

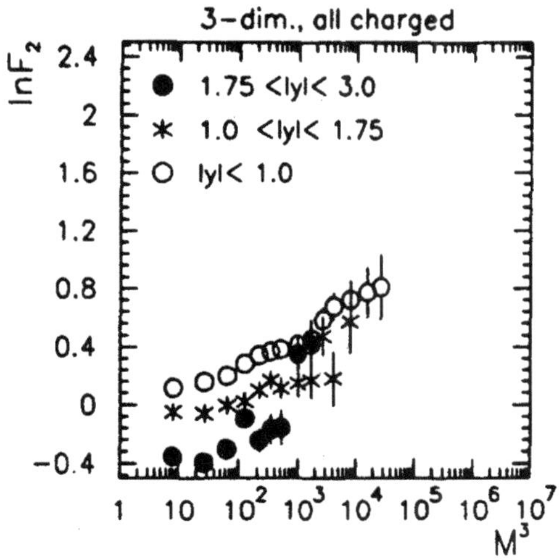
3-dim., all charged
lnF₂
1.75 <|y|< 3.0
1.0 <|y|< 1.75
|y|< 1.0
M³

Fig. 3

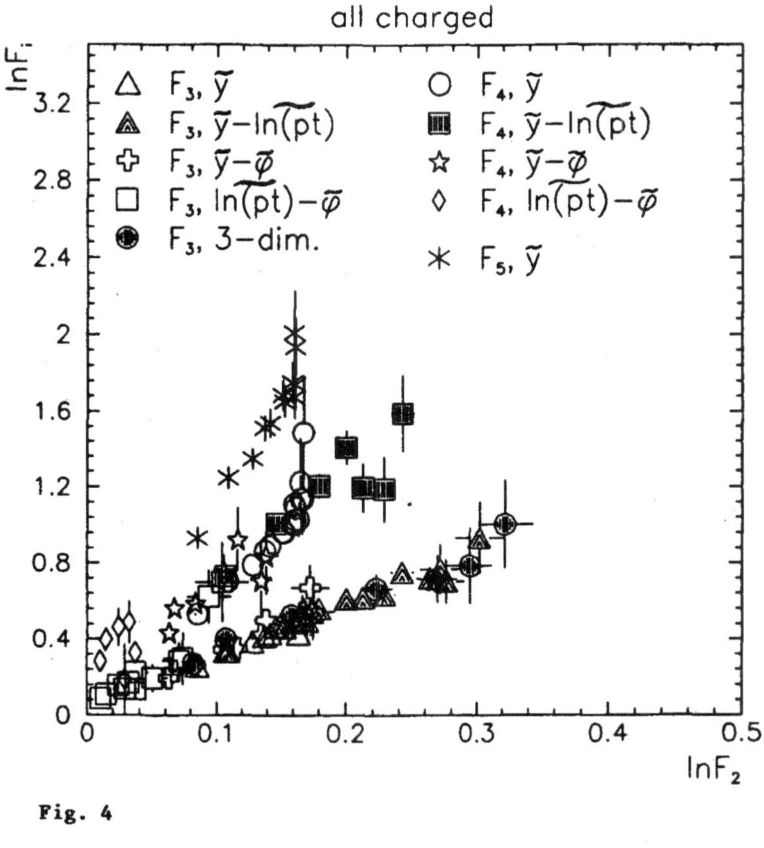
all charged
lnFᵢ
F₃, ỹ
F₃, ỹ−ln(pt)
F₃, ỹ−φ̃
F₃, ln(pt)−φ̃
F₃, 3−dim.
F₄, ỹ
F₄, ỹ−ln(pt)
F₄, ỹ−φ̃
F₄, ln(pt)−φ̃
F₅, ỹ
lnF₂

Fig. 4

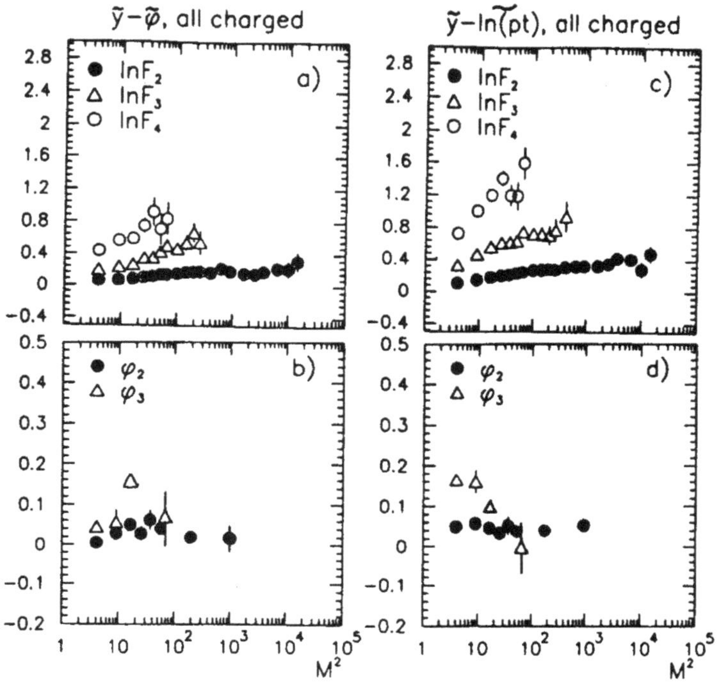

ỹ−φ̃, all charged
ỹ−ln(p̃t), all charged
lnF2
lnF3
lnF4
φ2
φ3
a)
b)
c)
d)
M²

Fig. 5

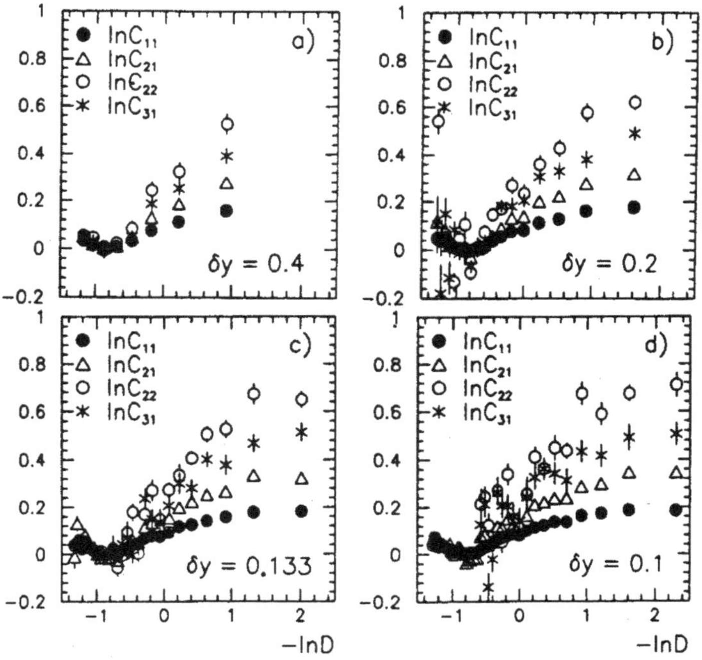

lnC11
lnC21
lnC22
lnC31
a)
b)
c)
d)
δy = 0.4
δy = 0.2
δy = 0.133
δy = 0.1
−lnD

Fig. 6

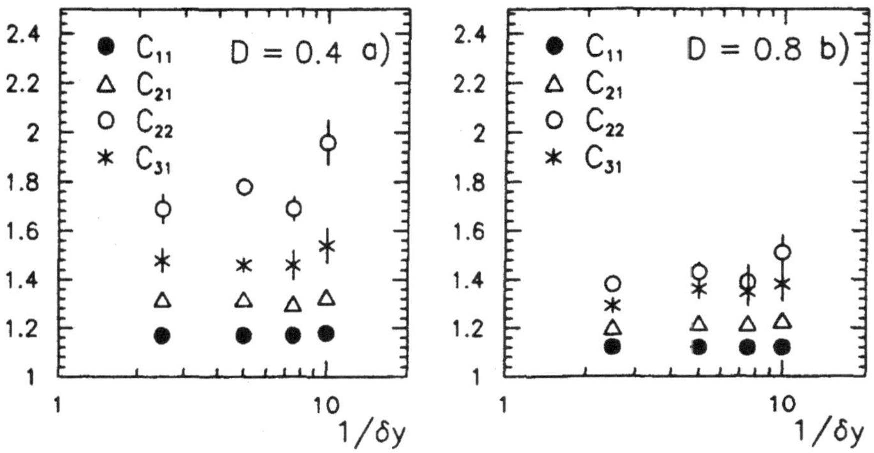
D = 0.4 a)
D = 0.8 b)
C_{11}
C_{21}
C_{22}
C_{31}
$1/\delta y$

Fig. 7

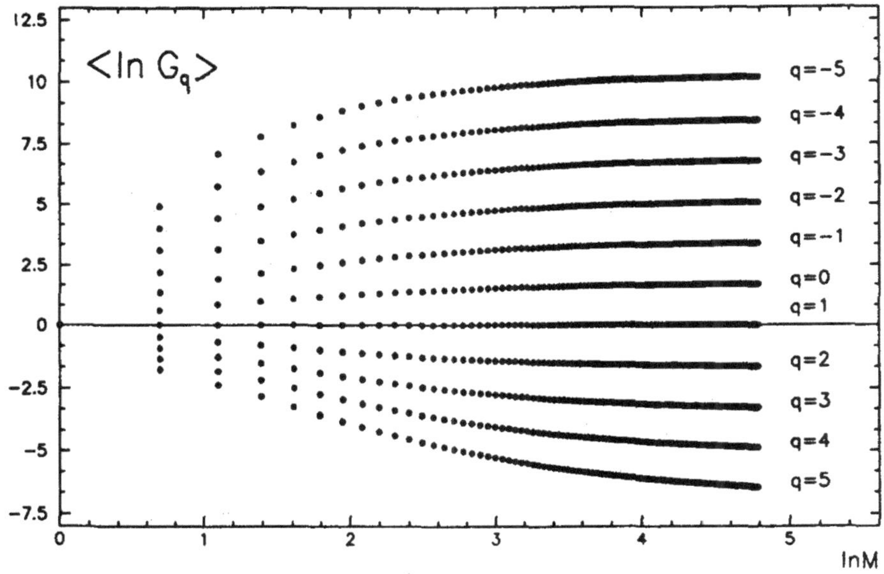
<ln G_q>
q=-5
q=-4
q=-3
q=-2
q=-1
q=0
q=1
q=2
q=3
q=4
q=5
lnM

Fig. 8

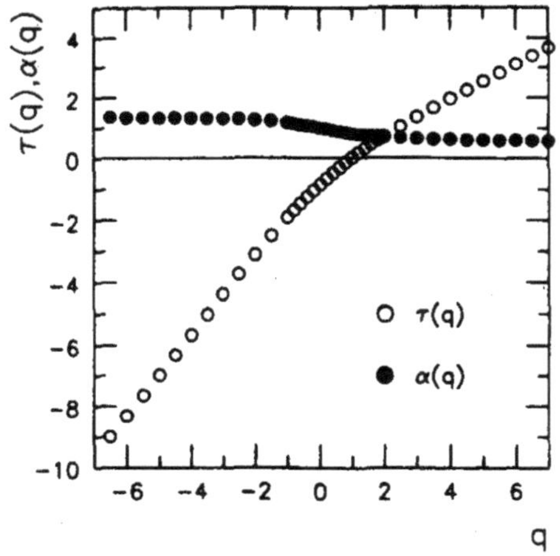

Fig. 9

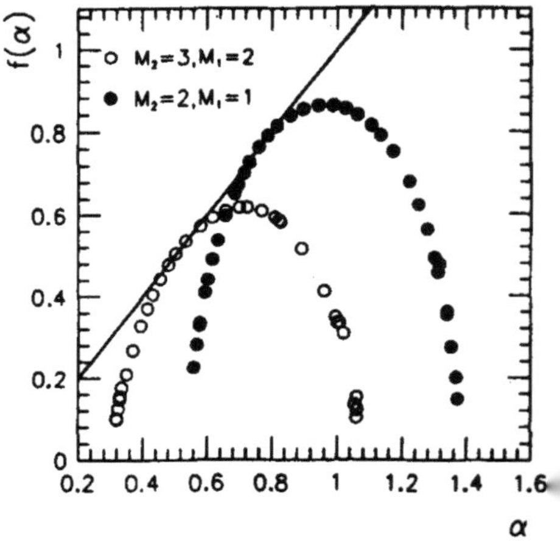

Fig. 10

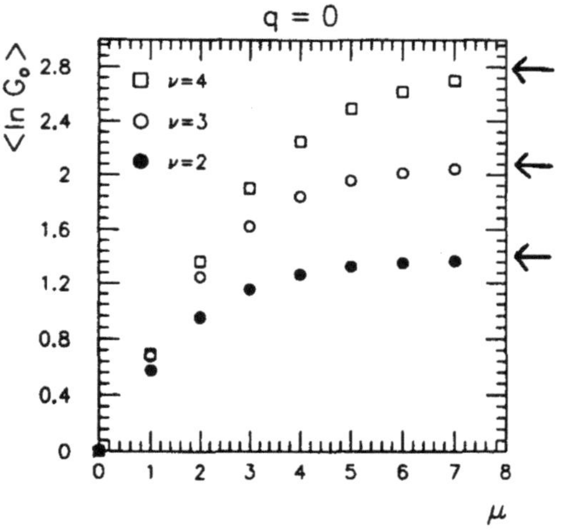

Fig. 11

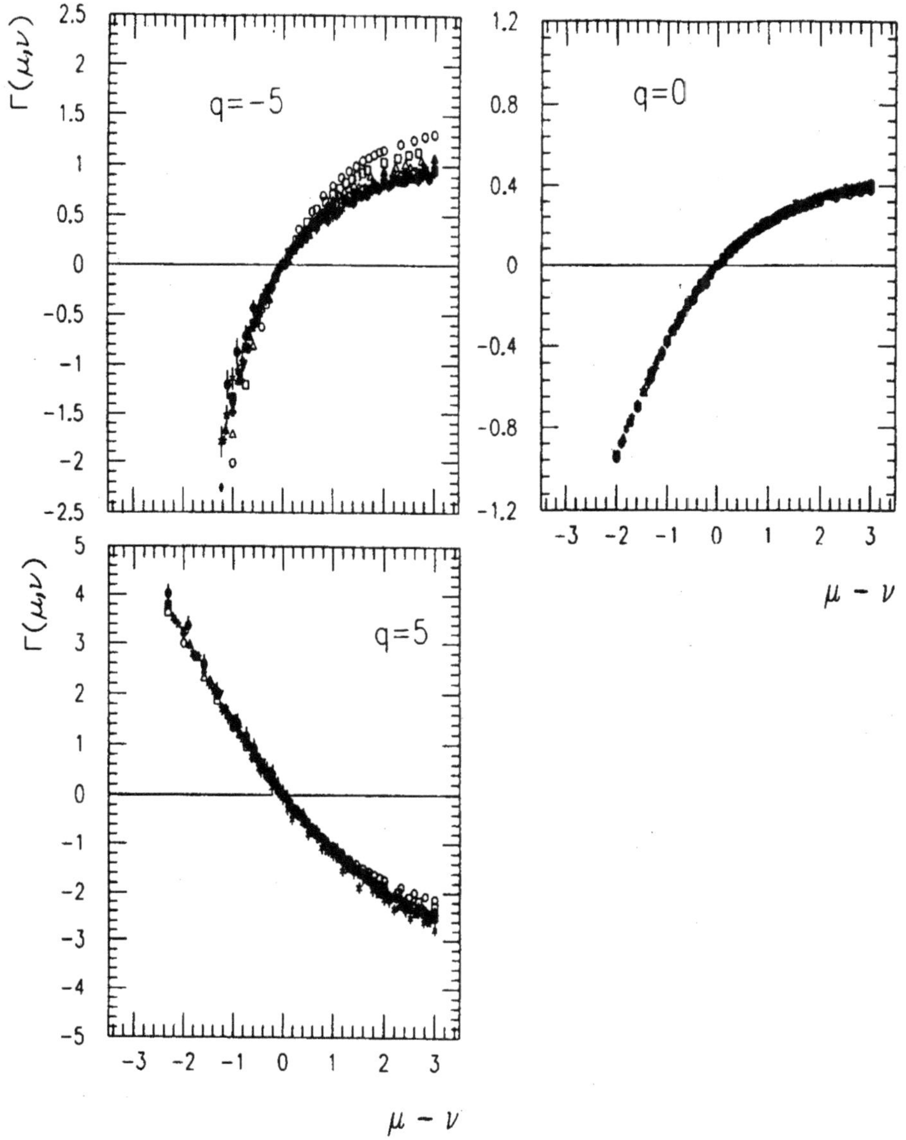
q=-5
q=0
q=5
Γ(μ,ν)
Γ(μ,ν)
μ - ν
μ - ν
2.5
2
1.5
1
0.5
0
-0.5
-1
-1.5
-2
-2.5
1.2
0.8
0.4
0
-0.4
-0.8
-1.2
5
4
3
2
1
0
-1
-2
-3
-4
-5
-3 -2 -1 0 1 2 3
-3 -2 -1 0 1 2 3

Fig. 12

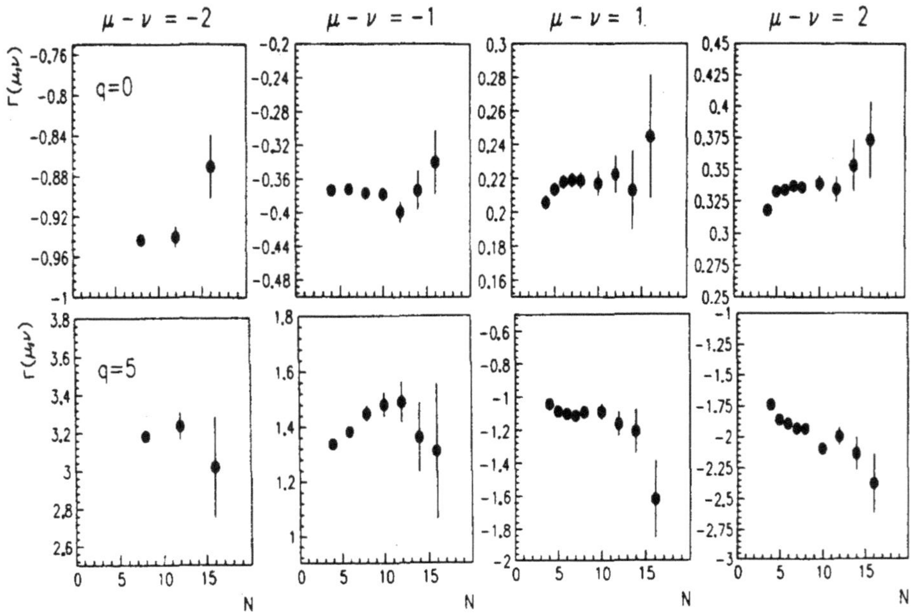

Fig. 13

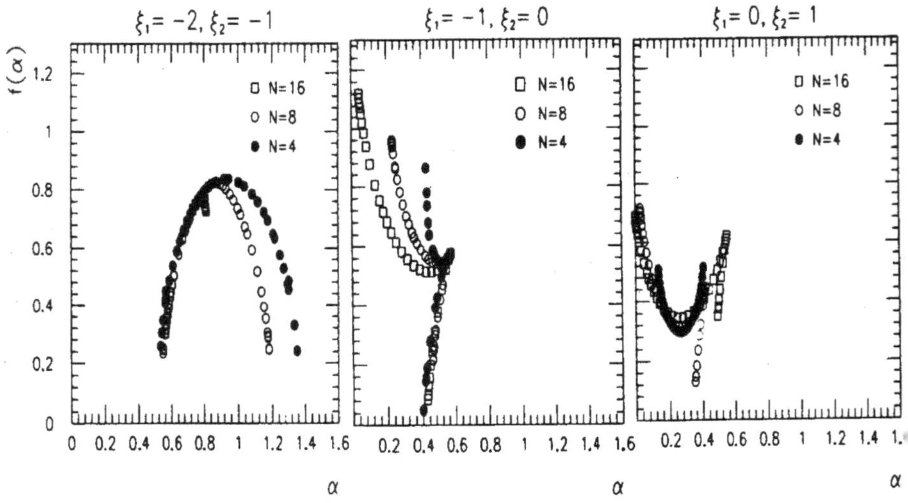

Fig. 14

RECENT EMU01-RESULTS ON FLUCTUATIONS AND INTERMITTENCY

Evert Stenlund
Div of Cosmic and Subatomic Physics, University of Lund,
Sölvegatan 14, S - 223 62 Lund, Sweden

and

M I Adamovich, M M Aggarwal, Y A Alexandrov, N P Andreeva, Z V Anzon, R Arora, F A Avetyan, S K Badyal, E Basova, K B Bhalla, A Bhasin, V S Bhatia, V G Bogdanov, V I Bubnov, T H Burnett, X Cai, I Y Chasnikov, L P Chernova, M M Chernyavski, G Z Eligbaeva, L E Eremenko, A S Gaitinov, E R Ganssauge, S Garpman, S G Gerassimov, J Grote, K G Gulamov, S K Gupta, H H Heckman, H Huang, B Jakobsson, B Judek, S Kachroo, F G Kadyrov, G S Kalyachkina, E K Kanygina, M Karabova, G L Kaul, S P Kharlamov, T Koss, S A Krasnov, V Kumar, P Lal, V G Larionova, V N Lepetan, L S Liu, S Lokanathan, J Lord, N S Lukicheva, S B Luo, L K Mangotra, N A Marutyan, N V Maslennikova, I S Mittra, S Mookerjee, H Nasrulaeva, S H Nasyrov, V S Navotny, G I Orlova, I Otterlund, H S Palsania, N G Peresadko, N V Petrov, V A Plyushchev, W Y Qian, R Raniwala, S Raniwala, N K Rao, V M Rappoport, J Ravina, J T Rhee, N Saidkanov, N A Salmanova, L G Sarkisova, V R Sarkisyan, G S Shabratova, T I Shakhova, D Skelding, K Söderstrom, Z I Solovieva, S C Strausz, J F Sun, L N Svechnikova, K D Tolstov, M I Tretyakova, T P Trofimova, U Tuleeva, S Vokal, H Q Wang, Z Q Weng, R J Wilkes, G F Xu, D H Zhang, P Y Zheng, S I Zhokova, D C Zhou

EMU01-collaboration

ABSTRACT

In this talk we will summarize the recent results concerning non-statistical fluctuations obtained by the EMU01 collaboration. In our one-dimensional analysis we have established a $1/\langle\rho\rangle$-scaling and furthermore found that higher order particle correlations are negligable in heavy ion interactions. In two dimensions the intermittency effect is very much larger, but so is the influence of γ-conversion. In fact, our estimates of γ-conversion, based on MC-calculations, are consistent with our experimental results from the two-dimensional analysis. This indicates that the particles forming dense clusters in η, which gives the main contribution to the signal in a one-dimensional analysis, are rather independent in the other dimension, φ.

1. INTRODUCTION

The topic of non-statistical fluctuations in the particle densities has attracted a lot of recent interest. Great efforts have been devoted to the description of these phenomena, but more theoretical work is still needed for the foll understanding of these effects. In the meantime the experiments produce a lot of data showing different aspects of these fluctuations, as can be seen in these proceedings. In this talk we will summarize the recent experimental achievements performed by the EMU01-collaboration. This experiment utilizes two different emulsion techniques; ordinary emulsion stacks, exposed parallel to the emulsion plates, and emulsion chambers, irradiated perpendicular to the plates. Some of the chambers are equipped with thin target foils of gold. The measurements in the chambers are performed with semi-automatic systems. In the conventional stacks the resolution in pseudo-rapidity is better than 0.1 units in the central region, whereas in the chambers a resolution of 0.013 units can be obtained. More details on the experiment are published elsewhere[1].

The second chapter summarizes the results obtained in the one-dimensional analysis, this dimension being along the pseudo-rapidity axis.

Chapter 3 contains results from the two-dimensional analysis (pseudo-rapidity and azimuth). Some emphasis will be put on γ-conversion since we find that the electron-pairs easily penetrate the filter provided by using a two-dimensional procedure.

Chapter 4 contains the conclusions.

2. FLUCTUATIONS IN ONE DIMENSION

2.1. Stochastic Emission

Due to the many sources expected in heavy-ion collisions, one can expect that most of the correlations between the produced particles will be washed out. Let us consider a pseudo-rapidity interval $\Delta\eta$, within which we look at the smaller window $\delta\eta$. If we assume that correlations between particles can be neglected and that the shape of the pseudo-rapidity distribution is independent of the multiplicity (or centrality), we obtain that $F_2(\eta,\delta\eta) \equiv \langle n\cdot(n-1)\rangle/\langle n\rangle^2$ is independent of the size and location of the window. F_2 is generally larger than unity due to the width of the global multiplicity distribution. Whenever $\langle n\rangle\cdot(F_2-1)^{1/2}$ is large enough we find[2]

$$\sigma_2 \simeq \langle n\rangle\cdot\sqrt{F_2-1} + \frac{1}{2\cdot\sqrt{F_2-1}} \quad (1)$$

i e we have a linear relationship between σ_2 and $\langle n\rangle$ for different windows. Both the intercept and the slope are determined by the value of F_2. Similar expressions are obtained for higher moments[3]. For instance, it is obtained that $\sigma_3 \equiv \mu_3^{1/3}$ depends linearly on $\langle n\rangle$, where now both the intercept and the slope are given by F_2 and F_3.

The relationship between σ and $\langle n\rangle$ has been tested for our

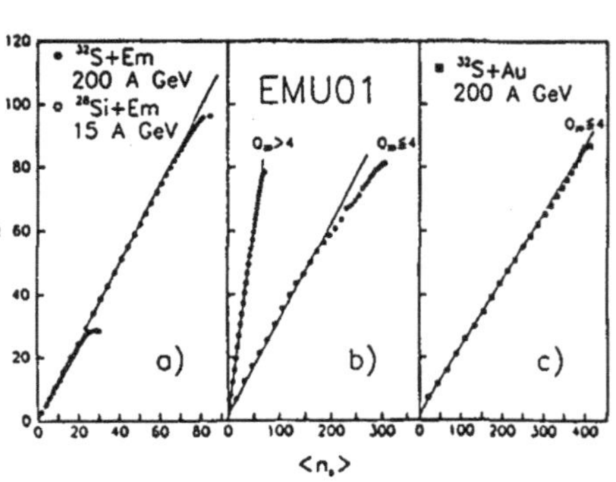

Fig 1. Demonstration of the linear relationship between σ_2 and $\langle n_s\rangle$ for some different data samples.

different data samples for η-windows centered around mid-rapidity. For all the samples (minimum bias as well as central selections) we find a nice linear behaviour as long as the windows are narrow enough to exclude the fragmentation regions[2,3]. Some examples are illustrated in fig 1.

These results clearly indicate that this kind of stochastic emission works well in the central region and that the correlation between the produced particles are well hidden in the combinatorial background. Thus the intermittency effect is, in a global sence, only a tiny effect, and is normally not seen, unless special tools are utilized in the analysis.

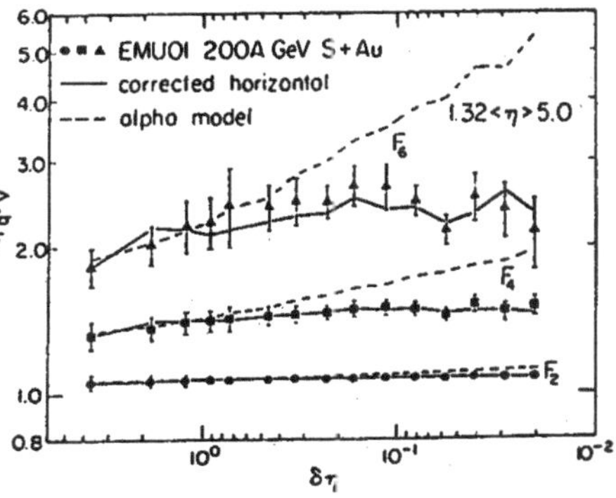

Fig 2. Scaled factorial moments calculated with two different metods and an MC-simulation using a model with built in intermittency.

We will return to these ideas in section 2.3. where the error estimates on the intermittency indices are discussed.

2.2 Scaled Factorial Moments

The most commonly used methods for studying non-statistical fluctuations are based on two papers by Białas and Peschanski[4]. These methods are based on the notion of scaled factorial moments and differ in the procedures of normalization and how to average over events and over different regions of phase space. One method, horizontal averaging, calculates the moments for each event and then averages over the events. A second method, vertical averaging, first calculates the moments for a given η-bin in all events and then averages over all bins. Horizontal averaging is normally followed by a correction for the variation of particle densities as a function of η, here referred to as Fialkowski correction[5]. The moments can either be normalized to the whole event sample or to the individual events (individual bins for the vertical averaging). The moments are studied as a function of the size of the chosen η-bin and the variation of the moments with varying δη may indicate an intermittent behaviour.

The very good resolution of our chambers allows us to study a region which is normally not accessible for other fixed target experiments. In fig 2 we show our results for central ^{32}S+Au events[6]. The moments are calculated by both vertical averaging and corrected horizontal averaging and give basically the same results. Also shown is a model calculation with built in intermittency, based on the α-model by Białas and Peschanski. The Fritiof model[7] shows no variation with δη (not shown in the figure). Our conclusions are that for $\delta\eta \geq 0.1$ the moments rise with decreasing δη, but for smaller values of δη the moments stay essentially flat, indicating a smallest inherent scale for the particle production. The rise for $\delta\eta \geq 0.1$ is consistent with other experimental findings[8].

The method used in this talk for calculating the moments is

$$F_q = M^{q-1} \cdot \sum_{m=1}^{M} \frac{k_m \cdot (k_m - 1) \cdot \cdots \cdot (k_m - q + 1)}{\langle N \rangle^q} \tag{2}$$

which is then averaged over all events and corrected for the shape of the pseudo-rapidity distribution[5]. Here M is the number of $\delta\eta$-bins and k_m is the multiplicity in bin m. $\langle N \rangle$ is the average multiplicity in the whole window under consideration. We will furthermore concentrate on the central region here defined as $|\eta - \eta_{cm}| < 1$. For all studied moments (up to and including the 6th order) a linear behaviour between $\ln(F_q)$ and $\ln(\delta\eta)$ is observed for our different data samples in the region $2 \geq \delta\eta \geq 0.1$[9,10]. For the highest energies the Fialkowski-correction is marginal.

2.3. Slopes of Scaled Factorial Moments

When lines are fitted to data in $\ln(F_q)$ versus $-\ln(\delta\eta)$ plots, one has to keep a few facts in mind. For a given moment the points are strongly correlated since the same sample is used, although with a different binning, for each point. This makes it very hard to estimate the error of the intermittency index, i e the slope of the line. Furthermore for the same reason the different moments are strongly correlated, which we will come back to later in this section. In order to study these correlations and to estimate the errors of the slopes we have devised a simple Monte Carlo method based on the stochastic emission picture in section 2.1. As input the experimental multiplicity and pseudo-rapidity distributions are used and the particles are randomly emitted. The M C sample is generated exactly as large as the experimental sample. For each experimental sample the generation is repeated N times and the standard deviation of the slope, ϕ_q, is estimated as

$$\sigma(\phi) \simeq \sqrt{\langle \phi^2 \rangle - \langle \phi \rangle^2} \tag{3}$$

with an uncertainty of $(2 \cdot N)^{-\frac{1}{2}}$. For N = 100 we have thus an uncertainty of about 7%. These error estimates can be considered as lower limits since the small contribution of non-statistical fluctuations tend to make them larger and this is accentuated for the higher moments.

If the denominator of eq (2) is taken outside of the sum we see that the definition of F_q contains a normalization factor times a sum. The normalization factor is the same for all events and the high multiplicity events give the main contribution to the sum. Thus the intermittency index is mainly determined by the high multiplicity tail when a broad multiplicity distribution is studied, again accentuated for the higher moments. This effect was investigated on our ^{16}O+Em sample at 200 A GeV. For the total sample, a multiplicity cut, $n > n_{cut}$, was applied and the results where compared as a function of n_{cut}. For small n_{cut} the slopes and the error estimates came out identical, whereas the intercepts decreased due to the changing normalization factor. A noticeable change in slopes (and errors) was only detected when less than 20% of the events remained. In heavy-ion experiments with widely varying multiplicities there is thus nothing to gain by treating full samples, it is rather preferred to divide the sample into subsamples with a limited multiplicity range or centrality cut.

If the produced particles from a heavy-ion interaction can be seen as a superposition particles comming from a varying number of independent subsources (cf section 2.1.), ϕ_q will be proportional to the inverse of the average particle density, i e $1/\langle\rho\rangle$. The results of a test of this scaling rule are shown in fig 3, where ϕ_2 is plotted versus $\langle\rho\rangle$ for different data samples[11]. As can be seen, data points from the same colliding system are rather well described by this scaling independent of energy and centrality cut. Normalized to the same particle density the largest system, however, shows a larger slope. When furthermore data from hadron induced interactions[12] are reanalyzed, the obtained slopes are close to the extrapolation for the oxygen-induced interactions. Also included in the figure is a data point from hadron-proton interactions at 250 GeV analyzed in a somewhat different way[13].

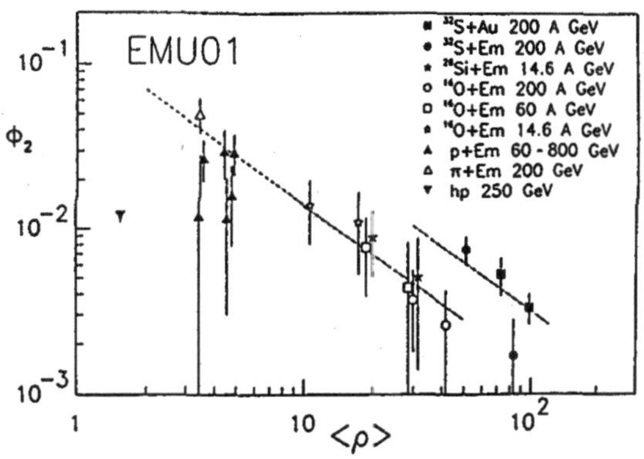

Fig 3. Intermittency index of the second order as a function of average particle density for different systems and centrality cuts.

Fig 4 suggests that also the η-dependence can be described by $1/\langle\rho\rangle$-scaling, although this has to be investigated further. The curve is obtained as the inverse of the average pseudo-rapidity density distribution. The parameter 0.315 is obtained from fig 3 for the Sulfur-induced interactions; no further parameters are introduced. Particles with $\eta \leq 1.3$ are not measured with the chamber technique.

Another scaling rule, which relates moments of different orders, is given by[4]

$$\phi_q = \binom{q}{2} \cdot \phi_2 \tag{4}$$

which should be obeyed if only two-particle correlations are significant. As we have seen, ϕ_2 is proportional to $\langle\rho\rangle^{-1}$. In a similar way higher order contributions to the slopes ϕ_q, drop faster with $\langle\rho\rangle$. Thus in heavy-ion interactions with a large number of produced particles we can expect higher order effects to die off quite fast due to the combinatoric background. Experimentally we do not see any significant deviation from the scaling-

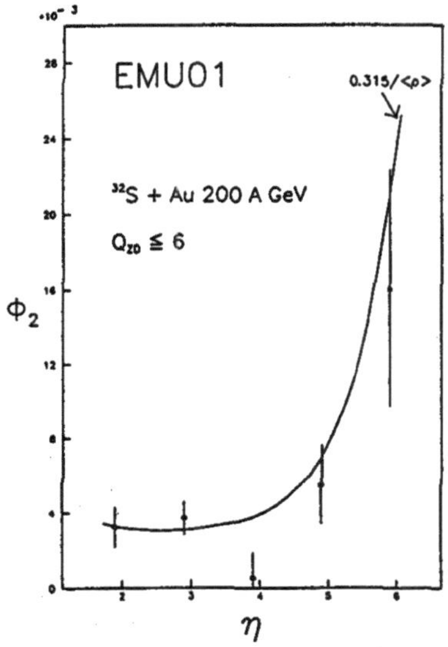

Fig 4. Intermittency index of the second order as a function of pseudo-rapidity.

rule given by eq (4), which indicates that the higher order contributions indeed are small. To obtain this result a great care was taken to eliminate the correlations between the different orders of the factorial moments. Here again the MC-procedure, based on stochastic emission, was utilized to study these correlations[9,11].

Białas and Hwa[14] have proposed a study of the anomalous fractal dimension d_q, related to ϕ_q by

$$d_q = \frac{\phi_q}{q - 1} \tag{5}$$

They have argued that, for a second-order phase transition, d_q should be independent of q. From what is discussed above about the validity of eq (4), we find that d_q is consistent with being proportional to q. It appears doubtful that the behaviour of d_q can be used as a probe for the quark-gluon plasma, since it requires that eq (3) should be substantially violated.

2.4. Uncertainties Due to Production of Electron Pairs

A possible source of systematic errors in the investigation is electron pairs produced either by a direct Dalitz decay or through γ-conversion. The influence on the slopes from both these processes has been carefully studied by Fritiof-simulations, where γ-conversion is added[15]. The results from the simulations are found to follow the same scaling rules as discussed in section 2.3. Dalitz production alone is found to be negligible and the combined effect can be parametrized as

$$\phi_{q,pair} \simeq 0.033 \cdot \binom{q}{2} \cdot \frac{p}{\langle\rho\rangle} \tag{6}$$

where p is the percentage of converted gammas. For all data samples the percentage is small enough to make this systematic error smaller than the quoted statistical errors in fig 3 and for the ^{32}S+Au samples, where p is of the order 3%, these corrections will lower the slopes by 20-30%.

2.5. Conclusions from the One Dimensional Analysis

We end this chapter with the following conclusions. In a limited $\delta\eta$-range (0.1 - 2) the intermittency-index, ϕ_q, is constant. For a given system ϕ_2 follows a $1/\langle\rho\rangle$-dependence, but normalized to the the same density, the largest systems show a larger slope. The data seems to be consistent with the scaling rule given by eq (3). Thus ϕ_q versus $\langle\rho\rangle$ will look similar to ϕ_2 versus $\langle\rho\rangle$ but with increased experimental errors. We also conclude that essentially all statistically significant information can be extracted from the moments of one specific order. To go beyond the information contained in moments of two different orders requires a substantial increase in statistics. The systematic errors introduced by electron-pairs influences the data, but can only partially explain the observed slopes. This is however not the case when the analysis is performed in two-dimensions as we will see in the next chapter.

3. FLUCTUATIONS IN TWO DIMENSIONS

3.1. Scaled Factorial Moments in Two Dimensions

The definition of the scaled factorial moments in two dimensions is quite straight forward; the only change in eq (2) is that now M is the number of $\delta\eta\delta\varphi$-bins.

In fig 5 a typical result from the two-dimensional procedure is shown, in this case F_2 for central ^{32}S + Au interactions at 200 A GeV. The $\delta\eta$-axis ranges from 2 down to 0.016 and the $\delta\varphi$-axis ranges from $2\cdot\pi$ to $0.016\cdot\pi$; in both directions one step corresponds to a factor of two more bins. Thus the innermost channel corresponds to ~16000 bins. The distribution has been Fialkowski-corrected along the η-direction but, as in the one-dimensional case, this correction is marginal. F_q for higher moments looks essentially the same (cf the one-dimensional case), but with our limited statistics the uncertainties become large.

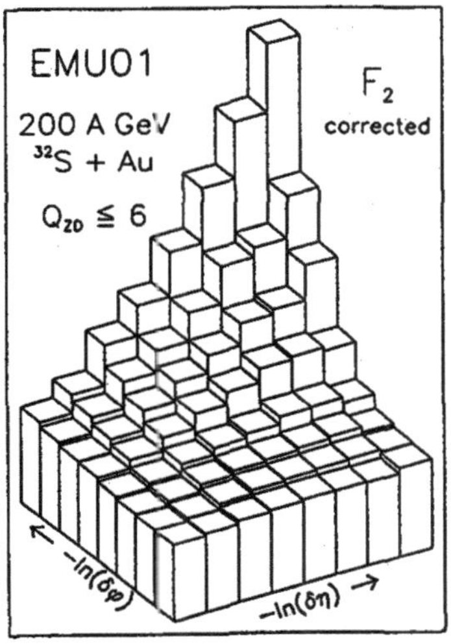

Fig 5. Scaled factorial moment in two dimensions (logarithmic scale).

We can now slice the two-dimensional histogram in fig 5 in different ways. It can be sliced along the main diagonal, sliced along the cross-diagonal keeping the total number of bins fixed, or it can be sliced perpendicular to one of the axes keeping either $\delta\varphi$ or $\delta\eta$ fixed.

Figure 6 a shows the outcome for slices along the $\delta\eta$-axis keeping $\delta\varphi$ fixed. The lines are fits in the region $M = 2^2 - 2^7$, where the

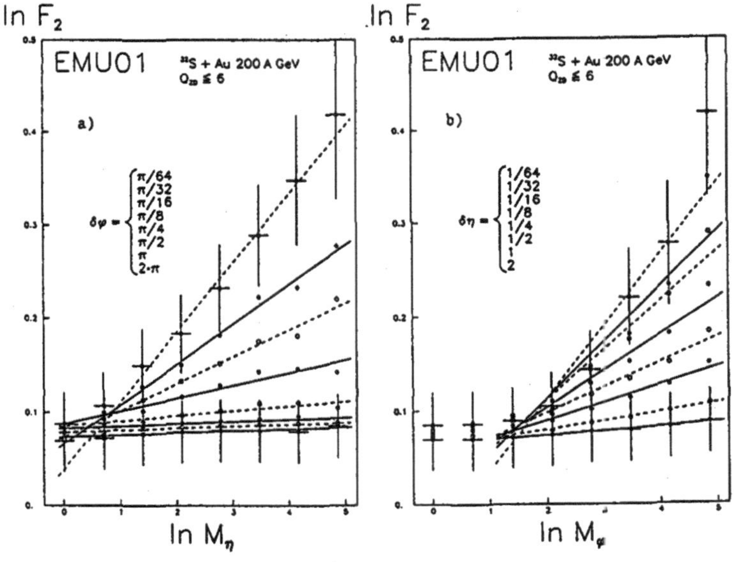

Fig 6. Slices along the axes of fig 5.

linearity is quite good. For decreasing $\delta\varphi$ we see that the slope fastly increases. In fig 6 b a similar behaviour is seen in the other dimension.

Figure 7 shows the slice along the main diagonal, which clearly indicates a stronger than linear dependence. Such a behaviour is expected from the behaviour in fig 6, as illustrated by the simple numerical example below.

linear ↑ (column) — parabolic ↗ (diagonal) — → linear (row)

	4	9	14	19	24
	3	7	11	15	19
↑	2	5	8	11	14
$\ln M_\varphi$	1	3	5	7	9
	0	1	2	3	4

$\ln M_\eta$ →

Along the cross-diagonal, fig 8, we find a rather flat spectrum, with a maximum for $M_\eta \simeq M_\varphi$. In fact, in a scenario with randomly oriented pairs of correlated particles, one would expect a maximum for

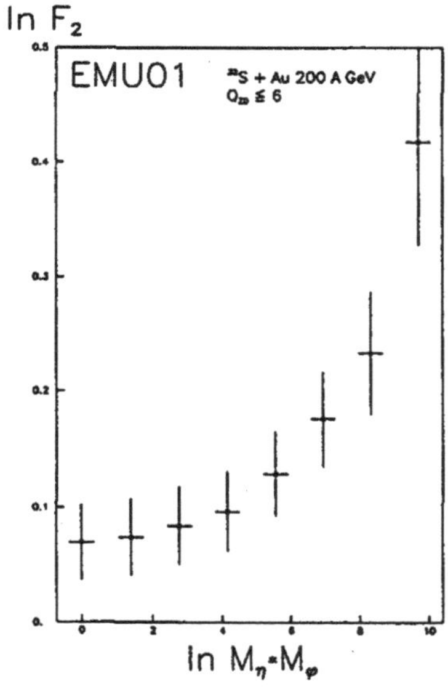

Fig 7. F_2 along the main diagonal of fig 5.

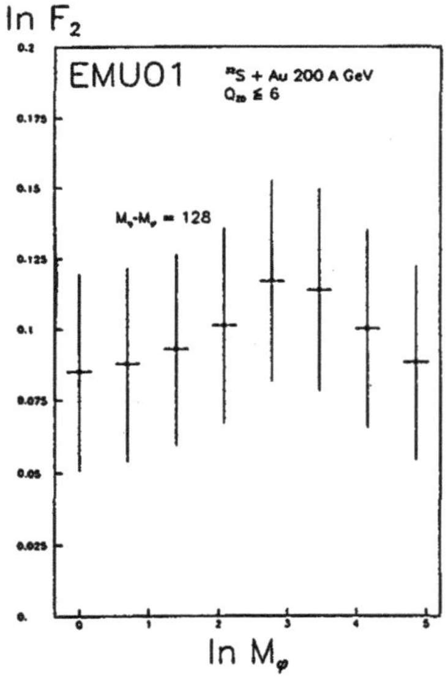

Fig 8. F_2 along the cross-diagonal of fig 5.

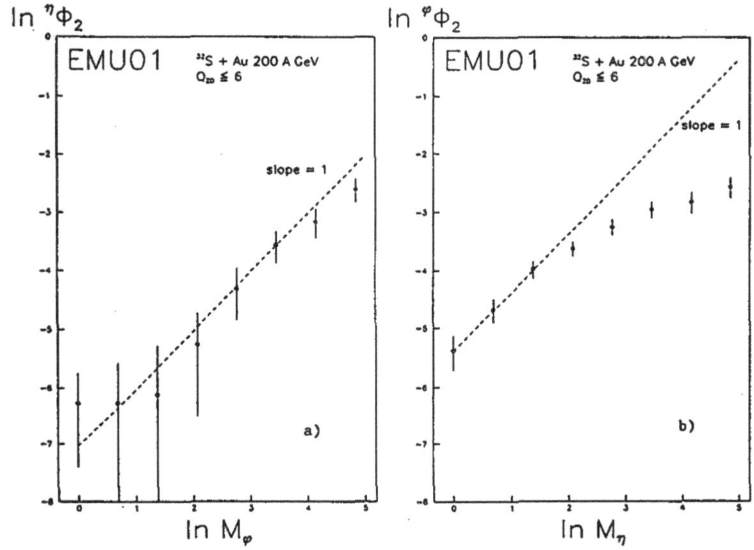

Fig 9. Slopes obtained in the η and φ dimensions from fig 6, as a function of the number of bins used in the other dimension, i e φ and η, respectively.

$\delta\eta = \delta\varphi$, corresponding to $\ln M_\varphi \simeq 3$. This indicates that what we see in our two-dimensional analysis may be attributed to some kind of pair-production, as we will see, to γ-conversion.

3.2 One-Dimensional Slopes along the Axes

In figs 9 a and b the slopes obtained from the fitted lines in figs 6 a and b are shown. As can be seen in the figures the data points essentially follows the dashed lines indicated in the figures. These lines are obtained if each new division doubles the slope. This is exactly the situation which appears if each subdivision of the events results in new subevents which are uncorrelated. This is the inverse situation to that discussed in section 2.3. in conection to the 1/<ρ>-rule. However, this trend is brooken when the number of subdivisions is large enough. What actually happens when the bins are smaller than the typical extension of an electron-pair from a γ-conversion, is that pairs of correlated particles are starting to end up in different bins, and the climb along the dashed line is hindered. The average opening angle of an electron-pair ($\simeq$ 0.07 rad) corresponds to $\ln M_\varphi$ = 4.5 in fig 9 a and to $\ln M_\eta$ = 3.35 in fig 9 b, which correspond well to where the

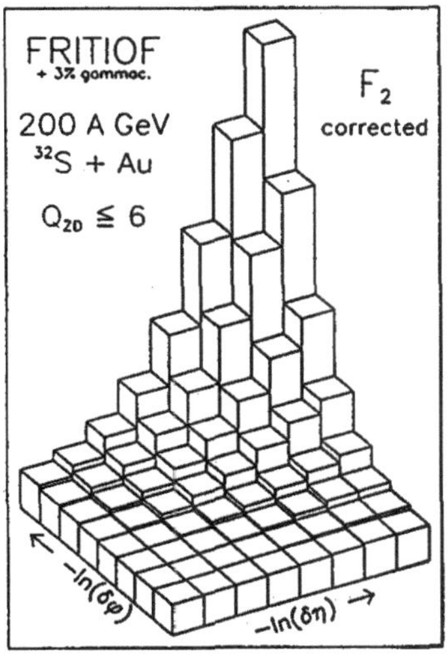

Fig 10. Scaled factorial moment in two dimensions for Fritiof with 3 % γ-conversion added (logarithmic scale).

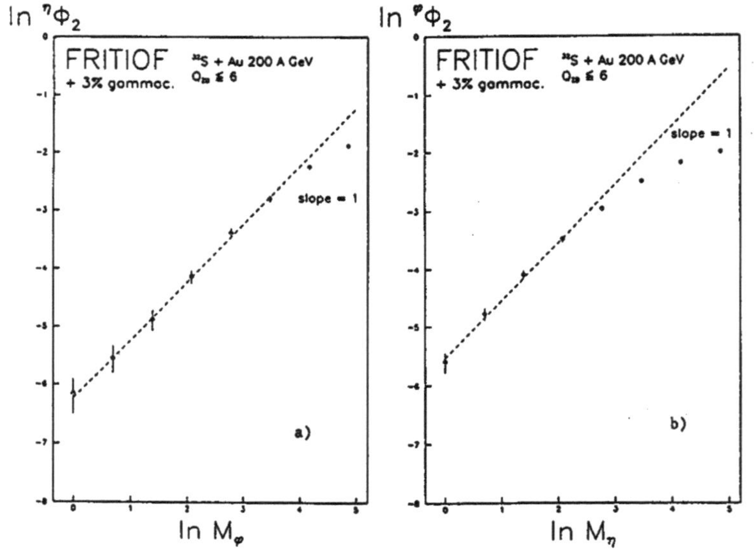

Fig 11. As fig 9 but for simulated data.

climbs along the lines are broken.

3.3 The Influence of γ-Conversion

The same MC-procedure as was used in section 2.4. is now applied to the two-dimensional analysis. For the ^{32}S + Au we estimate that 3 % of the produced gammas convert, the majority of them in the target gold-foil. Figure 10 shows the outcome of the MC-calculation analysed in two dimensions and it is in the essentials well corresponding to fig 5. The Fritiof model itself gives an essentially flat distribution. The same steps, which lead to figs 9 a and b, are now applied to the MC-data, resulting in figs 11 a and b. The comparison between the real data (figs 9) and the MC-data (figs 11) clearly shows that the intermittency effect drowns in the background when the two-dimensional procedure is applied.

The whole procedure has been repeated for a sample of central ^{16}O + Em interactions at 200 A GeV. Here the emulsion itself works as the target and the unresolved γ-conversion is estimated to be ≃ 1 %. Even here the obtained results indicate that the intermittency signal is completely drowned in the γ-conversion background.

3.4. Repeated Analysis in a Restricted Region of M

In section 3.2 the M-region used for the fitting was chosen to be 2^2 to 2^7. In order to study where the γ-conversion starts to kill the intermittency signal, we have repeated the two-dimensional analysis in the restricted region M = 2^0 to 2^4.

Figures 12 a and b show the results for the ^{32}S + Au data. As can be seen the MC-data still follows the dashed line, but the real data has a weaker dependence. At large ln M the two sets coincide. This clearly indicates that, when only one dimension is utilized, the

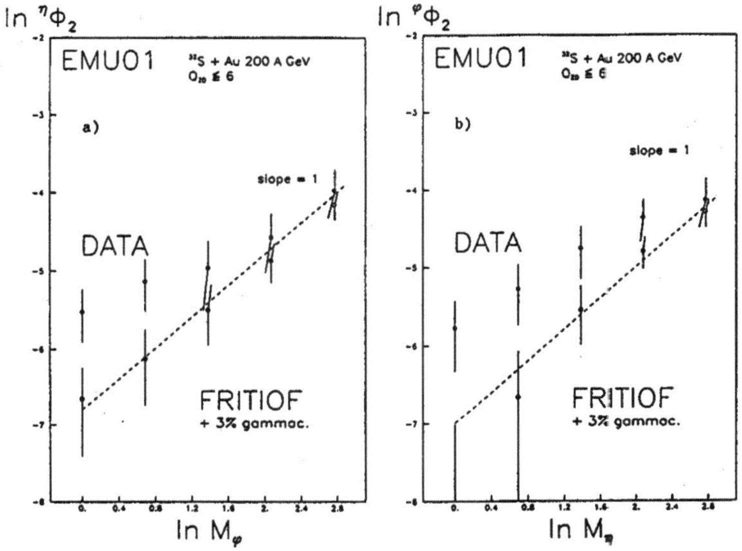

Fig 12. Slopes obtained in the restricted M-region.

experimentally observed effect is three times larger than the effect expected from γ-conversion. As soon as cuts are applied in both dimensions the net effect dies off. Essentially the same results are obtained for the ^{16}O + Em data.

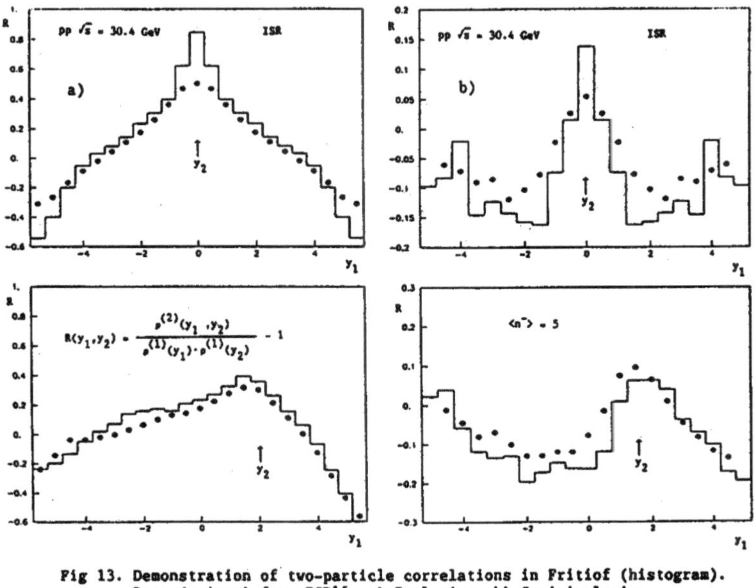

Fig 13. Demonstration of two-particle correlations in Fritiof (histogram). Data (points) from ISR[16]. a) Inclusive. b) Semi-inclusive.

3.5. Particle-Correlations in the Fritiof Model

As has been mentioned earlier (see sections 2.2. and 3.3) the Fritiof model seems to show no intermittency signal what-so-ever. One may thus ask whether there are any particle-correlations in such a model. The answer is yes, as illustrated in fig 13 a and b, which show two-particle correlations measured in pp-collisions at ISR-energies[16]. The data (points) and the results from the Fritiof calculation (histogram) agree qualitatively, although minor deviations are seen, especially in the semi-inclusive case.

3.6. Conclusions from the Two-Dimensional Analysis

We end this chapter with the following conclusions. In the two-dimensional analysis the data is strongly influenced by γ-conversion. The two-dimensional procedure actually works as a filter for the electron-pairs. This means, however, that the correlations which contribute most to the one-dimensional moments are not the same as the ones which contribute to the two-dimensional moments. It indicates that the bulk of the events responsible for the one-dimensional effect are of 'ring-type', i e strong correlations in η, diluted over the whole 2π in φ. This can be studied further by looking at the contribution of single events to the factorial moments, and such an investigation is planned in the near future.

4. SUMMARY

In this last chapter we would like to sum up our results obtained so far.

$1/\langle\rho\rangle$-scaling is observed for different interacting systems, for different centrality cuts, for different incident energies and possibly also for different η-regions. The same is to some extent also true for the subevents in the sence of section 3.2.

The data is consistent, within the statistical uncertainties, with the scaling expressed by eq (4), which relates indices of different orders.

One-dimensional results in η or in φ are essentially the same (cf fig 12).

The two-dimensional analysis is saturated by γ-conversion.

The total observed effect may by the sum of different physical phenomina like string fragmentation and hadronic decays, HBT (should influence the region $\delta\eta \sim 0.1 - 0.5$ depending on the radius), and γ-conversion.

5. ACKNOWLEDGEMENTS

The financal support from the Swedish Natural Science Research Concil is gratefully acknowledged. The authors thank A Oskarsson for valuable comments concerning this manuscript.

6. REFERENCES

1. S Garpman et al, Nucl Instr Meth A269(1988)134; M I Adamovich et al, Phys Lett 201B(1988)397; Phys Rev Lett 62(1989)2801.
2. M I Adamovich et al, Phys Lett 242B(1990)512.
3. M I Adamovich et al, Mod Phys Lett A6(1991)469.
4. A Białas and R Peshanski, Nucl Phys B273(1986)703; B308(1988)857.
5. K Fialkowski et al, Acta Physica Polonica B20(1989)639.
6. M I Adamovich et al, Phys Rev Lett 65(1990)412.
7. B Nilsson-Almquist and E Stenlund, Comp Phys Comm 43(1987)387.
8. See, for instance, R Holinski et al, Phys Rev Lett 62(1989)733.
9. E Stenlund et al, Proc Int Workshop on Correlations and Multiparticle Pruduction (CAMP) (LESIP IV), Marburg, F R Germany, May 14 - 16, 1990, pp 337. Eds: M Plümer, S Raha, R M Weiner (World Scientific).
10. M I Adamovich et al, Z Phys C 49(1991)395.
11. M I Adamovich et al, Phys Lett 263B(1991)539.
12. I Otterlund et al, Nucl Phys B142(1978)445; J Babecki, private communication.
13. I V Ajinenko et al, Phys Lett 222B(1989)306; 235B(1990)373.
14. A Białas and R Hwa, CERN preprint CERN-TH.5754/90(1990).
15. A Borzelino, Phys Rev 89(1953)1023.
16. W Bell et al, Z Phys C 22(1984)109.

Intermittency investigations in proton gold, sulphur gold and oxygen gold interactions at 200 GeV per nucleon

I. Derado
Max-Planck-Institut für Physik
München, Germany
NA35 Collaboration

Abstract

In the 3-dimensional intermittency analysis we observe intermittent behaviour in nucleus nucleus interactions in the second factorial cumulant moment and not in the normalised factorial moment. The intermittency slopes are large and comparable for different projectiles.

Nonstatistical density fluctuations of the multiplicity distributions in high-enery collisions have been observed in a large variety of experiments [1].

In this paper we report the results of the intermittency analysis in pAu, OAu and SAu reactions at 200 GeV per nucleon incident energy. In order to avoid possible contributions from positron-electron track contamination we used only negative particles. Special care was taken to avoid systematic errors due to double measurement of the same track.

In the analysis we considered only tracks in the central rapidity region (for pAu $\Delta Y = 4$; for OAu and SAu $\Delta Y = 2$). In this part of the phase space our acceptance was good (better than 80 per cent) and the measurement precision of the rapidity was ± 0.01. Therefore we did not apply any corrections to our data.

Fig. 1 shows the dependence of the normalised factorial moments F_q for the negative particle multiplicity distribution on the number M of 3-dimensional phase space boxes (rapidity, transverse momentum and azimuthal angle). We used cumulative variables [2] in order to correct for the non uniform distribution of these variables.

In order to have statically significant results we retained only those $F_q\,(M)$ for which at given M we had at least 10 multiples contributing. This cut restricted all F_q with rank q greater than 2 to $M < 1000$. Therefore we consider in the further analysis only the second rank moments. The F_2 moments for OAu and SAu reactions grow faster than the power law: (1) $F_2 = A_2 M^{\varphi_2}$. On the other hand for the factorial cumulant moment $K_2 = F_2 - 1$ we do observe a power law behaviour as is evident from Fig. 2.

In table 1 we give the results of the fit $K_2 = C_2 M^{\kappa_2}$

Table 1

	C_2	κ_2	$d\overline{N}/dy$	# of events
pAu	0.165±0.008	0.232 ± 0.017	2.5	1192
OAu	0.025±0.001	0.385±0.010	40	463
SAu	0.040 ±0.002	0.339±0.011	75	333

The pAu data are also consistent with a power law behaviour (1) for the normalised factorial moment F_2 with $A_2 = 1.077 \pm 0.031$ and $\varphi_2 = 0.059 \pm 0.012$.

It is interesting to point out that in the case of a cascading fragmentation model one expects the power law for the F moments and for the ISING model one expects a power law in the K moments [3].

REFERENCES

[1] See these proceedings "Ringberg Workshop on Multiparticle Production, Ringberg Castle, June 25-28, 1991"

[2] W. Ochs, Phys. Lett. **B247** (1990) 101,
A. Białas and I. Seixas, Phys. Lett. **B250** (1990) 161,
A. Białas and M. Gazdzicki, Phys. Lett. **B252** (1990) 483

[3] See e.g. J. Wosiek, Nucl. Phys. **B** (Proc. Suppl.) 9 (1989) 640

Figure captions

1.) Log-log plot of the factorial moments F_q versus M, the number of boxes in $y \times PT \times \Phi$ for Oxygen+Gold collisons at 200 GeV/nucleon

2.) Log-log plot of the factorial cumulant moment K_2 vs. M. The solid line is a fit to the function $C_2 M^{\kappa_2}$. Oxygen-Gold collisions at 200 GeV/nucleon

TABLE 1

The values of C_2, κ_2 from the fit to the expression $K_2 = C_2 M^{\kappa_2}$. $\frac{d\overline{N}}{dy}$ is the average number of negative particles per unit of rapidity in the central region and the last column is the number of events for the given reaction.

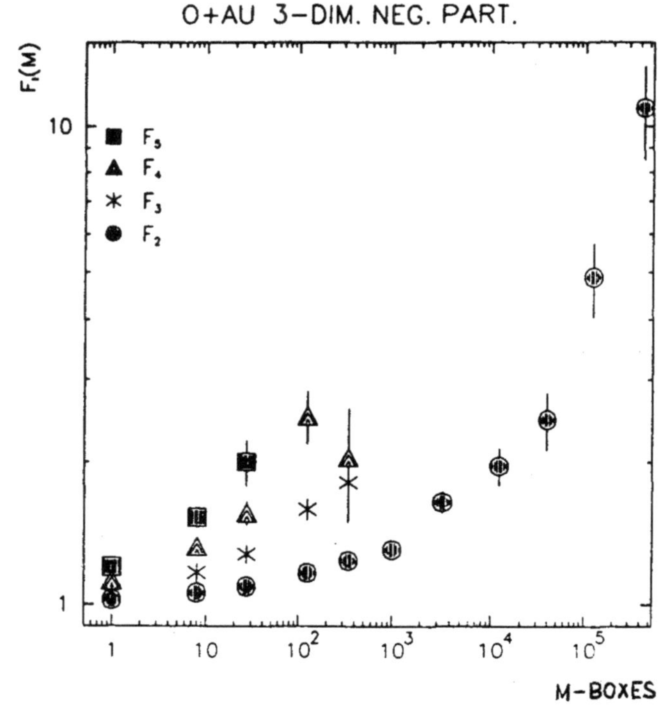
O+AU 3-DIM. NEG. PART.
F_i(M)
F_5
F_4
F_3
F_2
10
1
1
10
10^2
10^3
10^4
10^5
M-BOXES

Fig. 1

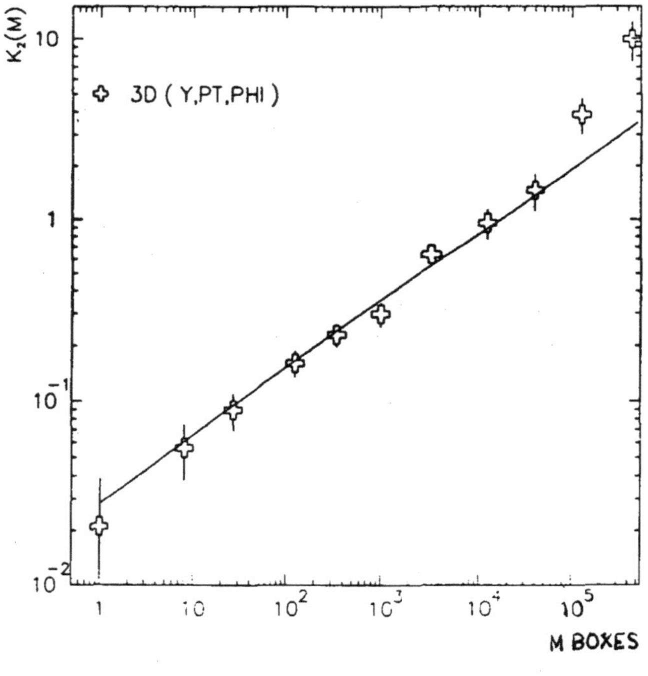
K_2(M)
3D (Y,PT,PHI)
10
1
10^-1
10^-2
1
10
10^2
10^3
10^4
10^5
M BOXES

Fig. 2

Entropy, dimensions and other multifractal characteristics of multiplicity distributions

M. PACHR, V. ŠIMÁK*, M. ŠUMBERA, I. ZBOROVSKÝ

Nuclear Physics Institute, Czechoslovak Academy of Sciences,
CS 250 68 Řež near Prague, Czechoslovakia
*) Institute of Physics, Czechoslovak Academy of Sciences,
Na Slovance 2, CS 180 40 Prague, Czechoslovakia

ABSTRACT

Entropy scaling of multiplicity distributions is shown to be a special case of more general set of multifractal measures. We have performed an analysis of multiplicity distributions data from hadron-hadron and electron-positron interactions.

1. Recently a large number of experimental data accumulated in hadron-hadron and electron-positron interactions stimulated a new development in multiparticle dynamics. A new methods introduced originally for description of chaotic dynamical systems, diffusion-limited growth of structures and other nonlinear phenomena[1] have been also applied to the spectra of particles created in high-energy collisions (for a review and extensive list of referencies see[2]). Pioneer work in multiparticle dynamics in terms of multifractal variables have been carried out in referencies[3]. Several authors have studied intermittent behaviour of particle distributions in a small rapidity bins. The exponential increase of a scaled factorial moments with decreasing window width was interpreted a signal of multifractal property of (a chaotic) dynamics of multiparticle production[4].

We have suggested[5] to study the multiplicity distribution of particles in terms of its information entropy. The scaling properties of entropy in rapidity windows together with its linear increase in full phase with maximum rapidity of produced hadrons $Y_{max} = \sqrt{s}/m_\pi$ were found. Using data on charged hadron multiplicity distributions P_n from hadron-hadron inelastic interactions it was demonstrated that for $\sqrt{s}$ > 20 GeV the information entropy

$$S = - \Sigma P_n . \ln P_n \qquad (1)$$

stays constant when divided by Y_{max}.

The Rennyi order-q information entropy[1]

$$Iq = \frac{1}{1-q} \ln \Sigma (Pn)^q \tag{2}$$

generalizes the information entropy S as follows from relation $I_1 = S$. The experimental values of ratios $Dq = Iq/Y_{max}$ for q = 1,3 and ∞ are presented on Fig.1a for pp[6] and $p\bar{p}$[7] data. For q > 1 they exhibit the same type of energy independent behaviour as observed earlier for q=1. Analogical plot (Fig.1b) for data from e^+e^- collisions[8] generalizes observed regularity to lepton–lepton collisions.

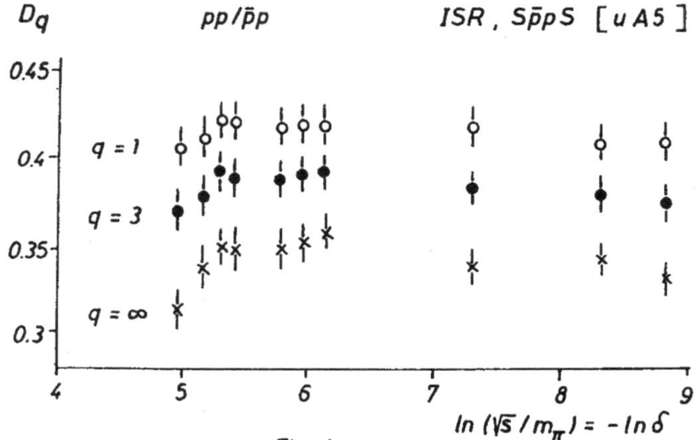

Fig. 1a

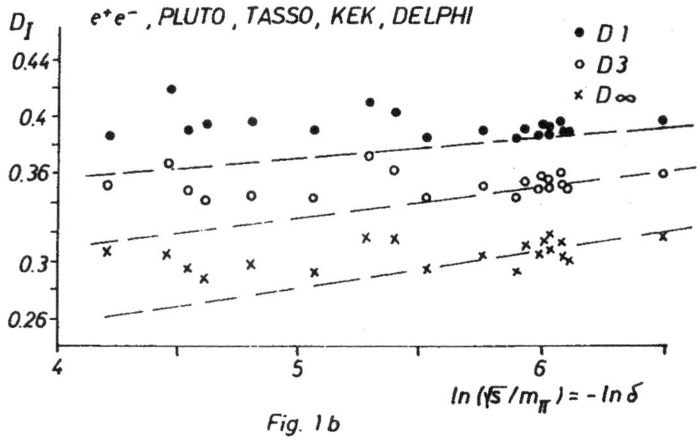

Fig. 1b

Fig.1. Generalized fractal dimensions Dq as a function of Y_{max} for pp/$\bar{p}$p(a) and ee (b) data.

Energy averaged values of Dq's for both data samples are summarized in Table 1.

Table 1. Generalized dimensions

D_q	q=1	q=2	q=3	q=4	q=5	q=∞
$pp/p\bar{p}$	.413 ±.002	.392 ±.002	.381 ±.002	.374 ±.003	.370 ±.003	.339 ±.004
e^+e^-	.393 ±.002	.365 ±.002	.352 ±.002	.344 ±.002	.338 ±.002	.303 ±.002

2. Let us now translate these facts into the language of fractal geometry as used when studying the growth structures[1]. The fractal with overall extent $\sqrt{s}$ is composed of $N = \sqrt{s}/m_\pi$ discrete sites each of size m_π. To the n-th site we assign a probability Pn. If the quantity $\Sigma(Pn)^q$ scales with improving relative resolution $\delta=1/N$ like

$$Z(q,\delta) = \sum (Pn)^q \sim \delta^{-\tau(q)} \tag{3}$$

for δ small enough one usually speaks about multifractality of distribution Pn [1]. Thus previously found energy and hence δ-independence of quantities Dq serves as a direct proof of multifractality of multiplicity distributions. These quantities (sometimes called generalized dimensions[1]) can be related to the function $\tau(q)$ by:

$$Dq = \frac{\tau(q)}{1-q} \tag{4}$$

It is interesting to note that the multifractality may be looked for even in the case when functional dependence of the relative resolution δ on essential parameter of the problem under study (e.g. on $\sqrt{s}$ in our case or on some more complicated function of both $\sqrt{s}$ and impact parameter b for the case of nucleus-nucleus collisions) is a priory unknown. This can be done by checking the q-independence of the ratios

$$R(\delta_1,\delta_2) = \frac{\ln[Z(q,\delta_1)]}{\ln[Z(q,\delta_2)]} = \frac{\ln\delta_1}{\ln\delta_2} \tag{5}$$

with different functions $\delta(\sqrt{s}, b, \ldots)$.

Meaning of function $\tau(q)$ is more transparent after performing the Legendre transformation from independent variables τ and q to the variables α and f :

$$\alpha(q) = -\frac{d\tau(q)}{dq}$$
$$f(\alpha(q)) = q\alpha(q) + \tau(q) \tag{6}$$

Then $f(\alpha)$ is a fractal dimension of a subset composed from bins whose occupancy probability lies in interval

$$(P-dP, P+dP) = (\delta^{\alpha+d\alpha}, \delta^{\alpha-d\alpha})$$

and is connected with frequency $N(\alpha,\delta)d\alpha$ of the occurrence of various probabilities at resolution δ by relation:

$$N(\alpha,\delta)d\alpha \sim \delta^{-f(\alpha)} \tag{7}$$

As $\tau(q=1)$ is defined to be zero for normalized probability distributions, from (7) it follows that $f(\alpha)$ has a slope of unity where α and $f(\alpha)$ are equal to each other: $\alpha(q=1) = f[\alpha(q=1)]$. From explicit dependence of $\alpha(q)$ on P_n and δ:

$$\alpha(q) = \lim_{\delta\to 0} \frac{\sum P_n^q \ln P_n}{\left(\sum P_n^q\right) \ln \delta} \tag{8}$$

one obtains $\alpha(q=1) = D_1 = - S/\ln\delta$. For all other points $f(\alpha)$ is smallerthen α. Hence the total probability $N(\alpha,\delta)d\alpha\ \delta^{\alpha}$ accumulated in bins with $P \approx \delta^{\alpha}$ is maximal for the subset with scaling exponent $\alpha(q=1) = D_1$. In the other words with decreasing δ the total probability gradually concentrates on subsets with fractal dimension close to the entropy dimension $-S/\ln\delta$.

The $f(\alpha)$ curve for pp/$\bar{p}$p data is presented in Fig.2a, for e^+e^- in Fig.2b. For both reactions the scaling is best around the region $\alpha(q=1)$.

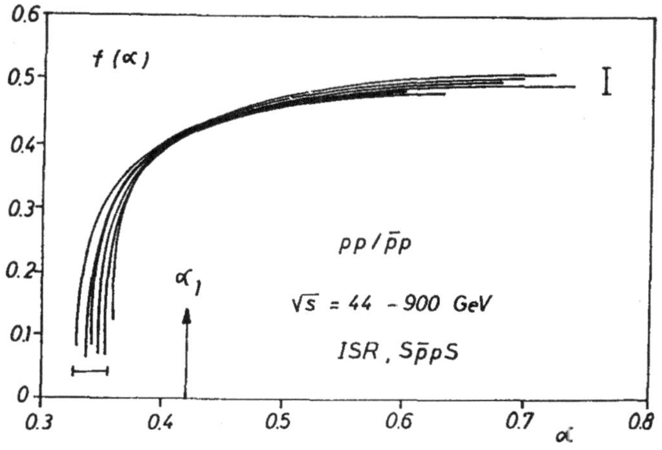

Fig. 2 a

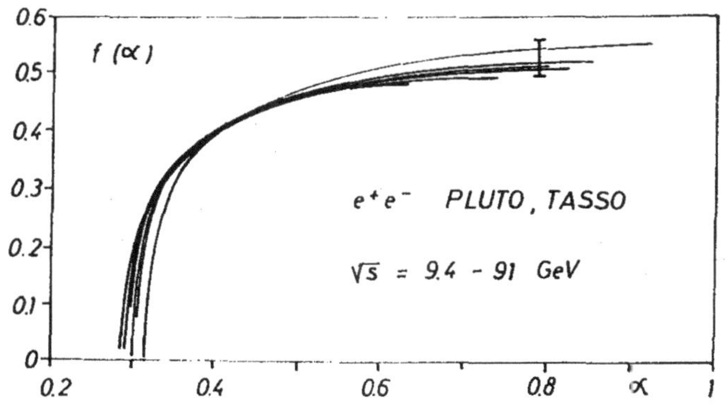

Fig. 2 b

Fig.2. $f(\alpha)$ curve for pp/$\bar{p}$p (a) and e^+e^- interactions (b). Statistical errors are indicated at the endpoints of curves.

3. In the intermittency studies multifractal behaviour is frequently interpreted as manifestation of some self-similar particle production mechanism[2]. In reference[9] a simple implementation of the iterative random cascade mechanism – scale invariant branching model[10] – was used. In its advanced Monte Carlo implementation the model can be applied to describe a lot of properties of the hadronic final states produced in

e^+e^- annihilation. However to discuss qualitatively our results on multifractality its skeleton form seems to be sufficient. The natural ansatz for the scale invariant kernel function (describing at each step of the cascade the mass distribution of hadronic cluster decay products) is a two body phase space. Charge conservation constraint is applied during the whole development of the hadronic cascade. Neutral or singly charged clusters are considered only.

We have generated 70000 MC events at several different energies starting from $\sqrt{s}$ = 10 GeV up to $\sqrt{s}$ = 100 TeV. Our results on Dq 's for energy range now accessible are presented by lines in Fig.1b. The model nothwistanding the fact that its agreement with the data seems to improve with increasing energy departs noticeably from the energy independent (multifractal) behaviour observed experimentaly. We have also tried other scale invariant kernel functions of very different shapes (pencil-like[9], uniform, etc.). Corresponding Dq's show only a weak dependence on the shape of the kernel function.

Let us now study the asymptotic behaviour of the scale invariant branching model. Up to 100 TeV convergence of Dq's is too slow to yield a convincing evidence for onset of the multifractality (Fig.3).

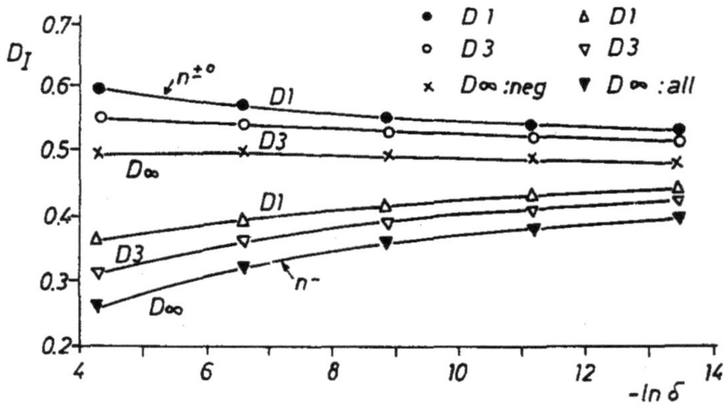

Fig. 3

Fig.3. Generalized fractal dimensions Dq calculated according to the scale invariant branching model[9].

However we can show that asymptotically considered branching model gives the energy independent Dq's . To this end let us recall[10] that in high energy limit the model exhibits both the KNO scaling of the multiplicity distribution and the power like behaviour $\langle n\rangle \sim s^{const.}$. After substitution $P_n = 1/\langle n\rangle \cdot \psi(z=n/\langle n\rangle)$ into Eq.(3) one gets:

$$\tau(q) = (q-1)\,\frac{\ln\langle n\rangle}{\ln\delta} - \frac{\ln\int \psi^q(z)dz}{\ln\delta} \tag{9}$$

where $\psi(z)$ is a well known KNO function. For $Y_{max} = -\ln\delta$ high enough the second term on the r.h.s. of Eq.(9) vanishes and (from Eq.(4)) multifractality follows. However this multifractality is a trivial one because all D_q's degenerate to a single value. In the same limit (as follows from Eq.(6)) the $f(\alpha)$ curve shrinks into one point. Using language of the fractal geometry one may say that for the selfsimilar scale invariant branching model the underlying fractal is a homogenious one.

Experimental data from both $pp/\bar{p}p$ and e^+e^- collisions provide much wider spectrum of generalized dimensions. Let us now show how such a nontrivial multifractal behaviour restricts the energy evolution of the KNO function $\psi(z)$. From (8) we have:

$$P_{max} \sim \delta^{-\alpha(q=\infty)} \tag{10}$$

where $P_{max} = \max\{P_n\}$. Using the entropy bound[6]:

$$\ln\langle n\rangle + 1 \le S = -\alpha(q=1)\ln\delta \tag{11}$$

one obtains:

$$\psi_{max} = \langle n\rangle P_{max} \sim \left(\frac{\sqrt{s}}{m_\pi}\right)^{\alpha(q=1)-\alpha(q=\infty)} \tag{12}$$

From unequality $\alpha(q=\infty) \le \alpha(q) \le \alpha(q=-\infty)$ (which can be derived using Eq.(8))[1] it follows that the multifractality with nondegenerate spectrum of α's must inevitably violate the KNO scaling. Qualitatively this means that in addition to the widening of function $\psi(z)$ with increasing energy (as unambiguously observed in $pp/\bar{p}p$ collisions) its shape will be determined by more and more pronounced peak. It is interesting whether such nontrivial behaviour can be reproduced by other phenomenological approaches.

For e^+e^- collisions the situation is less clear. Available data do not yet allow us to distinguish between trivial and genuine multifractality. At finite energies an approximate KNO scaling may not be in disagreement with observed energy dependence of the Negative Binomial parameter k^{-1} (see the DELPHI paper in[8]). However asymptoticaly both regularities can not coexist - one has to keep in mind the lesson on precousious scaling seen in pp collisions at the ISR energy range[11].

Finaly we have used Monte-Carlo QCD-based generator HERWIG[12] to calculate $f(\alpha)$ curve and D_1, D_3 and D_∞ for corresponding set of e^+e^- data (Fig.4 a,b). The model $^\infty$within the errors reproduces observed multifractal behaviour.

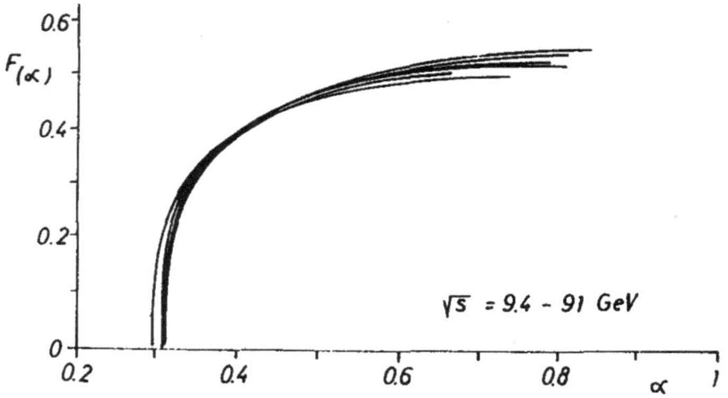

Fig. 4 a

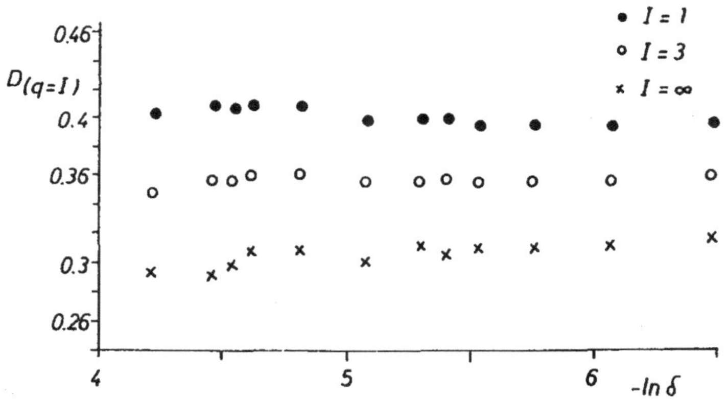

Fig. 4 b

Fig.4. HERWIG results for e^+e^- interactions: f(α) curve (a) and D1,D3 and D_∞ (b).

Compared to oversimplified scale invariant branching model HERWIG incorporates several ingredients which make it more realistic and hence provide a basis for a good description of the experimental data. Among other the effect of passing different flavour tresholds seems to gain on importance when going to still higher energies. Our Monte-Carlo study has shown that there should be an abrubt increase (~10%) in D_q 's when passing through $t\bar{t}$ treshold . Similar jump is observed in energy dependence of the Negative Binomial parameter k^{-1} and of all higher scaled factorial moments. Hence not only multifractality but also KNO scaling is expected to be vioalted near $t\bar{t}$ treshold. For even higher energies the situation is beyond the scope of our present understanding of the multiparticle production. Nevertheless we hope that the multifractal analysis of the multiplicity distributions will play a usefull role in future investigations.

References

1. T.C.Halsey et al.,*Phys.Rev.***A33**(1986)1141.
J.Feder, *Fractals*, (Plenum Press, New York and London, 1988).
G.Paladin and A.Vulpiani,*Phys.Rep.***156**(1987)147.
H.G.Schuster, *Deterministic chaos*, (VCH, Weinheim, 1988).
2. A.Bialas,"INTERMITTENCY '90", Preprint CERN-TH-5791/90 and to be published in Proc."Quark Matter '90", Menton, France 1990,*Nucl.Phys.***A** (Proc.Suppl.).
R.Peschanski,"Intermittency in particle collisions", Preprint CERN-TH.5891/90.
3. R.C.Hwa,*Int.J.Mod.Phys.***A4**(1989)481
R.C.Hwa,*Phys.Rev.***D41**(1990)1456
4. A.Bialas and R.Peschanski,*Nucl.Phys.***B273**(1986)703.
5. V.Šimák,M.Šumbera and I.Zborovský,*Phys.Lett.* **B206**(1988)159.
6. NA22 Collab., M.Adamus et al.,*Phys.Lett.***B177**(1986)239.
K.Alpgard et al.,*Phys.Lett.***B115**(1982)71.
7. UA5 Collab., G.J.Alner et al.,*Phys.Lett.***B138**(1984)304.
UA5 Collab., R.G.Ansorge et al.,*Z.Phys.***C43**(1989)357.
8. PLUTO Collab. {9.4,12.0,13.0,17.0,22.0,27.5,30.6 GeV} private communication.
TASSO Collab., W.Braunschweig et al.,*Z.Phys.***C45**(1989) 193; {14.0,22.0,34.8,43.6 GeV}.
AMY Collab.,H.W.Zheng et al.,KEK preprint 90-005, submitted to *Phys.Rev.***D** {50.,52.0,55.0,56.0,57.0,60.0, 60.8,61.4 GeV}.
DELPHI Collab.,*Z.Phys.***C50**(1991)185 {91 GeV}.
9. W.Ochs and J.Wosiek,*Phys.Lett.***B214**(1988)617.
10. W.Ochs,*Z.Phys.***C23**(1984)131.
11. W.Zajc,*Phys.Lett.***B175**(1986)219.
12. G.Marchesini, B.R.Weber,*Nucl.Phys.***B310**(1988)461.
T.Sjöstrand,QCD generators, in CERN yellow report "Z physics at LEP1", Vol.3, CERN 89-08.

CORRELATIONS AND FLUCTUATIONS IN MULTIHADRON FINAL STATES

PETER CARRUTHERS
Department of Physics, University of Arizona
Tucson, Arizona 85721 USA

ABSTRACT

We review the connections among rapidity density correlation functions, their associated cumulants and moments. The current status and significance of moment data, as a function of phase-space resolution, is reviewed. We note that the practice of slope-fitting to curved moment data can lead to misleading conclusion. The cumulant moments and correlators provide the best measure of true correlations. We show how to extract these quantities using identities similar in spirit to carrying out a phase shift analysis. Finally, we assess the status of the linked-pair correlation structure with negative binomial counts.

1. Introduction

Correlation functions are the tool of choice for the description of fluctuations in multihadron final states. These functions allow the computation of such observables as count probabilities, the rapidity gap distributions, bin averaged factorial moments, etc. In the case of scaling, a direct connection can be made to the fractal "correlation dimensions" characterizing the irregularity of the rapidity histograms.

The correlation functions themselves are very hard to measure beyond the second order. However their averages over suitable parts of phase space can be accurately found, in some cases, to fifth order. These data should be of importance in unraveling the mysteries of hadronization.

For historical reasons hadronic "intermittency" became identified with possible scaling of the bin-averaged factorial moments F_ρ for small bin size δy. By now it appears unlikely that such scaling occurs in one-dimensional rapidity data. The behavior of $F_\rho(\delta y)$ seems entirely compatible with saturation due to the existence of a finite correlation length. Whether or not scaling occurs in one or higher dimensions, the data on fluctuations are of critical importance. I suggest that we generalize the expression "intermittency" to mean the study of such fluctuations without regard to scaling behavior.

Most people believe in some kind of self-similar cascading dynamics of the pre-matter composed of partons. After a rocky start, various Monte Carlo cascade models now give a good account of the moment data. Recall, however, that originally the spurious difference between Monte Carlo and moment data led to the hope of new physics. In the case of F_2, which is <u>identically</u> given by integration over the <u>known</u> two-particle correlation function, this meant that the code had not even built in such a crucial observable as the two-particle correlation. This example should give pause in accepting black boxes of any kind as reliable physics. We believe that the magnitudes of the moment data are just as important as the local slopes. As we have learned here, there are many subtle effects that can confuse the moment data, especially with regard to slopes.

I also feel the necessity to criticize the practice of forcing straight-line "slopes" on moment data that exhibit obvious curvature. In many cases there is no unique slope whatsoever. This excessive concern with a small portion of the data set is like studying the tail of the dog while ignoring the rest of the dog. The dog, too, is important.

2. Identities Relating Moments and Correlation Functions

Consider measurements of rapidities s_i for events having n particles. Define idealized correlation densities[1]

$$\begin{aligned}
\hat{\rho}_1(y_1, \vec{s}) &= \sum_{i=1}^{n} \delta(y_1 - s_i) \\
\hat{\rho}_2(y_1, y_2; \vec{s}) &\equiv \sum_{i \neq j=1}^{n} \delta(y_1 - s_i)\delta(y_2 - s_j) \\
\hat{\rho}_3(y_1, y_2, y_3; \vec{s}) &\equiv \sum_{i \neq j \neq k=1}^{n} \delta(y_1 - s_i)\delta(y_2 - s_j)\delta(y_3 - s_k) \ .
\end{aligned} \tag{2.1}$$

The exclusion of overlapping indices is appropriate for systems having one species of particle (here the "species" of charged particle). The averaging over events and particle number is

$$\rho_n(y_1 \dots y_n) \equiv \sum_n P_n \int \prod_{j=1}^{n} Q_n(s_1, \dots, s_n)\hat{\rho}_n(y_1 \dots y_n; \vec{s}) \tag{2.2}$$

where P_n is the probability of finding n particles and $Q_n(s_1, \dots, s_n)$ is the joint probability of finding the n particle at positions $s_1 \dots s_n$. ρ_n is of course the normalized n-particle inclusive differential cross section. Integration of (2.2) over identical ranges of y_i (called Ω) clearly counts the number of particles in Ω according to

$$\begin{aligned}
&\int_\Omega \rho_1(y_1)dy_1 = \langle n(\Omega)\rangle \\
&\int_\Omega dy_1 \int_\Omega dy_2 \rho_2(y_1, y_2) = \langle n(\Omega)(n(\Omega) - 1)\rangle \\
&\int_\Omega dy_1 \int_\Omega dy_2 \int_\Omega dy_3 \rho_3(y_1, y_2, y_3) = \langle n(\Omega)(n(\Omega) - 1)(n(\Omega) - 2)\rangle \ .
\end{aligned} \tag{2.3}$$

The appearance of the factorial moment in Eq. 2.3 is in automatic consequence of the definitions 2.1. In this formulation no special discussion of statistical noise is needed to motivate factorial moments. Note that the absence of correlation defined in pth order as $\rho_p(y_1 \dots y_p) = \rho_1(y_1)\rho_1(y_2)\dots(y_p)$ implies Poissonian count statistics in Ω: $\langle n(n-1)\dots(n-p+1)\rangle = \langle n\rangle^p$ as is seen from Eq. 2.3.

Although translation invariance fails in hadronic physics, one finds $\rho_2 \to \rho_1\rho_1$ for large $|y_1 - y_2|$. The standard way to subtract uncorrelated background is via the cumulant C_2:

$$C_2(y_1, y_2) = \rho_2(y_1, y_2) - \rho_1(y_1)\rho_1(y_2) . \tag{2.4}$$

Consider as usual M bins of equal width δy, and normalize moments to the local population $\langle n_i\rangle = \rho_i \delta y$ for bin i

$$F_2^i \equiv \frac{\langle n_i(n_i-1)\rangle}{\langle n_i\rangle^2} = \int_{\Omega_i} \frac{dy_1 dy_2}{(\delta y)^2} \frac{\rho_2(y_1, y_2)}{\rho_i^2} \tag{2.5}$$

$$K_2^i \equiv \frac{\langle n_i(n_i-1)\rangle - \langle n_i\rangle^2}{\langle n_i\rangle^2} = \int_{\Omega_i} \frac{dy_1 dy_2}{(\delta y)^2} \frac{C_2(y_1, y_2)}{\rho_i^2} . \tag{2.6}$$

The usual bin-averaged factorial moment (here with local normalization) is $F_2 = \sum_{i=1}^M F_2^i/M$. The corresponding integration domain is the set of boxes along the diagonal shown in Fig. 1. It is related to the bin averaged cumulant factorial moment $K_2 = \sum_{i=1}^M K_2^i/M$.

$$F_2(\delta y) = 1 + K_2(\delta y) . \tag{2.7}$$

We exhibit this trivial equation as the prototype of the general case. Here no correlation ($K_2 = 0$ and Poissonian statistics) shows that $F_2 = 1$ in the absence of correlation.

Using well-known definitions of higher cumulant correlations allows the systematic removal of lower order "background" contributions to the F_P moments. One finds[3]

$$\begin{aligned} F_3 &= 1 + 3K_2 + K_3 \\ F_4 &= 1 + 6K_2 + 3\overline{(K_2)^2} + 4K_3 + K_4 \\ F_5 &= 1 + 10K_2 + 15\overline{(K_2)^2} + 10\overline{K_3K_2} + 10K_3 + 5K_4 + K_5 \end{aligned} \tag{2.8}$$

where the higher F_p and K_p moments are defined similarly to those of Eq. 2.5 and 2.6; the bar notation is an average of products bin moments. Eqs. 2.7-2.8 are identities which can be used phenomenologically to extract successively the highest cumulant factorial moment. This procedure is similar to decomposing a scattering amplitude by means of a phase shift analysis.

Note that the clean structure of Eqs. 2.8 depends on the use of local normalization. By now several prescriptions are available to deal with the non-uniformity of the single particle density ρ_1. We note interesting feature of the local (often known as "vertical") average, beginning with Eq. 2.6. For the negative binomial distribution one sees that $K_2^i = 1/k_i(\delta y)$. In fact for higher orders $K_P^i = (p-1)!/[k_i(\delta y)]^{p-1}$. Therefore to the extent that the negative binomial parameter k_i is independent of i, all bins are treated equally by this convention. (Although this bin-independence seems to be valid at collider energy, we have learned at this workshop of lower energy data where the contrary is true.)

3. Application of the Moment Identities to Data Analysis

The identities 2.7-2.8 can be used in several ways. First of all, we can sequentially determine the factorial cumulants K_p. These are the basic quantities sensitive to the correlations among the produced hadrons. Figs. 2-4 illustrate the procedure[3] for the UA1 experiment ($p\bar{p}$ collisions of 640 GeV c.m. energy). From measured F_2 data we can determine K_2 from Eq. 2.7. Inserting this value of K_2 into the first of Eqs. 2.8 determines K_3. The bottom curves of Fig. 2 show the difference between the measured F_3 and the value obtained keeping $1 + 3K_2$ only. Fig. 3 shows the resulting K_3 as a function of rapidity bin resolution δy. We note that for a given order p of the factorial moment F_p, the bulk of the data are numerically determined by lower order background contributions. For many experiments of current interest (NA22, KLM, EMUO1) with c.m. energy/particle of order 20 GeV the F_p moments are dominated by the two-particle moment K_2. This is shown in Table 1, where the NA22 moments[4] F_3, F_4, F_5. are compared with Eqs. 2.8 with $K_3 = K_4 = K_5 = 0$. For high resolution (small δy) this effect is particularly striking. As described elsewhere,[5,6] data involving nuclear targets and projectiles show little evidence for moments of higher order than K_2.

Our qualitative conclusions follow:

1. The cumulant moments K_p increase as δy decreases
2. For given s and δy, the K_p decrease with increasing p
3. For fixed p and δy, the K_p increase with c.m. energy $\sqrt{s}$
4. The numerical value of the factorial moment F_p is dominated by background correlations $K_{p'}$, $p' < p$
5. Increasing the complexity of the reaction drastically decreases the strength of the higher cumulants.

Although our detailed analysis has been restricted to the one-dimensional) rapidity variable, we expect these conclusions to generalize to higher dimensions. In any case the relation 2.7-2.8 are true in any number of dimensions. Generalization to more general integration domains can be found elsewhere.[7]

4. Models for the Higher Correlations

Many examples of the success of QCD-motivated event generators – all of the parton cascade variety – have been presented at this workshop. Offhand this would seem surprising because the values of the factorial moments are determined by higher correlations. Evidently the latter must have a simple structure. In fact the structure is not only simple in form (composed of two-particle correlations) but is closely related to the omnipresent negative binomial distribution. As mentioned at the end of Section 2, the connection of K_p, the cumulant correlation function C_p and the NBD parameter k for an interval δy are

$$K_p = \frac{(p-1)!}{k^{p-1}} \tag{3.1}$$

$$\frac{1}{k} = \int \frac{dy_1 dy_2}{(\delta y)^2} k_2(y_1, y_2) \tag{3.2}$$

$$k_2 = \frac{C_2(y_1, y_2)}{\rho_1(y_1)\rho_1(y_2)} \ . \tag{3.3}$$

Accepting the approximate success of the NBD, we can ask what is the simplest linked symmetric cumulant correlation function leading to Eq. 3.1. The answer is easily seen to be[8]

$$k_3 = \frac{C_3(1,2,3)}{\rho_1(1)\rho_1(2)\rho_1(3)} = \frac{A_3}{3}\left[k_2(1,2)k_2(2,3) + k_2(2,3)k_2(3,1) + k_2(3,1)k_2(1,2)\right] \tag{3.4}$$

$$k_4 = \frac{C_4(1,2,3,4)}{\rho_1(1)\rho_1(2)\rho_1(3)\rho_1(4)} = \frac{A_4}{12}\left[k_2(1,2)k_2(2,3)k_2(3,4) + \text{ perms}\right] \tag{3.5}$$

etc., with

$$A_p = (p-1)! \ . \tag{3.6}$$

The denominators in 3.4-3.5 are actually $p!/2$, counting the number of permutations appearing in the symmetrized correlation.

In our original proposal[8] we simply guessed the linked pair structure and fit the A_p to the F_p moments. Then DeWolf[9] pointed out that the fitted values were close to the NBD values of (3.6).

Although the linking formulas do not require translation invariance, they are much easier to treat in that case. To do so we change the box (bin) geometry to a strip geometry with equal volume. Using c.m. and relative coordinates

$$\begin{aligned} \eta &= \frac{1}{p}(y_1 + \ldots + y_p) \\ \zeta_{ij} &= y_j - y_i \ , \qquad\qquad i \neq j = 1 \ldots p \end{aligned} \tag{3.7}$$

we can write a typical contribution to F_3 as

$$\int \frac{d\eta}{M\delta y} \int_{-\delta y/2}^{\delta y/2} \frac{d\zeta_{12}}{\delta y} k_2(\zeta_{12}) \int_{-\delta y/2}^{\delta y/2} \frac{d\zeta_{23}}{\delta y} k_2(\zeta_{23}) \ . \tag{3.8}$$

Hence we find the results

$$K_3 = \frac{A_3}{3}\left(\int_{-\delta y/2}^{\delta y/2} d\zeta k_2(\zeta)\right)^2 \times 3 = \frac{A_3}{k^2(\delta y)}$$

$$K_4 = \frac{A_4}{12}\left(\int_{-\delta y/2}^{\delta y/2} d\zeta k_2(\zeta)\right)^3 \times 12 = \frac{A_4}{k^3(\delta y)} \tag{3.9}$$

and so forth.

There are some non-trivial numerical considerations in this analysis that will not be discussed here. We only mention that the bin (box) geometry is used for most experiments, while the strip geometry, used here as a good approximation to the box geometry has merit on its own as an approach to data analysis.[10] In our work[7] on "correlators" we have confirmed that good fits to moments of various kinds are not sensitive to details of the correlation function. For example good moments and correlators can be derived using (false) translation invariant correlation functions. Therefore it is of great interest to test the local structure of expressions like 3.4-3.5 in addition to the moments obtained by integration.

Figs. 3-4 show the predictions of Eqs. 3.4-3.5 for UA1 data. For lower energies (NA22) it seems that linked pairs do not give a good description of the moment data. We do not yet know if the problem lies with translation invariance or in the assumed linked structure. Although both the Monte Carlo codes and the linked pair correlations (with negative binomial coefficients) give correct moments in most cases involving hadronic collisions, it will be of great interest to see how either fares in comparison with accurate three-particle correlation data, for example.

Finally we mention that very similar correlation and count structures have been found in studies of galaxy correlation.[11] Excellent scaling behavior has been observed;[12] in addition the void probability obeys a nice scaling law[13] (derived from linking) which is indistinguishable from hadronic rapidity gap data.[14–15]

5. Acknowledgements

This research was supported by grants from the US Department of Energy, Divisions of High Energy and Nuclear Physics, and the Alexander von Humboldt Foundation.

6. References

1. Yu L. Klimontovich, *The Statistical Theory of Non-Equilibrium Processes in a Plasma*, (Pergamon, Oxford, 1967).
2. P. Carruthers, *Phys. Rev.* **A43** (1991) 2622.
3. P. Carruthers, H. C. Eggers and I. Sarcevic, *Phys. Lett.* **B258** (1991).
4. NA22 Collaboration: I. V. Ajinenko *et al., Phys. Lett.* **B222** (1989) 306; *ibid.*, **B235** (1990) 373.

5. I. Sarcevic, this workshop.
6. P. Carruthers, H. C. Eggers and I. Sarcevic, *Phys. Rev.* **C**, in press.
7. H. C. Eggers, P. Carruthers, P. Lipa and I. Sarcevic, *Phys. Rev.* **D**, in press.
8. P. Carruthers and I. Sarcevic, *Phys. Rev. Lett.* **63** (1989) 1562.
9. E. DeWolf, *Acta Phys. Pol.* **B21** (1990) 611.
10. P. Lipa, this workshop.
11. P. Carruthers, *Astrophys. J.*, in press.
12. T. Chmaj, W. Doroba and W. Słominski, *Zeitsch. Phys.* **C50**, (1991) 333.
13. J. N. Frŷ, R. Giovanelli, M. P. Haynes, A. L. Melott and R. J. Scherrer, *Astrophys. J.* **340** (1989) 11.
14. E. DeWolf, this workshop.
15. S. Hegyi, private communication.

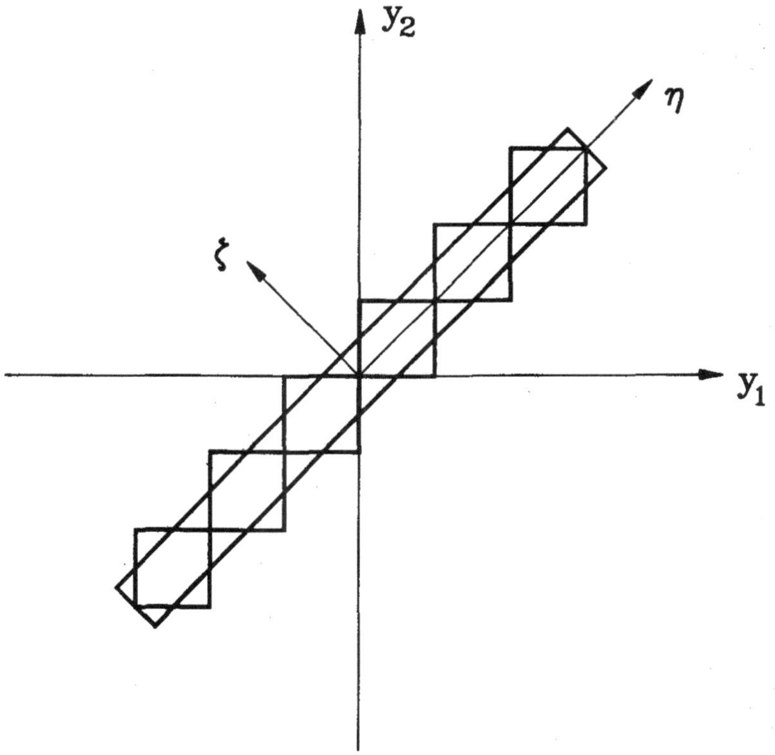

Fig. 1
The bins of width δy, labeled i in Eqs. 2.5-2.6 correspond to the square integration domains in this figure. We also indicate the strip domain, which is preferable for some purposed.

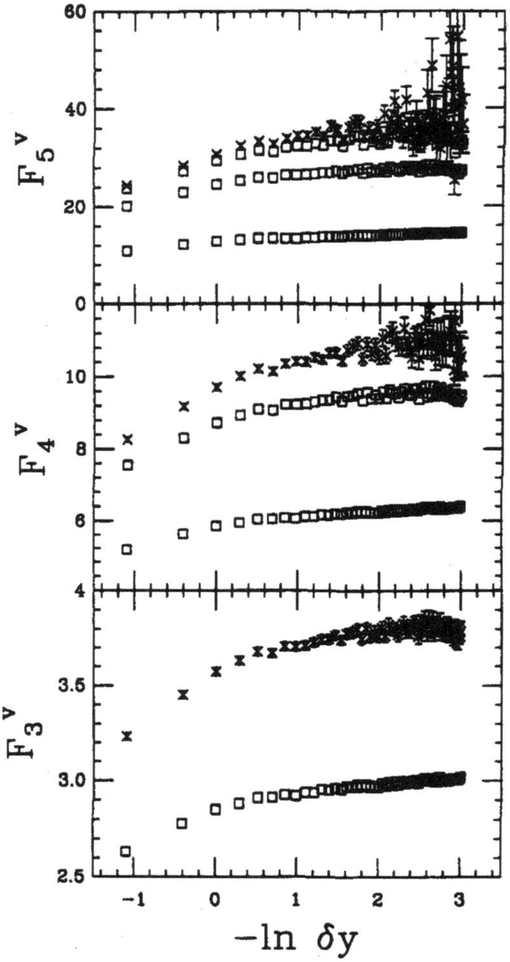

Fig. 2

This figure shows how the factorial moments F_p are built up from factorial cumulant moments K_p according to Eq. 2.8. In each case, the lowest squares are derived from the two-particle moment K_2, neglecting higher K_p. The next higher curve shows the effect of including the (experimentally determined) K_3, etc. We have used a linear scale for the F_p moments to make the percent contributions more visible.

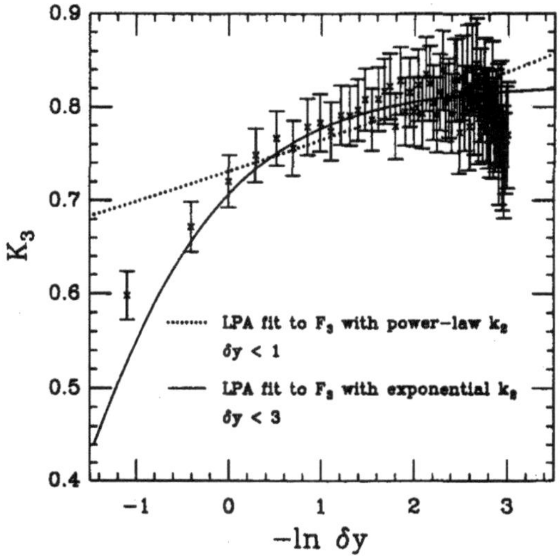

Fig. 3

The third order factorial cumulant moment $K_3(\delta y)$ has been determined from Eq. 2.8 for UA1 data. The data are compared to the linked pair ansatz (Eqs. 3.4-3.9) for the coefficient $A_3 = 1.84$ (exact negative binomial gives $A_3 = 2.0$).

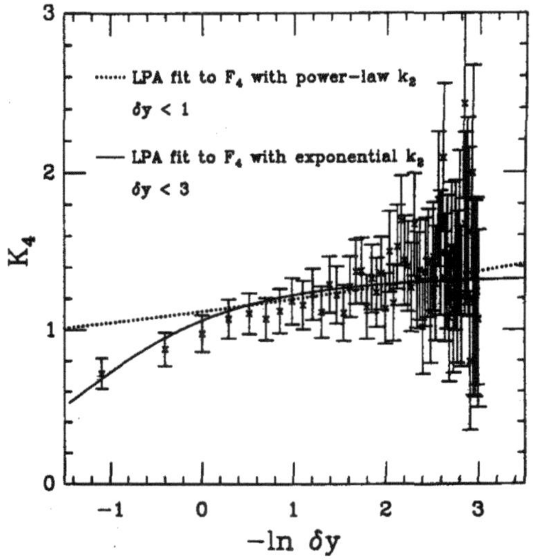

Fig. 4

The fourth order factorial cumulant moment $K_4(\delta y)$ has been evaluated as K_3 in Fig. 3. The best fit A_4/A_3 is 2.39 ± 0.07 as compared with 3.0 for exact negative binomial.

A STATISTICAL FIELD THEORY OF DENSITY FLUCTUATIONS IN HEAVY ION COLLISIONS

INA SARCEVIC*
Department of Physics, University of Arizona
Tucson, AZ 85721, USA

ABSTRACT

We construct a one-dimensional statistical field theory of density fluctuations, based on an analogy with Ginzburg-Landau theory of superconductivity which gives exponential two-particle cumulant correlations and no higher-order correlations, in agreement with all high-energy heavy-ion data. We discuss the extension of our model to higher dimension and the possibility of studying the phase transition of hadronic matter to a quark-gluon plasma in heavy-ion collisions. We give theoretical prediction for the probability distribution of producing n particles in high-energy heavy-ion collisons.

1. Introduction

In the last few years, we have witnessed an impresssive renewal of interest in studying the underlying dynamics of multiparticle production in ultrarelativistic heavy-ion collisions. In particular, the possibility of detecting the onset of a phase transition of hadronic matter to a quark-gluon plasma has created a new excitement.[1] Even though no conclusive prediction for the signature of quark matter has been found, the study of unusually large density fluctuations, recently observed in all high energy heavy-ion collisions has attracted considerable attention, especially as a means of revealing presently unknown dynamics of particle production.[2]

Several experimental groups have recently detected these fluctuations by measuring the bin-averaged factorial moments defined as

$$\begin{aligned} F_p(\delta y) &= \frac{1}{M}\sum_{m=1}^{M} \frac{\langle n_m(n_m-1)\ldots(n_m-p+1)\rangle}{\langle n_m\rangle^p} \\ &= \frac{1}{M(\delta y)^p}\sum_{m=1}^{M}\int_{\Omega_m}\prod_i dy_i \; \frac{\rho_p(y_1\ldots y_p)}{(\bar{\rho}_m)^p} \,, \end{aligned} \tag{1}$$

where n_m is the number of particles in a bin m, M is the total number of bins, δy is the rapidity bin size ($\delta y = Y/M$) and ρ_p is p-particle density correlation function. In all the experiments, the factorial moments are found to increase with decreasing bin size.[3] Presently there is no theory which explains the observed rise

* Humboldt Fellow (Max-Planck Institute for Physics, Munich)

of the moments and all the existing Monte Carlo programs for heavy-ion collisions fail to reproduce the data.[4]

We have previously emphasized the connection between the factorial moments and correlations[5] and in particular the fact that moments F_p ($p = 3, 4, 5$) have large combinatoric contributions from two-particle correlations.[6] Therefore, in order to examine the true higher-order correlations, we express F_p in terms of the bin-averaged cumulant moments

$$\begin{aligned}
F_2 &= 1 + K_2 \\
F_3 &= 1 + 3K_2 + K_3 \\
F_4 &= 1 + 6K_2 + 3(K_2)^2 + 4K_3 + K_4 \\
F_5 &= 1 + 10K_2 + 15(K_2)^2 + 10K_3K_2 + 10K_3 + 5K_4 + K_5 \ ,
\end{aligned} \tag{2}$$

where

$$K_p(\delta y) = \frac{1}{M(\delta y)^p} \sum_m \int_{\Omega_m} \prod_i dy_i \quad k_p(y_1 \ldots y_p) \tag{3}$$

and

$$k_2(1,2) = \frac{\rho_2(y_1, y_2)}{\langle \rho(y_1) \rangle \langle \rho(y_2) \rangle} - 1 \ ,$$

$$k_3(1,2,3) = \frac{\rho_3(y_1, y_2, y_3)}{\langle \rho(y_1) \rangle \langle \rho(y_2) \rangle \langle \rho(y_3) \rangle} - \sum_{perm}^{(3)} \frac{\rho_2(y_1, y_2)}{\langle \rho(y_1) \rangle \langle \rho(y_2) \rangle} + 2 \ ,$$

$$k_4 = \frac{\rho_4(y_1, y_2, y_3, y_4))}{\langle \rho(y_1) \rangle \langle \rho(y_2) \rangle \langle \rho(y_3) \rangle \langle \rho(y_4) \rangle} - \sum_{perm}^{(4)} \frac{\rho_3(y_1, y_2, y_3))}{\langle \rho(y_1) \rangle \langle \rho(y_2) \rangle \langle \rho(y_3) \rangle}$$

$$- \sum_{perm}^{(3)} \frac{\rho_2(y_1, y_2)\rho_2(y_3, y_4))}{\langle \rho(y_1) \rangle \langle \rho(y_2) \rangle \langle \rho(y_3) \rangle \langle \rho(y_4) \rangle} + \sum_{perm}^{(12)} \frac{\rho_2(y_1, y_2)}{\langle \rho(y_1) \rangle \langle \rho(y_2) \rangle} - 6 \ ,$$

$$k_5 = \frac{\rho_5(y_1, y_2, y_3, y_4, y_5))}{\langle \rho(y_1) \rangle \langle \rho(y_2) \rangle \langle \rho(y_3) \rangle \langle \rho(y_4) \rangle \langle \rho(y_5) \rangle} \tag{4}$$

$$- \sum_{perm}^{(5)} \frac{\rho_4(y_1, y_2, y_3, y_4))}{\langle \rho(y_1) \rangle \langle \rho(y_2) \rangle \langle \rho(y_3) \rangle \langle \rho(y_4) \rangle} + \sum_{perm}^{(10)} \frac{\rho_3(y_1, y_2, y_3)}{\langle \rho(y_1) \rangle \langle \rho(y_2) \rangle \langle \rho(y_3) \rangle}$$

$$- \sum_{perm}^{(10)} \frac{\rho_2(y_1, y_2)}{\langle \rho(y_1) \rangle \langle \rho(y_2) \rangle} + 4 \ .$$

Clearly, if there are no true, dynamical correlations, the cumulants K_p vanish.

2. Experimental Data

The KLM Collaboration has measured the so-called *horizontal* factorial moments defined as

$$F_p^h(\delta y) = \frac{1}{M}\sum_{m=1}^{M}\frac{\langle n_m(n_m-1)\ldots(n_m-p+1)\rangle}{\langle \bar{n}\rangle^p} , \tag{5}$$

where $\bar{n} = \sum_m n_m/M$, for proton and oxygen beams on emulsion target at 60 and $200GeV$/nucleon for oxygen and 200 and $800GeV$/nucleon for protons.[7] Since these moments are sensitive to the shape of the density distribution, they were corrected by the heuristic R factor

$$F_p^c(\delta y) = F_p^h/R(\delta y) , \tag{6}$$

where

$$R(\delta y) = \frac{\frac{1}{M}\sum_m \langle n_m\rangle^q}{(\frac{1}{M}\sum_m\langle n_m\rangle)^q} = \frac{1}{M}\sum_m(\langle n_m\rangle/\bar{n})^q . \tag{7}$$

This correction factor was introduced in order to exclude the contribution coming from the trivial fluctuations due to the non-flat shape of the rapidity distribution.[8] Although the expression (6) is not precisely the same as that connecting Eqs. (1) and (3), it is reasonable to use it to compare the exact Eqs. (2) with the KLM data.

In Fig. 1 we present cumulants K_2 for NA22[9] (π-p collision at $\sqrt{s} = 22GeV$), KLM[7] (proton, oxygen and sulfur beams on emulsion at $200GeV$ per nucleon) and EMU01 data[10] (sulfur on gold at $200GeV$ per nucleon). All these collisions are approximately at the same c.m. energy per particle. Both NA22 and EMU01 data are vertical moments defined by Eq. (1), while the KLM Collaboration gives horizontal (corrected) moments defined by Eqs. (6) and (7). We note that K_2 decreases in going from lighter to heavier projectiles, especially in the case of sulfur.

Higher-order cumulants for hadronic and nuclear collisions are shown on Figs. 2-4. The error bars on these figures are calculated for each point separately (i.e. we assume that all the points are independent/uncorrelated). Clearly, if one would want to fit the data points on Figs. 2-4 with a straight line, for example, the correlations between different points would have to be taken into account. For this particular analysis one needs the presently unavailable experimental correlation matrix.

While in hadronic collisions K_3 and K_4 are non negligible (K_3, for example gives up to a 20% contribution to F_3 at small δy), in nucleus-nucleus collisions, at the same energy, these cumulants are compatible with zero. This implies that there are no statistically significant correlations of order higher than two for heavy-ion collisions.[11] Thus, in high energy heavy-ion collisions, the observed increase of the higher-order factorial moments F_p is entirely due to the dynamical two-particle correlations. We find that this conclusion holds even in a higher-dimensional analysis. For example, the KLM Collaboration has done a two-dimensional analysis (in rapidity y and azimuthal angle φ) of the factorial moments.[7] We find that their measured two-dimensional cumulant K_2 is increasing with decreasing bin size $(\delta\varphi\delta y)$ faster than in the one-dimensional case, but the higher-order cumulants are still consistent with zero.[12] Preliminary NA35 data on O-Au at 200GeV/nucleon indicate that there are no true dynamical higher-order correlations present in *any* dimension. This can be seen on Figs. 5-7, where we present cumulant correlations K_3, K_4 and K_5 for O-Au in one dimension (rapidity y), two dimensions $(y - \varphi)$ and three dimensions $(y - \ell n\ p_T - \varphi)$.

3. A Statistical Field Theory of Density Fluctuations

In the previous section we have seen that particles produced in high-energy heavy-ion collisions exhibit only two-particle correlations, indicating that perhaps higher-order correlations are washed out by rescattering of the initially correlated particles. Presently there is no theory that describes this phenomena. Here we present our statistical field theory of density fluctuations which has these features. Our model is formulated in analogy with the Ginzburg-Landau theory of superconductivity.[13] We note that the large number of particles produced in ultrarelativistic heavy-ion collisions justifies application of the statistical theory of particle production. The formal analogy with the statistical mechanics of the one-dimensional "gas" was first pointed out by Feynman and Wilson[14] and later further developed by Scalapino and Sugar[15] and many others.[16] The idea is to build a statistical theory of the macroscopic observables, imagining that microscopic degrees of freedom are integrated out and represented in terms of a few phenomenological parameters, in analogy with Ginzburg-Landau theory of superconductivity. It is well known that in the case of superconductivity, a few years after Ginzburg and Landau proposed their theory, Bardeen, Cooper and Schrieffer formulated a microscopic theory from which one could derive G-L theory and all the phenomenological parameters. Thus, we will assume that our macroscopic theory of multiparticle production and density fluctuations in heavy-ion collisions will eventually be derived from a fundamental theory such as QCD.

While in the G-L theory of superconductivity the field (i.e. order parameter) represents superconducting pairs, in the particle production problem, the relevant variable is the density fluctuation. The "field" $\phi(y)$ is a random variable which depends on rapidity of the particle ($y = 1/2 \ln(E + p_{\parallel})/(E - p_{\parallel})$) and for simplicity we assume that its dependence on other variables, such as transverse momentum and azimuthal angle are integrated out. However we discuss later the possible extension of our one-dimensional theory to higher dimensions.

Even though particles produced in high-energy collisions need not be in thermal equilibrium, we can still introduce a functional of the field ϕ, $F[\phi]$, which plays a role analogous to the free energy in equilibrium statistical mechanics. In principle one should be able to derive this functional from the underlying dynamics.

As a starting point, we will present only a macroscopic theory of multiparticle production with phenomenological parameters.

We start by introducing our functional $F[\phi]$ in the Ginzburg-Landau form

$$F[\phi] = \int_0^Y dy \ \ [a\phi^2 + b(d\phi/dy)^2] \ , \tag{8}$$

Here Y is the maximum rapidity (i.e. $Y = ln(s/m)$) and a and b are the phenomenological parameters to be determined from the underlying dynamics. We note that this functional is a simple free field theory. The field $\phi(y)$ is identified with density fluctuation, i.e.

$$\phi(y) = \frac{\rho(y)}{\langle\rho(y)\rangle} - 1 \ , \tag{9}$$

so that $\langle\phi\rangle \equiv 0$. Here $\langle\rho(y)\rangle$ is the average density (averaged over all the events). The corresponding "partition function" of our one-dimensional model is given by

$$Z = \int \mathcal{D}\phi \ \ e^{-F[\phi]} \ . \tag{10}$$

All physical quantities can be obtained in terms of ensemble averages appropriately weighted by $F[\phi]$. For example, field correlations, which from our definition of the field are related to particle correlations, are given by

$$\langle\phi(y_1)\phi(y_2)...\phi(y_p)\rangle = \frac{1}{Z}\int \mathcal{D}\phi e^{-F[\phi]}\phi(y_1)\phi(y_2)...\phi(y_p) \ . \tag{11}$$

In particular, using our definition of the field (Eq. (9)), we find that the field correlations correspond to the following cumulant particle correlations k_p:

$$\begin{aligned}
\langle\phi(y_1)\phi(y_2)\rangle &\equiv k_2(1,2) \\
\langle\phi(y_1)\phi(y_2)\phi(y_3)\rangle &\equiv k_3(1,2,3) \\
\langle\phi(y_1)\phi(y_2)\phi(y_3)\phi(y_4)\rangle &\equiv k_4(1,2,3,4) + \sum_{perm}^{(3)} k_2(1,2)k_2(3,4) \\
\langle\phi(y_1)\phi(y_2)\phi(y_3)\phi(y_4)\phi(y_5)\rangle &\equiv k_5(1,2,3,4,5)
\end{aligned} \tag{12}$$

where the k_p's are defined by Eqs. (4).

From the expression for the functional $F[\phi]$ we can now calculate particle correlations. In order to do this, we expand $\phi(y)$ in a complete set of functions

$$\phi(y) = \sum_k \phi_k e^{i2\pi ky/Y} \ . \tag{13}$$

Then, functional $F[\phi]$ becomes

$$F[\phi] = \sum_k [a\phi_k^2 + b(2\pi k/Y)^2\phi_k^2] \; , \tag{14}$$

and

$$\begin{aligned} \langle\phi(y_1)\phi(y_2)\rangle &= \int \mathcal{D}\phi e^{-F[\phi]} \sum_k \sum_{k'} \phi_k\phi_{k'} e^{2\pi i(ky_1+k'y_2)/Y} \\ &= \sum_k \frac{e^{-2\pi ik(y_1-y_2)/Y}}{a+b(2\pi k/Y)^2} \\ &= \gamma e^{\frac{-|y_1-y_2|}{\xi}} \; , \end{aligned} \tag{15}$$

where $\gamma = \frac{1}{4\sqrt{ab}}$ and $\xi = \sqrt{\frac{b}{a}}$. Clearly, all the higher-order odd-power correlations vanish and even-power correlations can be expressed in terms of two-field correlations. In particular,

$$\begin{aligned} \langle\phi(y_1)\phi(y_2)\phi(y_3)\rangle &= 0 \\ \langle\phi(y_1)\phi(y_2)\phi(y_3)\phi(y_4)\rangle &= \sum_{perm}^{(3)} \langle\phi(y_1)\phi(y_2)\rangle\langle\phi(y_3)\phi(y_4)\rangle \\ \langle\phi(y_1)\phi(y_2)\phi(y_3)\phi(y_4)\phi(y_5)\rangle &= 0 \end{aligned} \tag{16}$$

From Eqs. (15), (16) and (12), we see that two-particle correlations are exponential and that there are no higher-order cumulant correlations (i.e. $k_3 = 0$, $k_4 = 0$ and $k_5 = 0$).

Assuming translational invariance, which seems to hold for the integrated cumulants,[17] we get the following cumulants K_p by integrating the cumulant correlations k_p over the appropriate rapidity region δy

$$K_2 = 2\gamma\xi^2[(\delta y/\xi) - 1 + e^{-\delta y/\xi}]/\delta y^2 \; , \tag{17}$$

and

$$\begin{aligned} K_3 &= 0 \\ K_4 &= 0 \\ K_5 &= 0 \; . \end{aligned}$$

Our result that there are no higher-order correlations is consistent with all experimental data (see Figs. 2-4). On Fig. 1 we compare our predictions for K_2 with the data (dashed lines). For NA22 we need $\gamma = 0.331$ and $\xi = 2.18$, for proton-emulsion $\gamma = 0.29$ and $\xi = 1.39$ and for KLM and EMU01 heavy-ion data, $\gamma = 0.13$ and $\xi = 1.28$ (for oxygen-em), $\gamma = 0.09$ and $\xi = 2.14$ (for sulfur-em), and $\gamma = 0.08$ and $\xi = 2.14$ (for sulfur-gold). This implies that both coefficients in our

Ginzburg-Landau functional ("mass" a and "kinetic coefficient" b) are increasing with the complexity of the system. The value of the correlation length ξ usually determines how far the system is from the critical point. In the case when $\xi \to \infty$ (or $a \to 0$), the system goes through the phase transition. Even though our fitted values for ξ do not indicate critical behavior for the system, one has to keep in mind that it is not even possible to study phase transitions in one dimension. In the three-dimensional case, using our G-L functional, we get Yukawa type two-particle correlations and no higher-order correlations, in agreement with multidimensional data (see Figs. 5-7). Whether phase transitions of hadronic matter to a quark-gluon plasma could be observed in density fluctuations in the three-dimensional case is presently under investigation.

So far we have only discussed correlations and factorial moments. Another interesting observable is the multiplicity distribution. To obtain this distribution theoretically, one needs to know the correlations to all orders. Luckily, in heavy-ion collisions there are no correlations higher than two-particle, so we can express all the factorial moments in terms of K_2

$$\begin{aligned} F_2 &= 1 + K_2 \\ F_3 &= 1 + 3K_2 \\ F_4 &= 1 + 6K_2 + 3{K_2}^2 \\ F_5 &= 1 + 10K_2 + 15(K_2)^2 \end{aligned} \tag{18}$$

or in general

$$F_p = p! \sum_{n_1 \geq n_2} \frac{\delta(p - n_1 - n_2)}{(n_1 - n_2)! n_2!} (K_2/2)^{n_2} . \tag{19}$$

The corresponding generating function defined as

$$G(z) = \sum_m z^m P_m = \sum_m \frac{(z-1)^m}{m!} F_m \bar{n}^m \tag{20}$$

or

$$\ell n G(z) = \sum_m z^m P_m = \sum_m \frac{(z-1)^m}{m!} K_m \bar{n}^m \tag{21}$$

is given by

$$G(z) = e^{\bar{n}(z-1) + K_2 \bar{n}^2 (z-1)^2/2} . \tag{22}$$

From the generating function, we can get the following probability distribution

$$\begin{aligned} P_n &= \frac{1}{n!} \frac{\partial^n}{\partial z^n} G(z)|_{z=0} \\ &= \frac{e^{(K_2 \bar{n}^2/2) - \bar{n}}}{n!} (K_2 \bar{n}^2)^{n/2} (-i)^n H_n([\frac{(K_2 \bar{n}^2 - \bar{n})^2}{2K_2 \bar{n}^2}]^{1/2}) . \end{aligned} \tag{23}$$

This distribution has two parameters, $\bar{n}$ and K_2 and it is narrower than the hadronic distribution at the same energy which can be very well described by the negative binomial distribution. This indicates that one can test our statistical theory which predicts no higher-order correlations, by directly comparing experimental data on distributions with our distribution given by Eq. (23). Our theory will be best checked by a direct comparison with, not only the first five moments of the distribution function presented in Section 2, but with the full distribution function for producing n charged particles in high-energy heavy-ion collisions.

4. Acknowledgements

The work presented here was done in collaboration with T. Elze, P. Carruthers and H. Eggers whom I also thank for many useful discussions. I am grateful to I. Derado, A. Bokan and G. Jancso for sharing their preliminary analysis of NA35 cumulant moments which are presented in this paper. I am also indebted to the Max Planck Institute for Physics in Munich for their hospitality while this manuscript was completed. This work was supported in part by the United States Department of Energy, Division of High Energy and Nuclear Physics and the Humboldt-Stiftung.

5. References

1. For a review see, Proceeding of *Quark Matter '90*, *Nucl. Phys.* **A525** (1991).
2. For a review see, *Intermittency in High Energy Collisions*, eds. F. Cooper, R. Hwa and I. Sarcevic (World Scientific, Singapore, 1991).
3. A. Bialas, in Proceeding of *Quark Matter '90*, *Nucl. Phys.* **A525** (1991) 345.
4. E. Stenlund, in Proceeding of *Quark Matter '90*, *Nucl. Phys.* **A525** (1991) 67.
5. P. Carruthers and I. Sarcevic, *Phys. Rev. Lett.* **63** (1989) 1562.
6. P. Carruthers, H. C. Eggers and I. Sarcevic, *Phys. Lett.* **B254** (1991) 258.
7. KLM Collaboration, R. Holynski et al., *Phys. Rev. Lett.* **62** (1989) 733; *Phys. Rev.* **C40** (1990) 2449.
8. F. Fialkowski, B. Wosiek and J. Wosiek, *Acta Physica Polonica* **B20** (1989) 639.
9. W. Kittel (NA22 Collaboration), in Proceedings of Santa Fe Workshop on *Intermittency in High Energy Collisions*, eds. F. Cooper, R. Hwa and I. Sarcevic (World Scientific, Singapore, 1991) p. 83.
10. EMU01 Collaboration, M. I. Adamovich et al., *Phys. Rev. Lett.* **65** (1990) 412.
11. P. Carruthers, H. C. Eggers and I. Sarcevic, University of Arizona preprint AZPH-TH/91-17, *Phys. Rev.* C, in press.
12. H. C. Eggers, Ph. D. Thesis, University of Arizona (1991).
13. V. L. Ginzburg and L. D. Landau, *Zh. Eksp. Teor. Fiz.* **20** (1950) 1064.

14. R. P. Feynman (unpublished); K. Wilson, Cornell Report No. **CLNS-131** (1970) unpublished.
15. D. J. Scalapino and R. L. Sugar, *Phys. Rev.* **D8** (1973) 2284; J. C. Botke, D. J. Scalapino and R. L. Sugar, *Phys. Rev.* **D9** (1974) 813; **D10** (1974) 1604.
16. P. Carruthers and I. Sarcevic, *Phys. Lett.* **B189** (1987) 442; N. G. Antoniou et al., *Phys. Rev.* **D29** (1984) 1470; *Phys. Rev.* **D14** (1976) 3578; I. Dremin and M. T. Nazirov, this volume.
17. H. C. Eggers, P. Carruthers, P. Lipa and I. Sarcevic, University of Arizona preprint AZPH-TH/91-15, *Phys. Rev.* **D**, in press.

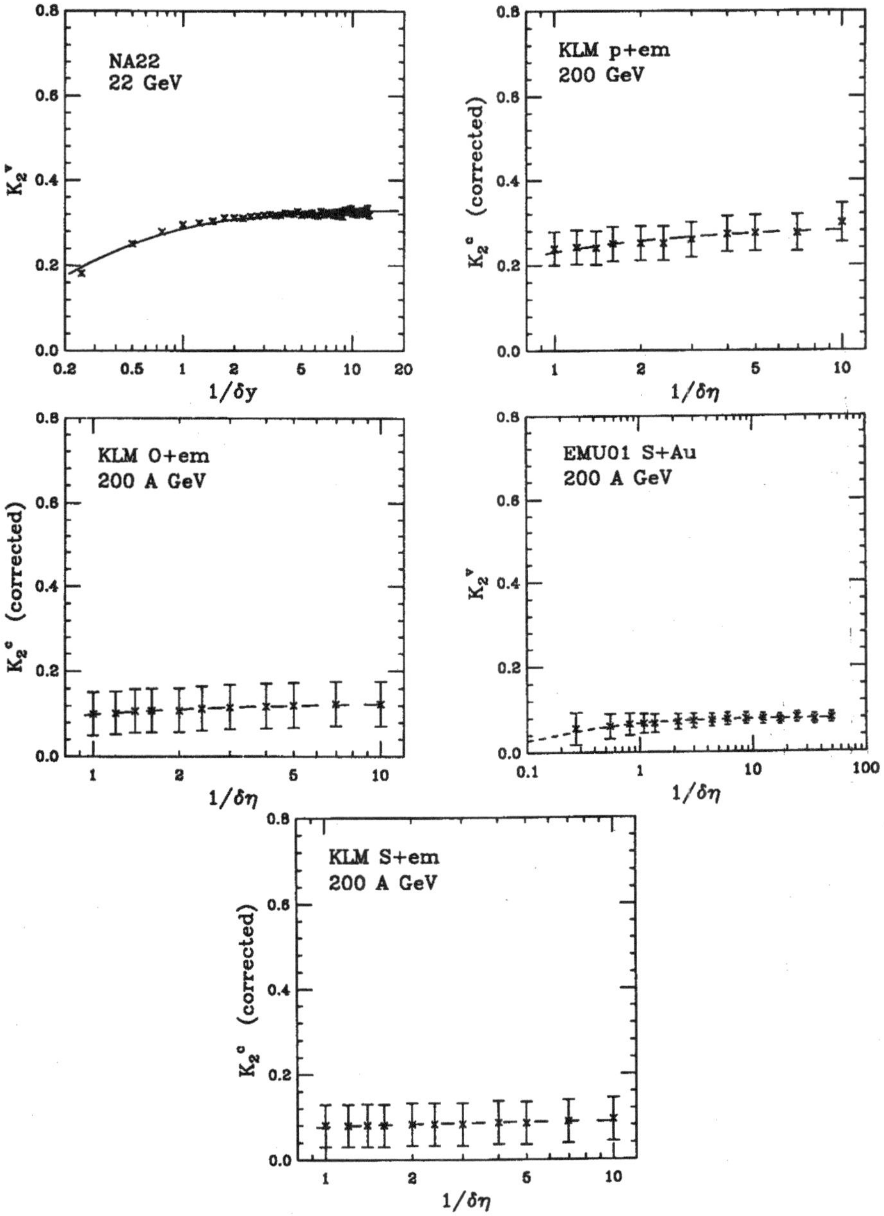

Fig. 1
Theoretical results for K_2 (dashed lines) compared with NA22, KLM and EMU01 data.

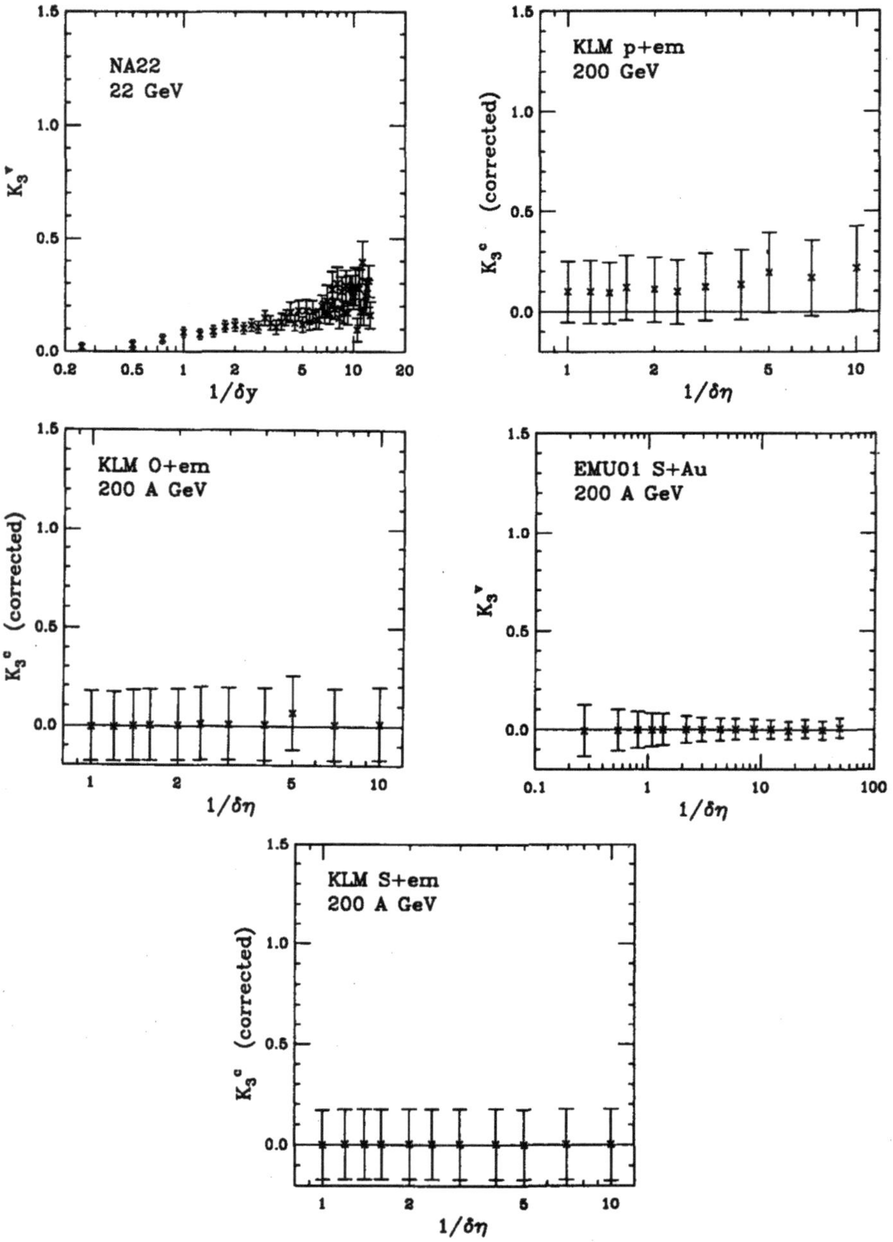

Fig. 2
The cumulant K_3 for NA22, KLM and EMU01 data.

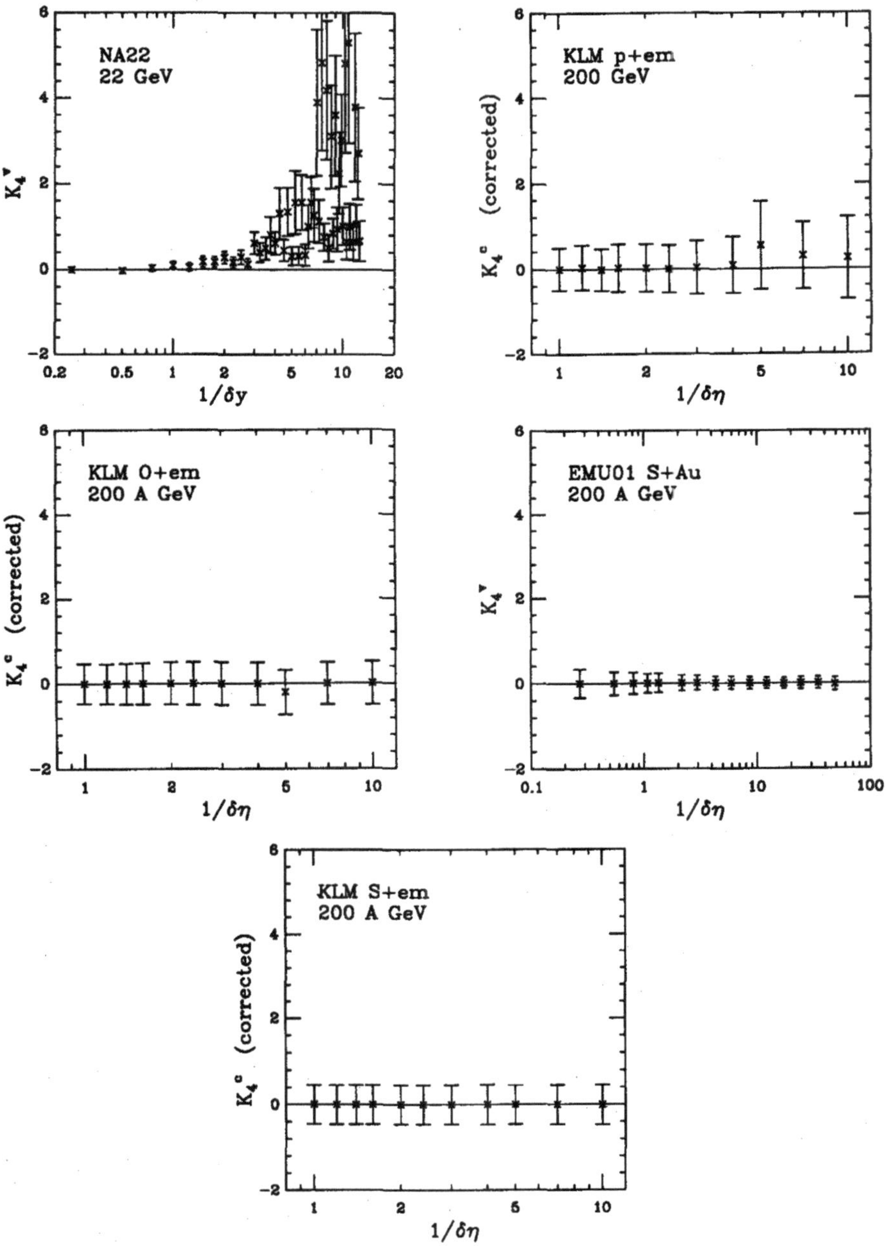

Fig. 3
The cumulant K_4 for NA22, KLM and EMU01 data.

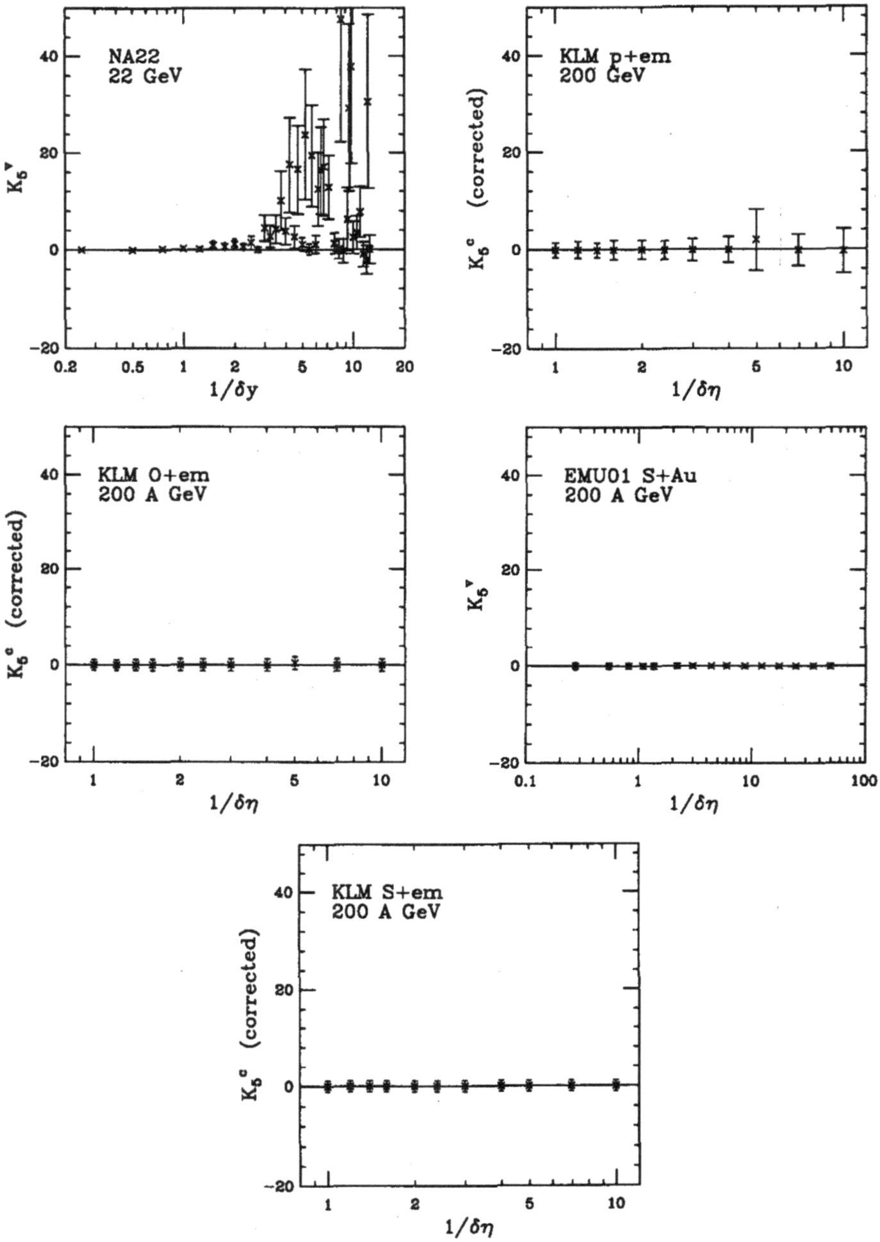

Fig. 4
The cumulant K_5 for NA22, KLM and EMU01 data.

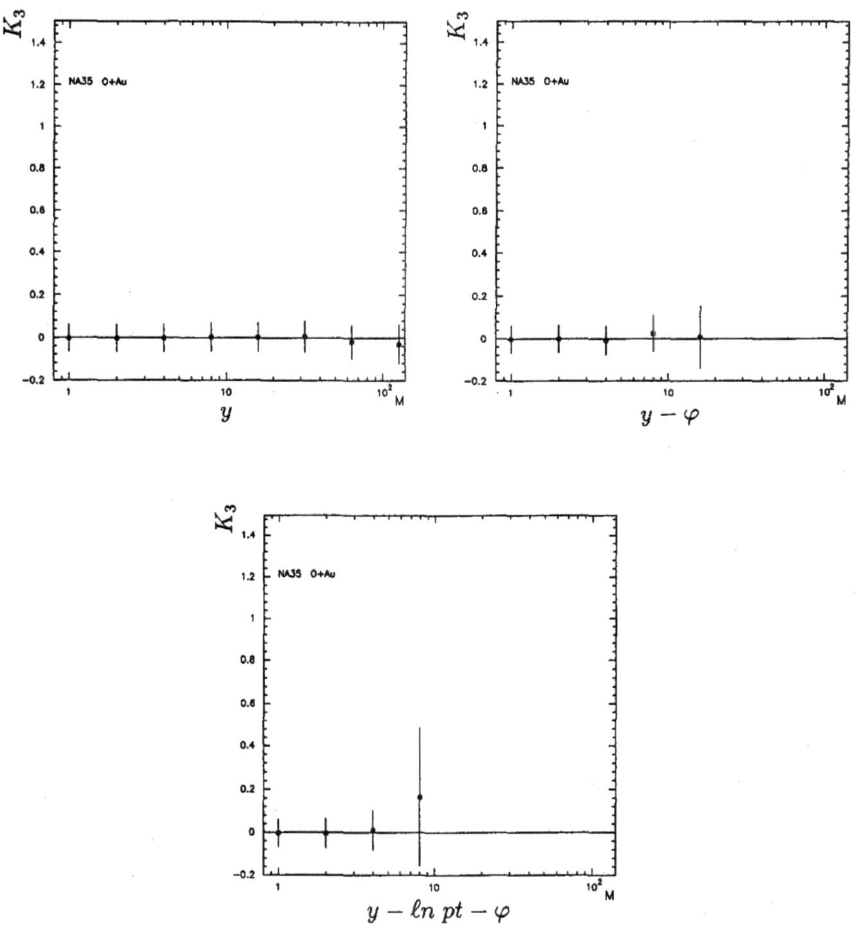

Fig. 5
The cumulant K_3 in one dimension (rapidity y), two dimensions (rapidity y and azimuthal angle φ) and three dimensions (rapidity y, transverse momentum p_T and azimuthal angle φ) for NA35 data.

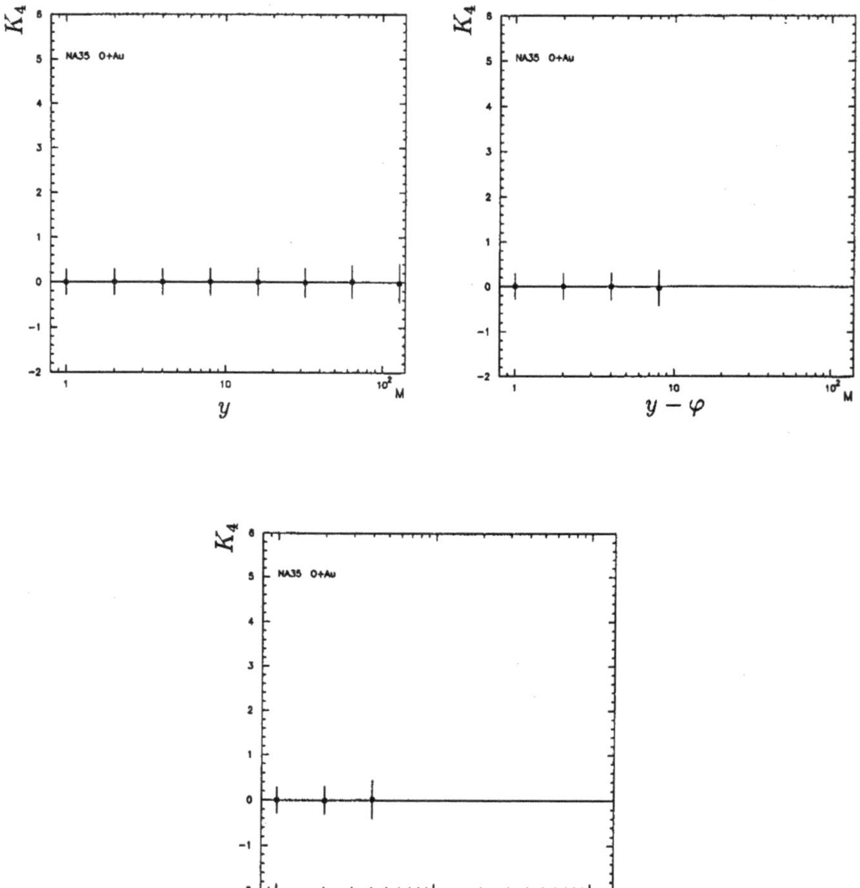

Fig. 6
The cumulant K_4 in one dimension (rapidity y), two dimensions (rapidity y and azimuthal angle φ) and three dimensions (rapidity y, transverse momentum p_T and azimuthal angle φ) for NA35 data.

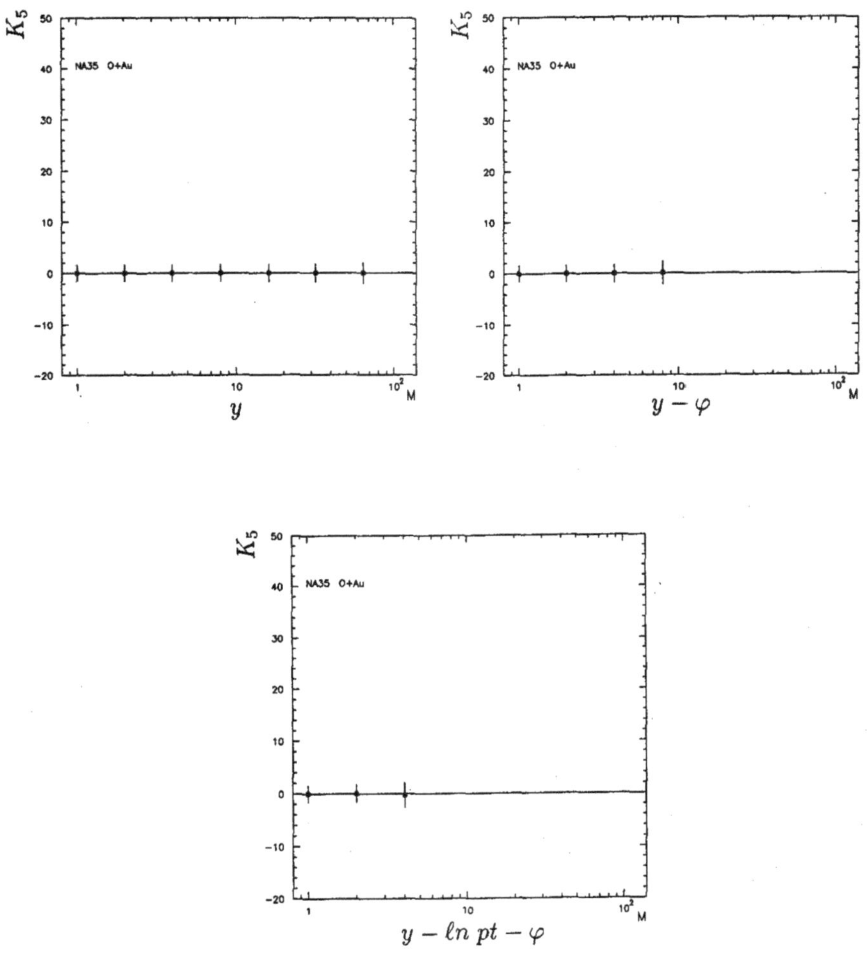

Fig. 7
The cumulant K_5 in one dimension (rapidity y), two dimensions (rapidity y and azimuthal angle φ) and three dimensions (rapidity y, transverse momentum p_T and azimuthal angle φ) for NA35 data.

INTERMITTENCY , CORRELATIONS AND CORRELATORS

E.A. De Wolf
Physics Department
Universitaire Instelling Antwerpen
B-2610 Wilrijk, Belgium

ABSTRACT

Unpublished NA22 data on the rapidity dependence of multiplicity distributions in small δy-intervals are used to discuss possible implications for intermittency studies. A simple Monte Carlo calculation, based on ISR correlation data, illustrates the effect of resonance effects on the increase of factorial moments with decreasing cell-size in $(\delta y, \delta\phi)$ space. The scaling behaviour of void-statistics in δy-intervals is discussed using NA22 data on hydrogen and nuclear targets, and JETSET 6.3 PS "data" for e^+e^- annihilations in an wide energy range. The NA22 correlator data are discussed in terms of a model with "standard" short-range order.

1 Introduction

At a Workshop devoted almost exclusively to "Intermittency", it is quite superfluous to elaborate on the importance of the subject for the field of multiparticle production in high energy collisions. This "old" chapter in high energy physics recovered some of its former vitality thanks to the ideas and ... persistence of Białas and Peschanski. Their greatest merit may well be to have recognized that fruitful ideas and techniques from many modern branches of physics that were largely ignored, could also be exploited, with proper modifications, to HEP.

The search for Intermittency is a search for self-similarity in the dynamics. At a more down-to-earth level, it is a study of multiplicity distributions (an all-time favorite) as a function of resolution-scale in momentum space, with particular emphasis on the high-multiplicity tail of the distribution. Technically, factorial moments of the multiplicity distribution turn out to be the most appropriate observables in view of their important property (well known in optics since the late fifties) of "noise-suppression". Factorial moments $\tilde{F}_r$ are integrals of r-particle inclusive densities (for identical particles). Intermittency studies are therefore equivalent, in principle, to studies of these densities as a function of the particle separation in phase space. Power-laws for factorial moments imply singular densities.

Given the strictly mathematical relationship between factorial moments and inclusive densities, it is not surprising that different experimental and theoretical groups tend to emphasize different aspects of the same physics-issue.

Some (most?) researchers direct their attention to the "initial problem": "Can we find evidence for power-law dependence of factorial moments on the resolution-scale?". To answer this question, the originally suggested techniques are used which usually imply global analyses, over a large part of phase space, of highly averaged factorial moments. This approach has led to the quite optimistic conclusion that intermittency in observed in "all types of collisions". Taking the experimental data (especially in ≥ 2 dimensions) at face value, such a conclusion seems unescapable, but is it the correct one?

A second (minority?) group stresses the importance of a thorough analysis of the inclusive particle densities, or rather of the genuine multiparticle correlations, in as great detail as possible. Their efforts are motivated in part by a suspicion that the present evidence for scaling might if fact be explainable by (admittedly less spectacular or less interesting) physical mechanisms known to exist since long. The two "schools" are represented at this Workshop and it is obvious that their complementary approaches will eventually converge to a better-balanced and deeper understanding of whole subject.

In Sects. 1 and 2, we present some results which address potential problems of factorial moment analyses due to the non-stationarity of the multiplicity distribution in different parts of phase space and the non-trivial shape of the 2-particle correlation function in more than one dimension. The well-known fact that multiplicity distributions in small phase space intervals are of approximate Negative Binomial type in almost all types of collisions, except perhaps for AA interactions, suggests a particular "hierarchy" of high-order correlations. Astrophysicists discovered that this property should reflect in a particular scaling-behaviour of the probability to observe zero particles—a void—in a given domain of phase space. In Sect. 3 we show evidence for "void-scaling" for NA22 data and for e^+e^- annihilations from JETSET simulations. Finally, in Sect. 4, we analyse factorial correlators in terms of a simple model for the higher-order rapidity correlation functions and conclude that these interesting quantities reveal no firm evidence for intermittent behaviour.

2 y-dependence of the multiplicity distribution

Mainly to reach sufficient statistical accuracy, factorial moments at a given resolution, say δy, are almost always[1] averaged over a considerable region of phase space, typically several units of rapidity. Previous studies learn, however, that multiplicity distribution in a given interval changes shape, depending on the rapidity-value around which δy is taken. The multiplicity distribution typically becomes narrower when moving outside the central region towards the fragmentation regions. Although this fact should not prevent detection of strong local fluctuations, it certainly

[1] For an exception see N. Lieske's talk at this meeting

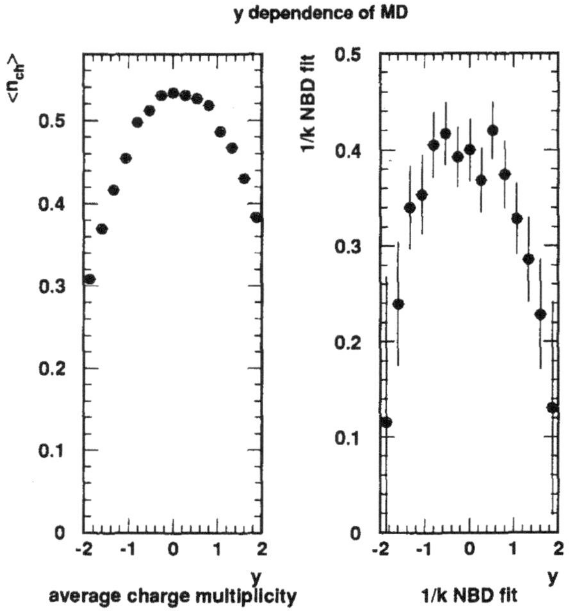

Figure 1: Rapidity dependence of the charge multiplicity distribution (MD): the average charge multiplicity and the parameter $1/k$ from a fit of the MD in a rapidity interval $\delta y = 4/15$ to the negative binomial. Data from NA22 [1]

influences the dependence of the factorial moments on resolution-scale and makes comparison to simplified (analytical) models hasardeous.

An example illustrating the above point is presented in Fig. 1. It shows the parameter values $<n>$ and $1/k$ from fits of a Negative Binomial distribution (NBD) to 15 NA22 multiplicity distributions as a function of the central rapidity value of the intervals of size $\delta y = 4/15$ [1]. The parameters show quite strong variation outside the "stable" region $-0.75 < y < 0.75$. In each interval, an acceptable fit to the NBD is found. The multiplicity distribution obtained by averaging the 15 individual distributions is, however, no longer well described by a NBD (not shown). In this case the NBD exceeds the data in the large-n tail.

The usually quoted NA22 factorial moment data for $\delta y = 4/15$ represent an average over differential factorial moments that vary significantly with y. The data in Fig. 1 are typical for the $1 + 2 + \cdots + 45 = 1035$ (!) multiplicity distributions that are obtained in the course of a (vertically averaged) factorial moment analysis. The experimental situation (regarding this point) is clearly more favorable at collider energies where the "stable η"-region is more extended. For e^+e^- annihilations, however, the situation is much more dramatic, as we learned from the ALEPH data, because of the deformation of single- and higher-order inclusive densities induced by the sphericity-axis (or thrust-axis) defining algorithms.

The non-stationarity of the multiplicity distributions thus raises a number of questions.

- From differential studies, we find that the Negative Binomial parametrization of the NA22 multiplicity distribution data remains good to acceptable down to $\delta y \sim 0.1$ (except for those intervals where the "famous" spike-event contributes). Under these circumstances, all higher-order factorial moments F_r ($r > 2$) are fixed by F_2 and contain no new information. In the light of the foregoing, it is, however, no surprise that the NBD-relations among factorial moments are much less well satisfied if the vertically averaged factorial moment data are used. From Fig. 1, we learn that more reliable information would be obtained if the analysis is restricted to the "stable" region $-0.75 < y < 0.75$.

- Little to nothing is known about the variation of the multiplicity distribution in two- or three-dimensional phase space, but it is likely to be still stronger than in rapidity space. How much does it affect, or contribute to, the approximate power-law dependence, now known to exist? Could it explain the stronger-than-power-law rise of the factorial moments in three dimensions?

- If the factorial moments depend strongly on the location of the "resolution-cell" in phase space (witness the ALEPH results), can one reasonably expect to get useful, easily interpretable and physically meaningful information from such a highly averaged experimental observable, except for the knowledge that, in some cases, some versions of some Monte Carlo models reproduce the data?

We feel that, statistics permitting, a new round of more differential factorial moment analyses could clarify some of the questions raised. This would also provide, with little additional effort, valuable information on the local structure (be it intermittent or smooth) on two- and higher-order particle correlations and on the phase space dependence of the multiplicity distribution. Such data may eventually turn out to be at least as relevant as information on intermittency slopes from highly averaged data.

3 To higher dimensions

Factorial moment analyses in one dimension reveal, with few exceptions, a gradual flattening of the moments with decreasing resolution-scale, which is not attributable to detector- or statistical effects. This feature is a sign for a finite correlation length

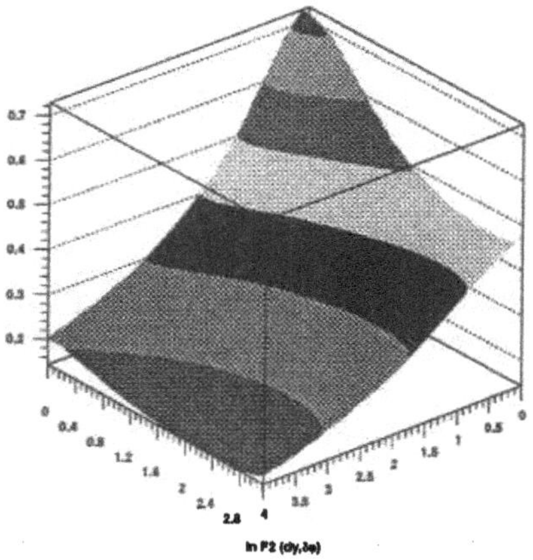

Figure 2: $\ln F_2$ versus $(\delta y, \delta\phi)$ calculated from the cluster Monte Carlo of ref. [2]. For details see text

and predictions of analytical models based on "standard" non-singular short-range correlations have no great difficulty to describe the data, in particular at collider energies [3–5].

Recently is has become clear from the work of Ochs [6] and Białas and Seixas [7] that true intermittency may "live" in higher dimensions. Projection onto lower-dimensional phase space would then explain the bending. Data now available seem to confirm this hypothesis. However, what is lacking is a reliable "conventional" phenomenological analysis in higher dimensions similar to what was done in one dimension. From this one might learn whether the data show additional effects besides those expected from a finite correlation length, and in particular whether a singular correlation function is not only sufficient but also necessary to explain the empirical "power-law like" behaviour in two and three dimensions.

Whereas data on rapidity correlations exist for many experiments, experimental information in 2- or higher dimensions is still a rare commodity. NA22 has presented some data at this meeting which indicate that the 2-particle correlation function changes considerably as a function of (y, ϕ) or when low- and higher p_T particles are selected. Although models such as FRITIOF fail to reproduce the data in detail, there is no evidence for "dramatic" features that would point to singular-like behaviour.

Quite precise data from an ISR streamer chamber experiment [2] at $\sqrt{s} = 23$ and 53 GeV exist on the $(\Delta\eta, \Delta\phi)$-dependence of the 2-particle correlation function

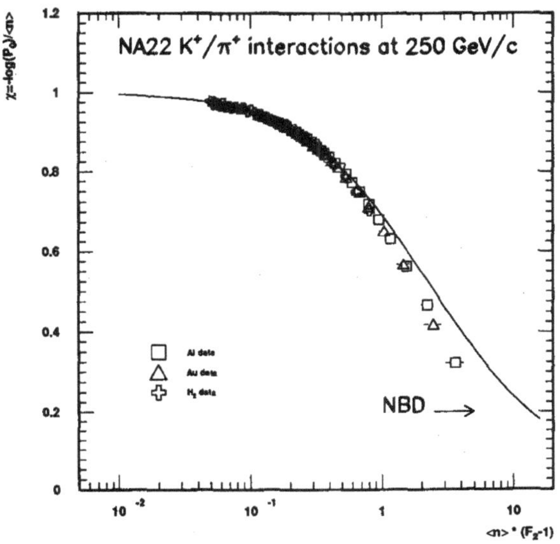

Figure 3: Test of the void-scaling relation for NA22 data [1]

which contain valuable information[2]. For example, Fig. 5 in [2] shows that the 2-dimensional correlation function has a 2-bump structure at small and large $\Delta\phi$. At small $\Delta\phi$ the pseudo-rapidity correlation length ξ is approximately twice smaller than at large $\Delta\phi$. Smaller ξ means faster increase of the factorial moments as $\delta\eta \to 0$. A two-dimensional factorial moment analysis of these data would therefore reveal a significantly larger (local) slope than in the 1-dimensional case. The authors of [2] qualitatively explained their observations in terms of resonance effects (ρ, ω, η) using a simple cluster model Monte Carlo.

In Fig. 2 we plot results for $\ln F_2$ using the same Monte Carlo procedure as in [2]. The results are plotted as a 2-dimensional "cumulative" distribution in $(\delta y, \delta\phi)$. F_2 shows a maximum in the smallest $(\delta y, \delta\phi)$-bin. The 1-dimensional δy dependence is given by the intersection of the surface with the vertical plane at $\delta\phi = \pi$. The dependence as found in a "standard" 2-dimensional factorial moment analysis is given by the intersection of the vertical diagonal plane through the points $(\delta y = 4, \delta\phi = \pi)$ and $(\delta y = 0, \delta\phi = 0)$. Clearly, F_2 rises much faster along the diagonal than along the δy direction.

From this exercise, which is mainly of pedagogical interest, we learn that a cluster model with a reasonable mixture of only a few resonances, has no difficulty in producing strongly increasing factorial moments even though the 1-dimensional δy

[2]See also [8] for a further interesting example of 2D data

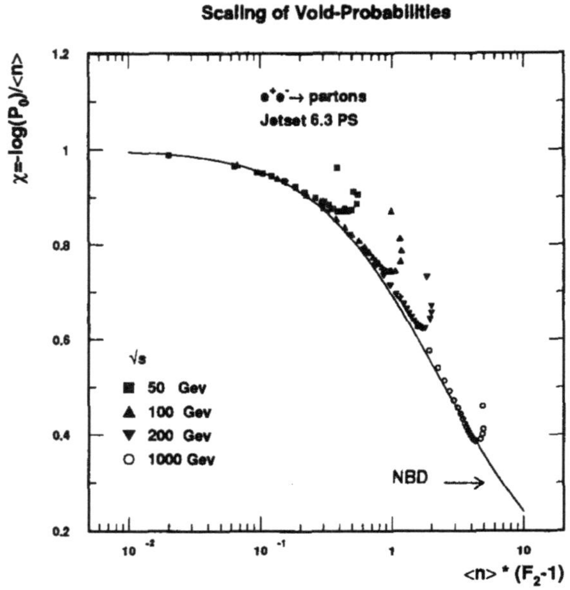

Figure 4: Test of the void-scaling relation for $e^+e^- \to$ partons at various c.m. energies.

projection shows quite early saturation. We find that the ratio of 2D to 1D slopes is about 2.5, close to what is measured by NA22.

We conclude that a proper assessment of the presence of intermittency in 2D and 3D analyses needs a detailed "conventional" counter-analysis. This requires, above all, improved and updated data on two- and higher-order correlations in e.g. (y, p_T, ϕ)-space.

4 Statistics of Voids

In intermittency studies, attention is focussed on the tail of the multiplicity distribution in decreasing domains of phase space. With finite statistics this leads to serious biases in the estimates of the factorial moments. Obviously, in small intervals, the number of counts is very frequently zero. The probability $P_0(\delta y)$ of finding no particles in some δy-interval—a void—is therefore measured with quite good accuracy!

The statistics of voids is attracting much attention in astrophysics as an alternative to the N-point galaxy correlations (see [9,10] and refs. therein).

Define the factorial moment ($\tilde{F}_r(\delta y)$) generating function, $Q(s)$, for a phase space cell of size δy:

$$Q(s) = \sum_{n=0}^{\infty}(1-s)^n P_n = 1 + \sum_{r=1}^{\infty}\frac{(-s)^r}{r!}\tilde{F}_r, \tag{1}$$

where P_n is the multiplicity distribution. The factorial cumulants K_r (integrals of the irreducible correlation functions $K_r(1,2,\ldots,r)$ over the r-cube of linear size δy) appear in the power expansion of $\ln Q(s)$:

$$\ln Q(s) = \sum_{r=1}^{\infty}\frac{(-s)^r}{r!}K_r. \tag{2}$$

From (1-2) it is seen that $P_0(\delta y)$ is given by $Q(s)$ evaluated at $s=1$ and:

$$\ln P_0 = \sum_{r=1}^{\infty}\frac{(-1)^r}{r!}K_r. \tag{3}$$

Consequently, P_0 involves the cumulants of all orders and should allow a sensitive test of models for higher-order correlations. This is true, in particular, for the so-called "hierarchical clustering scheme", a generalized form of the linked-pair *ansatz* [4].

The hierarchical clustering scheme was originally introduced to describe correlations of galaxies and galaxy clusters. Its implications have been extensively studied [11]. The general hierarchical (reduced) N-point correlation function $k_N(1,2,\ldots,N)$ is written as [12,13]:

$$k_N = \sum_{\text{trees}} Q_{N,a} \sum_{\text{labels}} \prod_{\text{edges}}^{N-1} k_2(A,B); \tag{4}$$

a product over $N-1$ edges linking N particles A,B, summed over relabelings of the particles A, B, and summed over all distinct connected N-point tree graphs. The quantities $Q_{N,a}$ are a priori unknown correlation amplitudes associated with tree graphs of type a. From (4) it follows [9] that the quantity $\chi = -\ln P_0/ <n>$ is a function only of the combination $<n>$ (F_2-1); $<n>$ is the average multiplicity and F_2 the normalized factorial moment of order 2 in the phase space domain under study. Consequently, χ scales as a function of $<n>$ (F_2-1) provided (4) holds. Some special cases are of interest. For a Poisson distribution $\chi \equiv 1$; for a NBD $\chi = (k/<n>)\ln(1+<n>/k)$. The latter expression shows that the NBD is a member of the class of discrete "hierarchical" multiplicity distributions.

In Fig. 3 we show a test of the void-scaling relation for NA22 π^+- and K^+-beam data at $\sqrt{s}=22$ GeV in interactions on hydrogen, and on Al and Au nuclear targets. As input we used (bin-averaged) multiplicity distributions for a given δy interval, obtained in previous intermittency studies [14,15]. From these P_0, $<n>$ and $<F_2>$ were derived.

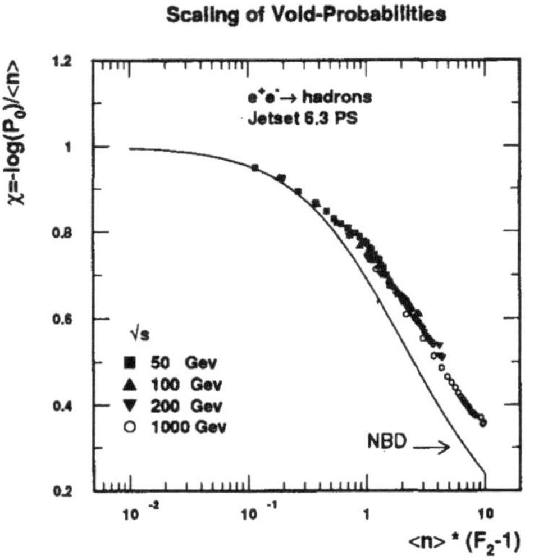

Figure 5: Test of the void-scaling relation for $e^+e^- \to$ stable hadrons at various c.m. energies.

From Fig. 3 we note that the three data sets clearly cluster around a common curve, close to that expected for a Negative Binomial. The data approach the Poisson limit for the smallest δy intervals, follow the NBD curve at intermediate δy, but deviate from it for $\delta y > 1$.

Encouraged by the success of the scaling prediction, we have repeated the test for e^+e^- annihilations using the JETSET 6.3 Parton Shower Monte Carlo with default parameters [17–19]. Fig. 4 gives the results at $\sqrt{s} = 50, 100, 200$ and 1000 GeV at the parton level, for partons with rapidities (w.r.t. the sphericity axis) $|y| < y_{cut}$ and y_{cut} in the range 0.05 to 2.5 in steps of 0.05. The data show remarkable scaling over this wide energy range for y_{cut}-values small (typically $y_{cut} < 0.7$ at 100 GeV) compared to the FWHM of the rapidity spectrum. Scaling violations become stronger the wider the δy-interval, indicating that the hierarchical-clustering predictions for P_0 are no longer valid[3]. Note that the Monte Carlo data cluster around the Negative Binomial curve. This behaviour agrees with the expectation [20] that the soft parton multiplicity distribution is approximately of NBD-type.

In Fig. 5 we show analogous results for stable hadrons. The "data" still exhibit scaling for not too large δy but now deviate much more strongly from the NBD

[3]Note that the derivation of the scaling-law assumes approximate stationarity of all cumulant correlation functions over the considered δy-interval. This assumption is certainly unjustified if δy is of the same order, or larger, than the typical correlation length

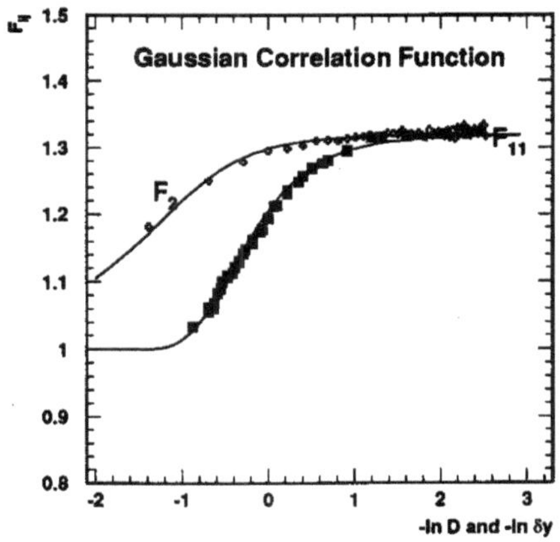

Figure 6: Factorial correlator F_{11} versus D ($\delta y = 0.2$) and factorial moment F_2 versus δy, for Gaussian-shaped two-particle correlation functions with parameters $\gamma_2 = 0.319$, $\xi = 1.556$ and $D = 0.2$. Data from [16]

curve. The "curls" seen in Fig. 4, are no longer present but irregularities remain due to the data points for the largest δy-values.

The above results, especially those in Fig. 4, imply that the higher-order correlations are consistent with a hierarchical structure, for partons located in a phase space cell of the order of one unit in rapidity (or less) at LEP energies. Given the complexity of the shower-evolution, this property can only be approximate at best. On the other hand, one might speculate that the mere complexity of many-parton final states, with their many degrees of freedom, could allow a relatively simple, statistical-type description of the correlations, as so often is encountered in other domains of many-body physics.

We have also examined the void-probabilities in 2-dimensional $(\delta y, \delta\phi)$ phase space cells[4]. We find that void-scaling remains satisfied for ($\delta y < 1$, $\delta\phi < \pi/2$). Interestingly, the scaling curves are now identical at the parton and at the stable hadron level and thus offer a further (rather unexpected) example of parton-hadron duality.

It would be of interest to pursue further studies on experimental data and on QCD parton showers to check the limits of validity of the linked-pair correlation structure. One should ultimately be able to elucidate in mathematical terms how these properties come about and how they are linked to basic features of the shower

[4]These results will be published elsewhere

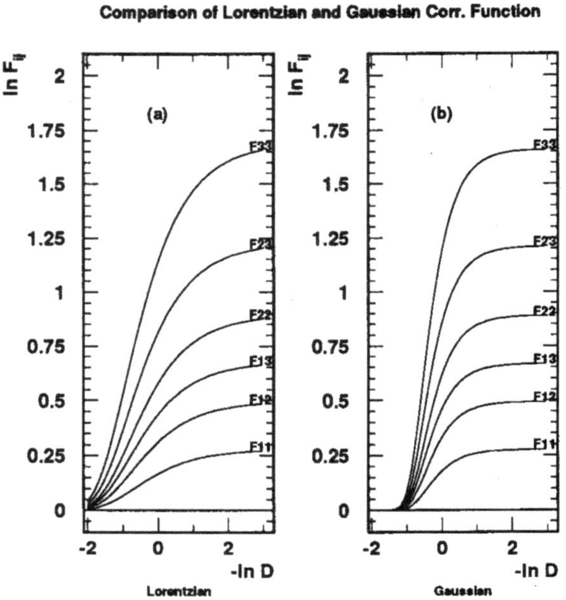

Figure 7: Factorial correlators as a function of interval-distance D for Lorentzian and Gaussian two-particle correlation functions with parameters $\gamma_2 = 0.32$, $\xi = 1.77$ and $\delta y = 0.1$.

and hadronization mechanism.

5 Factorial Correlators

The usual univariate factorial moments $\tilde{F}_r \equiv <n^{[r]}>=<n(n-1)\ldots(n-r+1)>$ characterize multiplicity fluctuations in a single phase space cell and thus reflect only local properties. More information is contained in the correlations between fluctuations (within the same event) in two or more cells. This has lead to consider multivariate factorial moments. For two cells they are defined as:

$$\tilde{F}_{pq} = <n^{[p]}n'^{[q]}>, \tag{5}$$

where n (n') is the number of particles in cell 1 (cell 2). A normalized version of the two-variate correlator F_{pq} is discussed in [21] and defined as:

$$F_{pq} = \frac{<n^{[p]}n'^{[q]}>}{<\tilde{F}_p><\tilde{F}_q>}. \tag{6}$$

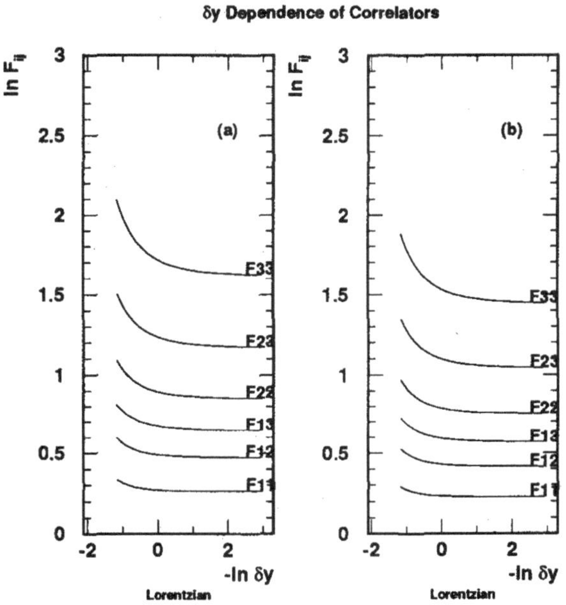

Figure 8: Factorial correlators as a function δy for Lorentzian two-particle correlation function with parameters $\gamma_2 = 0.32$, $\xi = 1.77$ and D-bin of 0.2 (a) and 0.4 (b)

N-fold factorial moments are a familiar and widely used tool in radio- and radar physics and in quantum optics [22]. There they relate to simultaneous measurement of photo-electron counts detected in N counting time-intervals, or in N space points, which yield a probability distribution $P_N(n_1, n_2, \ldots, n_N)$. In the simplest case of two cells, $\tilde{F}_{11} = <nn'>$ is related to the autocorrelation function of the electromagnetic radiation field. It obeys the Siegert-relation [22] for very short counting times, whatever the statistical properties of the field.

Factorial moments and factorial correlators are strongly related quantities. In terms of inclusive densities one has:

$$\tilde{F}_{pq} = \int_{\Delta_1} dy_1 \ldots dy_p \int_{\Delta_2} dy_{p+1} \ldots dy_{p+q}\, \rho_{p+q}(y_1, \ldots, y_p; y_{p+1}, \ldots, y_{p+q}), \qquad (7)$$

where ρ_{p+q} is the inclusive density of order $p+q$. The integrations are performed over two arbitrary (possibly overlapping) phase space cells Δ_1 and Δ_2, separated by a "distance" D.

For $\Delta_1 = \Delta_2$ or $D = 0$ (7) gives the correct definition of $\tilde{F}_2$ whereas (6) is equal to $<n^2>$ and misses the so-called "shot-noise" term $- <n>$.

The Białas-Peschanski "theorem" on Poisson-noise suppression is valid in general for multivariate factorial moments, as well as for multivariate factorial cumulants[5]. An early proof may be found e.g. in [23] which generalizes a well-known result by L. Mandel for the univariate case [24].

From (7) follows that F_{11} is directly derivable from measured two-particle correlation functions or from appropriate analytical parametrizations. Higher-order correlators involve higher-order correlation functions which are unknown in general.

In the following we use a specific *ansatz* for the higher-order correlation functions and therefore need to express the multivariate factorial moments in terms of multivariate factorial cumulants. Such expressions are quite lengthy and omitted here (examples may be found in [25]). They are derived from the relation:

$$1+\sum_{l=1}^{\infty}\frac{(-s)^l}{l!}\tilde{F}_l=\exp\left(\sum_{l=1}^{\infty}\frac{(-s)^l}{l!}K_l\right), \tag{8}$$

for the single-interval case by identifying in the Taylor expansion of the exponential the coefficients of the same powers of s on the left and right-hand side of the above equation. In the two-variate case, one considers two arbitrary non-overlapping intervals and uses the relation:

$$\sum_{l=0}^{\infty}\sum_{m=0}^{\infty}\frac{(-s)^l(-t)^m}{l!\,m!}\tilde{F}_{lm}=\exp\left(\sum_{l=0}^{\infty}\sum_{m=0}^{\infty}\frac{(-s)^l(-t)^m}{l!\,m!}K_{lm}\right), \tag{9}$$

where $\tilde{F}_{00}\equiv 1$ and K_{00} is defined equal to zero.

In the following we parametrize the two-particle (rapidity) correlation function $k_2(y_1,y_2)=K_2(y_1,y_2)/\rho(y_1)\rho(y_2)$ as

$$k_2(y_1,y_2) = \gamma_2 e^{-|y_1-y_2|/\xi} \quad \text{Lorentzian} \tag{10}$$

or

$$k_2(y_1,y_2) = \gamma_2 e^{-(y_1-y_2)^2/\lambda^2} \quad \text{Gaussian} \tag{11}$$

where λ is chosen such that the two functions have the same FWHM.

Fig. 6 shows NA22 data [16] for $F_{11}(D)$ versus the bin-distance D together with $F_2(\delta y)$. The curves are obtained from a Gaussian parametrization of the NA22 two-particle correlation function and clearly describe the data well. With the Lorentzian form, we can describe $F_2(\delta y)$ but not F_{11} (not shown).

Factorial correlators are considered to be important as an additional tool in the search for scaling-laws. In the α-model [21], F_{ij} has a power-law dependence on D: $F_{ij}\sim(\Delta/D)^{\phi_{ij}}$, where Δ is the total rapidity interval. In addition, F_{ij} is independent of δy and one further derives the relation:

$$\phi_{ij}=\phi_{i+j}-\phi_i-\phi_j=ij\phi_2, \tag{12}$$

[5]Multivariate factorial cumulants are integrals of the type (7) with inclusive densities replaced by irreducible correlation functions

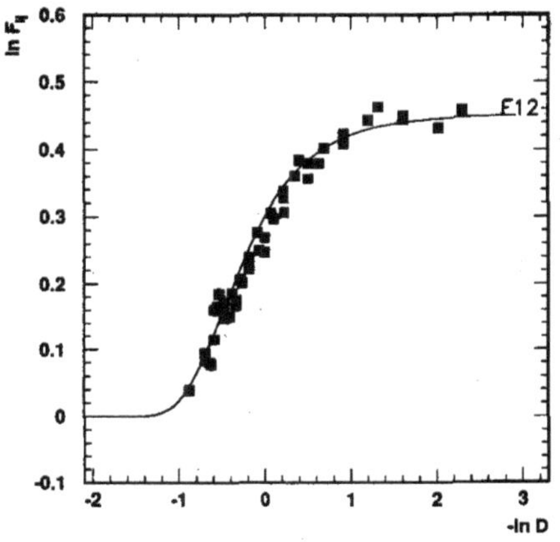

Figure 9: Factorial correlator F_{12} versus D (bin-width $\delta y = 0.2$) for Gaussian two-particle correlation function with parameters $\gamma_2 = 0.319$, $\xi = 1.556$. Data from NA22 [16]

where the first equality is due to the α-model and the second one follows from the lognormal approximation for long cascades.

To derive analytical predictions for higher-order correlators, we adopt a linking-ansatz described earlier in [5]. We write the normalized factorial cumulant functions $k_r(1, 2, \ldots, r)$ in terms of a "coherence function" $G(y_1, y_2)$ as follows:

$$
\begin{aligned}
k_2(y_1, y_2) &= |G(1,2)|^2, \\
k_3(y_1, y_2, y_3) &= G_2(y_1, y_2)\, G_2(y_2, y_3)\, G_2(y_3, y_1) + \text{cycl. perm.} \\
k_4(y_1, y_2, y_3, y_4) &= G_2(y_1, y_2)\, G_2(y_2, y_3)\, G_2(y_3, y_4)\, G_2(y_4, y_1) + \text{cycl. perm.} \qquad (13) \\
& \ldots
\end{aligned}
$$

This *ansatz* predicts a multiplicity distribution in a cell of size δy that approaches a Negative Binomial for $\delta y \ll \xi$. In [5] it is proven that this type of linking is numerically very close to the linked-pair assumption of Carruthers and Sarcevic [4]. The former, however, is based on a Random Phase Approximation scheme [26, 27] and is valid for (complex) stochastic fields with Gaussian fluctuations.

Fig. 7 shows the results for factorial correlators in the above scheme with parameters appropriate to the NA22 data. With a Lorentzian correlation function, the F_{ij} increase strongly with $-\ln D$ up to the smallest D-values. The rise is slower for the Gaussian parametrization but the local derivatives (slopes) still differ from zero in the experimentally accessible range. We find (not shown) that the slopes increase nearly linearly as a function of $i \times j$, as also predicted by the α-model, eqn.(12).

Fig. 8 shows the correlators for fixed D as a function of the bin-size δy. They are nearly independent of δy for $\delta y < 1$. Strict δy-independence is a property of random (multiplicative) cascade models. From Fig. 8 we learn that the δy-independence, experimentally confirmed by the NA22 analysis [16], also follows from a model with "standard" short-range order and non-singular correlation functions, and does not require intermittency.

Fig. 9 shows, as an example, a comparison with the NA22 data on F_{12} which tests the linking-*ansatz* for the 3rd order correlation function. The agreement is excellent. Similar results hold for still higher orders but the experimental errors become too large to allow strong conclusions.

Before drawing too optimistic conclusions, it should be pointed out that "our" formalism assumes strict stationarity of the correlation functions. This approximation is not well justified at NA22 energies. A careful analysis of the issue of non-stationarity by the Arizona group [28] shows, however, that a more complete calculation surprisingly gives results very similar to ours. For a completely satisfactory test, data at much higher energies are eagerly awaited.

The extension of single-variate factorial moment studies to the multivariate case is bound to offer better insight into the complicated structure of the correlations in multihadron final states. However, the expectation that correlators would help in clarifying the issue of intermittent behaviour is not borne out by the present data. Simple but reasonable assumptions about higher-order correlation functions, which use the experimental 2-particle correlations as input, have no difficulty in reproducing the behaviour of factorial correlators as measured by NA22.

Acknowledgements. *It is a great pleasure to thank the organisers of this Workshop, I. Derado, W. Ochs and N. Schmitz for a most enjoyable and fruitful meeting. I further wish to thank F. Botterweck and M. Charlet (Nijmegen) for generous help with the NA22 data and numerous discussions.*

References

[1] M. Charlet, private communication

[2] K. Eggert et al. *Nucl. Phys.*, **B86**:201, 1975.

[3] A. Capella, K. Fiałkowski, A. Krzywicki *Intermittency patterns in hadron production preprint BNL-43078/Orsay LPTHE 89/21*, 1989.

[4] P. Carruthers and I. Sarcevic *Phys. Rev. Lett.*, **63**:1562, 1989.

[5] E.A. De Wolf *Acta Phys. Pol*, **B21**:611, 1990.

[6] W. Ochs and J.Wosiek *Phys. Lett.*, **B232**:271, 1989.

[7] A. Białas and J. Seixas *Phys. Lett.*, **B250**:161, 1990.

[8] N.N. Biswas et al. *Phys. Rev. Lett.*, **37**:175, 1976.

[9] J.N. Fry *Astrophys. J.*, **306**:358, 1986.

[10] J.N. Fry et al. *Astrophys. J.*, **340**:11, 1989.

[11] P.J.E. Peebles *The Large Scale Structure of the Universe*, Princeton Univ. Press, 1980

[12] A.J.S. Hamilton, J.R. Gott III *Astrophys. J.*, **331**:641, 1988.

[13] J.N. Fry *Astrophys. J.*, **279**:499, 1984.

[14] NA22 Collaboration: I.V. Ajinenko et al. *Phys. Lett.*, **235B**:373–378, 1990.

[15] NA22 Collaboration: F. Botterweck et al. *Z. Phys. C, Particles and Fields*, **51**:37–47, 1991.

[16] NA22 Collaboration: I.V. Ajinenko et al. *Phys. Lett.*, **B258**:487–492, 1991.

[17] T.Sjöstrand *Comp. Phys. Comm.*, **39**:347, 1986.

[18] M.Bengtsson and T.Sjöstrand *Comp. Phys. Comm.*, **43**:367, 1987.

[19] M.Bengtsson and T.Sjöstrand *Nucl. Phys.*, **B289**:810, 1987.

[20] E.D. Malaza, B.R. Webber *Nucl. Phys.*, **B267**:702, 1986.

[21] A. Białas and R. Peschanski *Nucl. Phys.*, **B273**:703, 1986.

[22] B. Saleh *Photoelectron Statistics* (Springer Verlag, Berlin, 1978)

[23] C.D. Cantrell *Phys. Rev.*, **A1**:672, 1970.

[24] L. Mandel *Proc. Phys. Soc.*, **72**:1037, 1958.

[25] M.G. Kendall and A. Stuart *The Advanced Theory of Statistics*, Vol. 1 C. Griffin and Co., London 1969

[26] J. Karczmarczuk *Nucl. Phys.*, **B78**:370, 1974.

[27] J. Karczmarczuk *Acta Phys. Pol.*, **B9**:233, 1978.

[28] H.C. Eggers et al. *Correlations and Factorial Correlator data*, Arizona preprint AZPH-TH 91-15

UNIVERSAL INTERMITTENCY SLOPES?

K. FIAŁKOWSKI

Max-Planck-Institut für Physik
– Werner-Heisenberg-Institut –
P.O.Box 40 12 12, Munich (Fed. Rep. Germany)

Institute of Physics, Jagellonian University
30-059 Krakow, Poland

ABSTRACT

The 3-dimensional data on factorial moments in phase space bins show remarkable similarity for various processes. In particular, they seem to be dominated by a power-like dependence of second-order cumulant on the binsize with nearly universal exponent values of 0.4 – 0.5. Some future investigations are suggested and discussed.

1. Introduction

In the last few years a lot of attention has been paid to the investigations of multiplicity fluctuations in small phase-space bins. Following the suggestion of Białas and Peschanski[1], many experimental groups have investigated the dependence of scaled factorial moments on the bin size[2]. In many processes an increase above the predictions of standard MC models have been found for small bins. However, the linear dependence on log-log scale ("intermittency") expected for the random cascade models was often found compatible with data only in a narrow range of very small bin sizes. The fit parameters were found to depend significantly on the process investigated, variables chosen and details of the data processing.

It has been pointed out repeatedly[3] that in the standard multiparticle production models the correlation structure is better reflected by the Mueller correlation coefficients[4] (cumulants[5]) than by the factorial moments, which contain the "background" terms from uncorrelated emission. Obviously, if for very small bins cumulants grow power-like, so do the factorial moments, and the exponent values can be extracted from either. However, for the moderate values of moments and limited range of bin size, the difference between the behaviour of these quantities may be very important. In particular, it has been noted that the exponent values extracted from the factorial moment analysis may be much smaller than the asymptotic values for very small bins[6].

It has been advocated[7] that the real test of the possible power-dependence of moments will be the investigation of 3-dimensional phase-space bins, since integrating over some variables may obscure the underlying power-like singularities in the correlation functions (densities). In this workshop, the first preliminary data for the multiplicity moments in 3-bins have been presented[8-11], using the common choice of variables corrected for non-flat single distributions[12]. We show in the next section that whereas the scaled factorial moments seem to grow faster than any power of the bin number, the cumulants show the good agreement with the power behaviour for the full range of bin sizes (in particular, if a constant term corresponding to the long-range constant correlations is subtracted). Moreover, the similar values of exponent for the second-order correlation is found for different processes: μ p, π/K p, pAu, 0Au and SAu collisions. Unfortunately, statistics is not high enough to allow for significant tests of similar universality for higher order correlations. Nevertheless, we feel that this observation should influence future investigations of intermittency in multiple production and its theoretical interpretation. The possible lines of further investigation are shortly discussed in the last section and some conclusions are presented.

2. Data analysis

Let us remind first the 3-dimensional preliminary data on factorial moments presented in this workshop fom EMC[9], NA22[10] and NA35[11] experiments. As seen in Figs. 1a-1e, they all show the marked upward curvature. Moreover, as one can immediately see from the curves, they seem to follow quite well a simple behaviour

$$F_2 \simeq 1 + f_N + C(M^3)^{\varphi_2} \tag{1}$$

where the values of F_N (0 for EMC; 0.16 for NA22; .20, 0, 0 for p, 0, S NA35 data) corresponding to long-range component are estimated just from the distribution in full phase-space region considered and are significant only for the first few points. The values of C and φ_2 were not really fitted, but simply estimated from two points on the smooth curve interpolating the data. The really surprising thing is that very similar φ_2 values of .45÷5 seem to describe quite well all five processes, except of SAu data, for which some strange break is seen. Moreover, also the C values are quite similar; only for heavy ions lower values are needed.

This simple exercise suggests, that the power behaviour occurs indeed for the cumulant f_2, and may be observed in the factorial moment only when $F_2 \gg 1$. Thus one should really try to fit the "intermittency slopes" fitting $ln f_2 = ln(F_2 - 1)$ to the straight line

$$ln f_2 = ln c + \varphi_2 ln(M^3) \tag{2}$$

In two cases, where F_2 is significantly higher than 1 for largest phase-space region considered (NA22 and pAu NA35 data), one should subtract and plot $ln(f_2 - f_N)$. In Fig. 2 a corresponding plot is shown for EMC μp data[9]. One finds indeed very satisfactory linear behaviour in the whole range of bin sizes. Similar results were found for "positive only" and "negative only" data of EMC[13] and for NA35 data[14]. This seems to refute the suspicions , that the "intermittency" signal is due to some background effects (Dalitz pairs, γ conversion) or experimental problems (track doubling), since the effect of similar size and similar bin dependence is seen in three widely different experiments and different data selections. Nevertheless, further studies are certainly needed to compare in detail the correlations of like-sign and unlike-sign pairs. Unfortunately, the Ochs-Białas-Gazdzicki variables[12] are not suitable for such a study if the single-particle distributions for positive and negative particles differ (as is the case for μp data). The higher order moments at present statistics do not allow to decide, if higher order cumulants show a similarly simple power dependence and to estimate the slopes. The μp data, e.g. are perfectly compatible with $f_3 \alpha (M^3)^{\varphi_3}$ with $\varphi_3 \simeq 2\varphi_2$, but, as seen in Fig. 1a, the available range of M is not even sufficient to check if lnF_3 plot indeed curves up, as lnF_2 does.

Finally, let us note that we have not discussed the $3D$ e^+e^- CELLO data[8] which were unfortunately presented in figures too small to decide if the agreement with Eq. 1 is significant. It would be very interesting to check for e^+e^- data both in PETRA and LEP energy range.

3. Conclusions and outlook

We have seen that all the preliminary data on factorial moments in 3 dimensions presented at this workshop show surprising simplicity and similarity, if analyzed in terms of cumulants. By subtracting (when necessary) the long-range term (which may be achieved by subtracting simply the F_2 value for full phase-space), one should get approximately linear log-log plots for the whole range of bin sizes. Obviously, this does not replace the fitting procedure. We suggest that all the data should be fitted to formula (1), and, if the fit quality is really good, one should analyse quantitatively the obtained values of fit parameters. Let us note here few remarks concerning these parameters:

1. f_N can be practically read out of few first points and does not affect the fit for large M. If one just sets all $f_N = 0$, trying to fit directly the power dependence for the cumulant f_2, only few first points in NA22 and pAu NA35 data will deviate from the fit. Moreover, the value of f_N in NA22 data is practicaly the same as one fitted to the analysis of 1-dimensional data[3a)] and predicted theoretically 10 years earlier[15], so f_N hardly can be called a free parameter.
2. Intriguingly close values of φ_2 used to draw the lines in fig. 1 should be most carefully fitted, taking into account all the error correlations. If indeed the universality observed here is confirmed, it may suggest a common origin of intermittency (q-g plasma droplets? long tail of the distribution of hadronization times?) for very different processes.
3. The values of C parameters are of utmost importance to test the collective character of singular correlations in ion collisions suggested by our analysis. If one applies independent collision model and estimates the effective number of collisions as the ratio of particle density to that of N-N collision, one expects that the C value of pAu should be 1/4, for 0Au 1/20, and for SAu 1/30 of the value of NA22[11]. It does not seem to be compatible with Fig. 1, but certainly detailed fitting may change the results.

4. Finally, let us comment on values of φ_2 suggested by the lines in Fig. 1 and observed slopes in Fig. 2. They are surprisingly large and do not seem to follow from any existing model. For large M equal for all variables the size of a bin $\Delta = \delta y \delta\varphi \delta(lnp \perp)$ (proportional to M^{-3}) is related to the invariant four-momentum difference squared Q^2 by a simple relation

$$\Delta \simeq [-Q^2/ < p_\perp^2 >]^{3/2}$$

Thus if indeed $\varphi_2 = .4 \div .5$ for $\Delta \to 0$, it suggests a $(Q^2)^{\div 6 \div .75}$ behaviour of the correlation function at $-Q^2$ values below $< p_\perp^2 >$. (Let us remember that the Mueller-Regge picture predicts actually such a behaviour, but for $-Q^2 \gg m_n^2!$). It would be certainly a challenge for all the models to account for such correlations.

Concluding, we have found a remarkably universal regularity in the moments describing the multiplicity distributions in three-dimensional bins. The second cumulant is dominated by a term increasing power-like for decreasing bin size. Its exponent has very similar values of .45-.5 for all the types of collisions analysed. It seems to open quite new perspective in the investigation of intermittency in multiple production.

ACKNOWLEDGEMENTS

I would like to thank W. Ochs and all the theory group at MPI, Munich, for the warm hospitality during the post-workshop visit at Munich, which has allowed me to prepare this note. Fruitful discussions with I. Derado, J. Figiel, R. Hwa, G. Jancso, W. Ochs and N. Schmitz are appreciated. I am extremely grateful to the MPI group for the free access to their unpublished data.

REFERENCES

1. A. Białas and R. Peschanski, *Nucl. Phys.* **B 273** (1986) 703.
2. See e.g. A. Białas, *Nucl. Phys.* **A 525** (1991) 345c: B. Buschbeck, Intermittency, a short experimental review, Vienna preprint HEPHY-PUB 554/91, May 1991, to be published in the proceedings of XXVth Recontre de Moriond (High energy hadronic interactions): W. Kittel, Intermittency, a review of experimental results, Nijmegen preprint HEN 335/90, and references therein; Intermittency in High Energy Collisions, eds. F. Cooper, R. Hwa and I. Sarcevic.
3. a) A. Capella, K. Fiałkowski and A. Krzywicki, *Phys. Lett.* **B 230** (1989) 271,
   b) P. Carruthers and I. Sarcevic, *Phys. Rev. Lett.* **63** (1989) 1562,
   c) P. Carruthers, H.C. Eggers and I. Sarcevic, *Phys. Lett.* **B 254** (1991) 258.
4. A.H. Mueller, *Phys. Rev.* **D 4** (1971) 150.
5. P. Carruthers and C.C. Shih, Int.J.of Mod. Phys. **A2** (1987) 1447.
6. K. Fiałkowski, Acta Phys. Polon. **B22** (1991).

7. W. Ochs, *Phys. Lett.* **B 247** (1990) 101, A. Białas and J. Seixas *Phys. Lett.* **B 250** (1990) 161.
8. O. Podobrin, CELLO, contribution to this workshop.
9. N. Schmitz, contribution to this workshop.
10. W. Kittel, NA22, contribution to this workshop.
11. I. Derado, NA35, contribution to this workshop.
12. A. Białas and W. Gazdzicki, *Phys. Lett.* **B 252** (1990) 483, W. Ochs, Z. Phys. **C50**, 339 (191).
13. J. Figiel, private communication.
14. I. Derado, private communication.
15. A. Capella and A. Krzywicki, *Phys. Rev.* **D 18** (1978) 4120.

FIGURE CAPTIONS

Fig. 1 Second factorial scaled moment F_2 averaged over bins as the function of number of bins in log-log scale for a) μp collisions[9], b) π/Kp collisions[10], c) pAu, d) 0Au, and e) SAu collisions[11]. Solid lines represent Eq. 1, with parameter values a) $F_N = 0, C = 0.025, \varphi = 0.45$; b)$F_N = 0.16, c = 0.02, \varphi = 0.45$; c) $F_N = 0.2, C = 0.02, \varphi = 0.45$; d) and e) $F_N = 0, C = 0.01, \varphi = 0.5$.

Fig. 2 Second cumulant $f_2 = F_2 - 1$ averaged over bins as the function of number of bins in log-log scale for μp collisions[9].

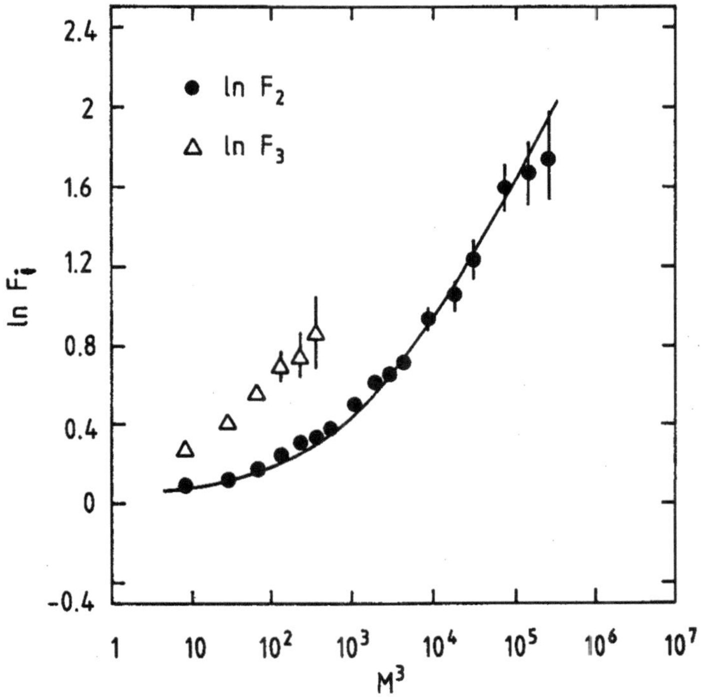

FIG. 1a

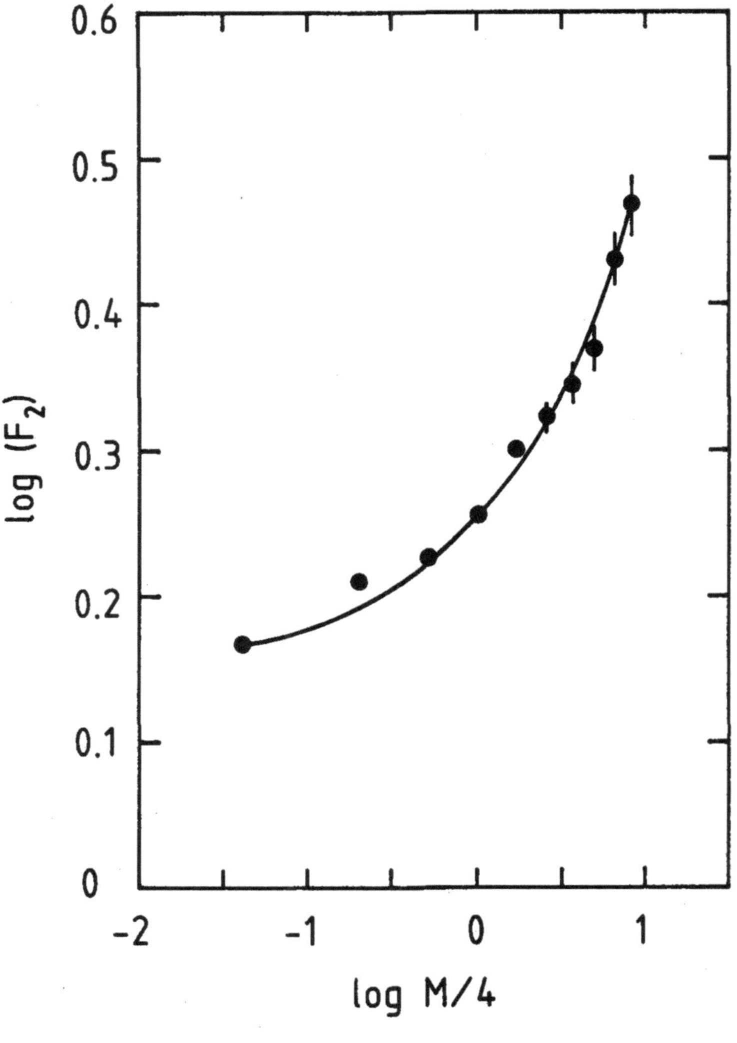

FIG. 1b

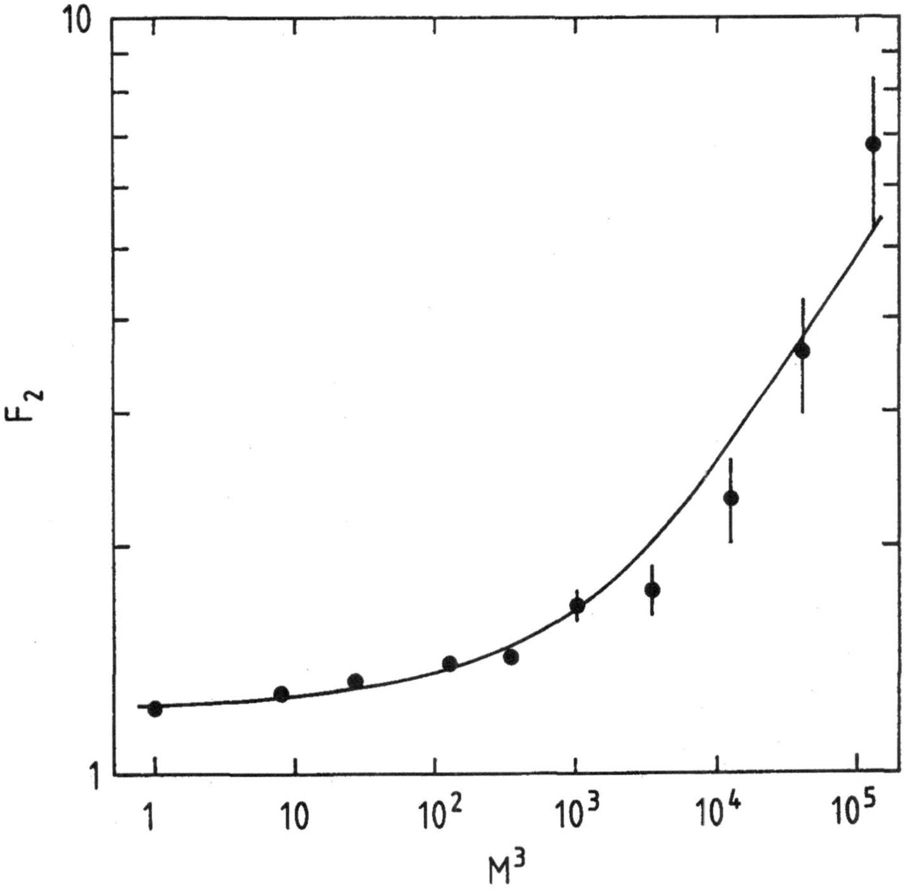

FIG. 1c

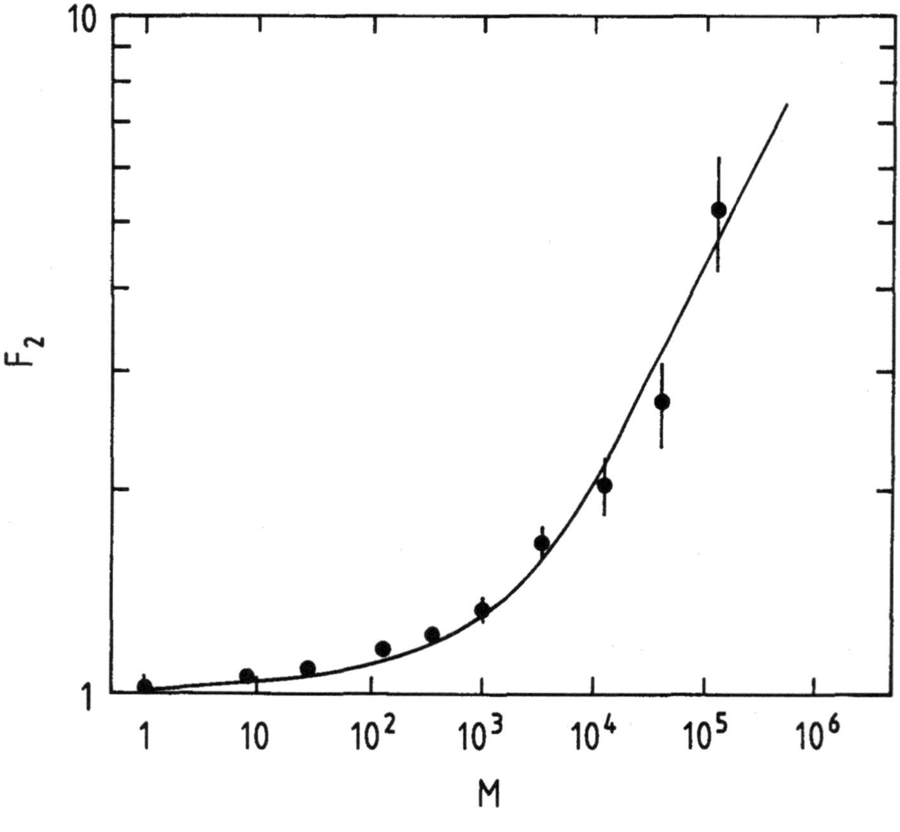

FIG. 1d

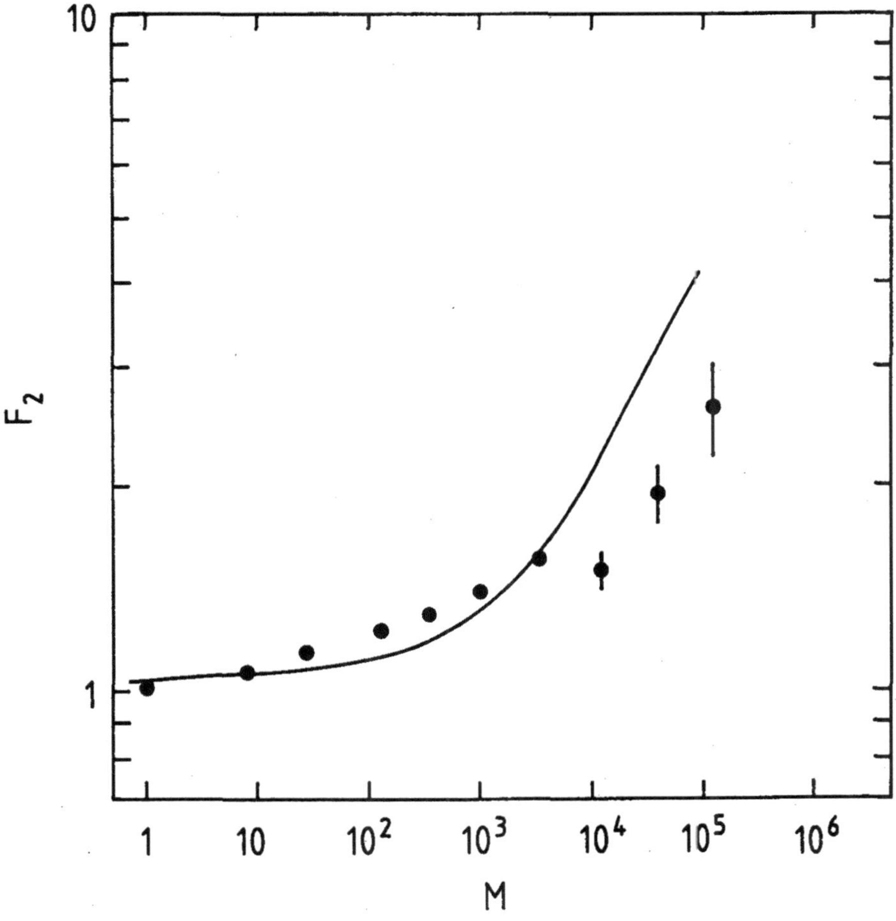

FIG. 1e

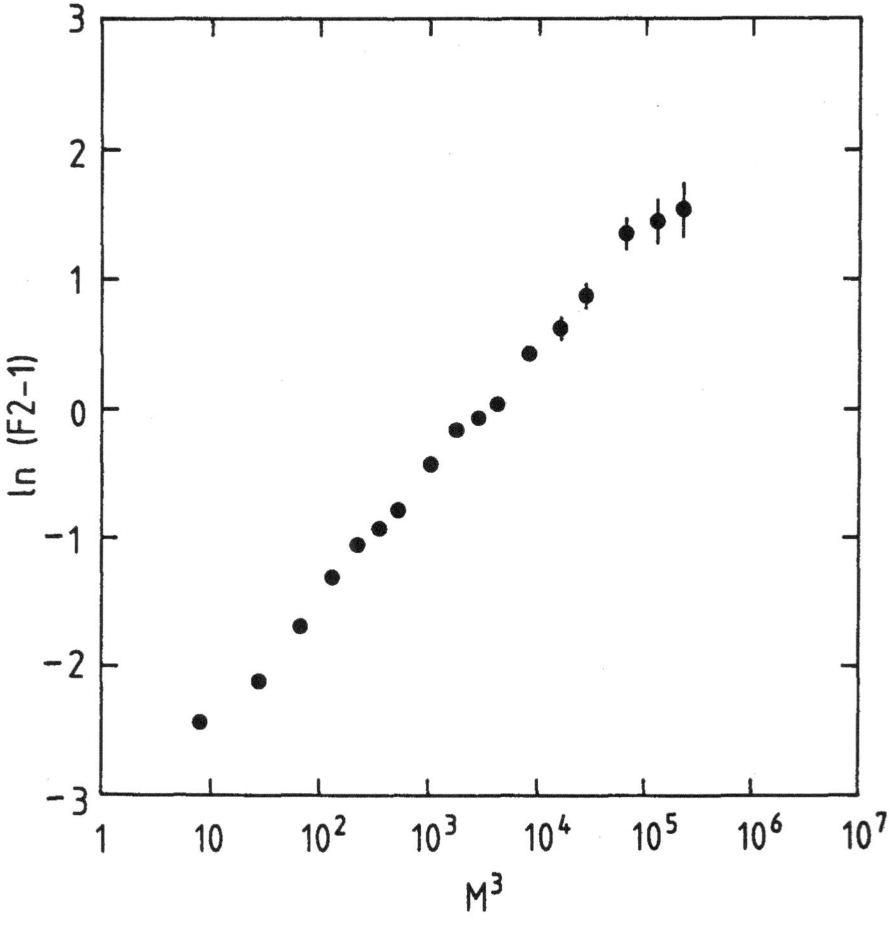

FIG. 2

INTERMITTENCY IN MULTIPLE PRODUCTION AND THE SINGULARITIES OF CORRELATION FUNCTIONS

K. FIAŁKOWSKI
Institute of Physics, Jagellonian University
30-059 Krakow, Poland

ABSTRACT

It is pointed out that power-like singularities in the correlation functions are expected to lead to power-like increase of factorial moments only in the limit of bin size going to zero. For available range of δ the power-like fits may be successful, but the fitted exponent is then usually much smaller than the the singular term in the correlation function.

It has been a common knowledge since few years that the power-like increase of factorial moments F_q for the bin size δ decreasing to zero

$$lnF_q \approx \phi_q ln\delta + a_q \qquad (\delta \to 0) \tag{1}$$

corresponds to the power-like singularity in the q-particle correlation function C_q, and thus in the q-particle distribution ρ_q. In particular, for the 1-dimensional case of a single kinematical variable X (rapidity, Φ angle etc.) and for the second order moments we have

$$\begin{aligned} F_2(\delta) &\equiv [\int_\delta\int dx_1dx_2\rho_2(x_1,x_2]/[\int_\delta d^x\rho_1(x)]^2 \equiv \\ &\equiv 1+[\int_\delta dx_1dx_2C_2]/[\int_\delta dx\rho_1(x)]^2 \equiv 1+f_2(\delta) \end{aligned} \tag{2}$$

Thus indeed in the limit $\delta \to 0$ $F_2(\delta)$ as well as the cumulant $F_2(\delta)$ behave as given by Eq.1 if ρ_2 and C_2 contain a singular term

$$C_2^s(x_1,x_2) \simeq \bar{C}_2^s|x_1-x_2|^{-\varphi_2} \tag{3}$$

However, for most of the existing 1-dimensional data $F_2 - 1 \ll 1$, which means that the uncorrelated term $\rho_1(x)\rho_1(x)$ dominates in

$$\rho_2(x_1,x_2) \equiv \rho_1(x_1)\rho_1(x_2) + C_2(x_1,x_2) \tag{4}$$

In fact, C_2 may have as well more terms than just C_2^s. In any case, however for $F_2 - 1 \equiv f_2 \ll 1$ one may have to a good approximation

$$lnf_2 \simeq -\phi_2 ln\delta + b_2 \tag{5}$$

but there seems to be no reason why Eq.1 should hold. While in some data the lnF_q vs $ln\delta$ plot is indeed not really linear except in very limited range of δ (which is usually attributed to the dominance of other terms than C_2^s in C_2), some data show excellent linear behaviour over quite wide range of δ, e.g. the UA1 data for ϕ variable[1]. We will show here that this apparently surprising effect may result simply from small values of ϕ_2. However, the coefficient found when fitting Eq.1 to the data will not be equal to the singularity exponent in Eq.3, and usually it should be much smaller. Indeed, let us assume that C_2 is dominated by C_2^s with a constant $\bar{C}_2^s$ much smaller than $[\rho_1(x)]^2$ (assumed to be constant).

Then we find

$$\begin{aligned} lnF_2(\delta) &\approx ln[1 + \frac{\bar{C}_2^s}{\rho^2}\delta^{-\phi_2}] = ln[1 + \frac{\bar{C}_2^s}{\rho^2}e^{-\phi_2 ln\delta}] \\ &\approx ln[1 + \frac{\bar{C}_2^s}{\rho_1^2}(1 - \phi_2 ln\delta)] \approx \frac{\bar{C}_2^s}{\rho_1^2}(1 - \phi_2 ln\delta) \end{aligned} \tag{6}$$

This has the form of Eq.[1], but the coefficient in front of $-ln\delta$ is not ϕ_2, but $\bar{C}_2^s\phi_2/\rho_1^2$. Since $\bar{C}_2^s/\rho_1^2 \ll 1$ has to be assumed to keep $F_2 - 1 \ll 1$, the fitted "intermittency slope" from Eq.1 will be much smaller than the "true" value ϕ_2, which one would find only for very small δ, corresponding to very big values of F_2. A slightly more realistic formula including the long-range correlations was found to describe the UA1 data quite well[2]. Obviously, in the intermediate rante of δ when $-\phi_2 ln\delta$ is no longer much smaller than 1, lnF_2 will not be linear in $-ln\delta$ and should develop an upward curvature. However, this may be obscured by the presence of other terms in C_2, such as conventional non-singular short-range correlations[3,4]. One can only state that if lnF_q is linear for moderate $-ln\delta$ it start curving up when higher values of $-ln\delta$, are available (or in higher dimensions). The situation is even more involved for higher order moments, where many terms contribute. We refer to[2] for further remarks concerning this subject.

Summarizing, we have shown that the linear dependence of lnF_2 on $-ln\delta$ observed already for $F_2 - 1 \ll 1$ is compatible with a power-like singularity in C_2 being responsible for this increase of F_2. However, the "intermittency slope" fitted to this dependence may be then much smaller than the exponent in the singular term of C_2, which determines the asymptotic behaviour when $\delta \to 0$. It is tempting to suspect that the big differences observed between the slope values in different processes may result from different $\bar{C}_2^s/\rho_1^2$ values, and that the singularity exponent may be more universal. It is also interesting if the upward curvature will appear when the data for smaller bins (or higher dimensions) become available.

REFERENCES

1. UA1 coll., C. Albajar et al., *Nucl. Phys.* **B 345** (1990) 1.
2. K. Fiałkowski, Acta Phys. Polon. **B22** (1991).
3. A. Capella, K. Fiałkowski, A. Krzywicki, *Phys. Lett.* **B 230** (1989) 148.
4. P. Carruthers, I. Sarcevic, *Phys. Rev. Lett.* **63** (1989) 1562.

INTERMITTENCY AND CORRELATIONS IN HARD AND SOFT COLLISIONS COMPARED

WOLFGANG OCHS
Max-Planck-Institut für Physik und Astrophysik
Föhringer Ring 6, W-8000 München 40, Fed. Rep. of Germany

ABSTRACT

We discuss the expectations from cascade models for particle correlations in variable phase space intervals. Data suggest a rather universal pattern for multiplicity moments. Several simplified models and their predictions for the shape of multiplicity distributions are compared. Finally we discuss the multiplicity dependence of moments in connection with the phenomenology of parton hadron duality.

1. Introduction

In the studies of multiplicity fluctuations in phase space intervals of variable size it is an important question whether there is a fractal structure of these fluctuations ("intermittency"[1]). One speaks of intermittency in multiparticle production if the factorial moments F_q of a multiplicity distribution in a phase space interval of size Δ follows a power law

$$F_q(\Delta) \sim \Delta^{-\varphi_q}. \tag{1}$$

The new results presented at this workshop show that the moments in a 3-dimensional analysis rise unlimited for decreasing Δ, a power law (1) however can be seen in general only in limited ranges of scales but not in the full range of scales. Another important question is whether the results can be understood in terms of the current models for multiparticle production. Here the answer is largely confirmative for the e^+e^- annihilation reaction, contrary to the other processes investigated.

In this contribution we will discuss how the higher moments F_q are related to F_2. A power like relation[2] has been suggested from the analysis of models and this relation seems to be valid in the full range of scales. Furthermore, this structure is rather universal for all processes which may help to further constrain models in the full variety of collision processes. Then we investigate several simple forms for the multiplicity distributions derived from cascade type models which yield the observed correlation structure (results from recent work with K. Fiałkowski and I. Sarcevic). Finally, we adress some observed discrepancies between experimental data and models for low multiplicity events and a possible interpretation in terms of an extended version of the hypothesis of local parton hadron duality[3].

2. Phenomenology of higher order factorial moments F_q vs. F_2

2.1. Observation on cascade models.

In realistic selfsimilar cascade models the moments F_q do not show a power behaviour in the 1D projection into the rapidity space. After an initial rise for large δy there is a tendency of flattening of the moments for $\delta y \to 0$ and the shape of all moments F_q is similar. It has been found that the moments follow a modified power law[2]

$$F_q(\delta y) \sim (g(\delta y))^{\varphi_q} \tag{2}$$

with arbitrary function g (intermittency is then obtained for $g(\delta y) = 1/\delta y$). One can express $g(\delta y)$ by $F_2(\delta y)$ and then obtains the relation

$$\log F_q = c_q + \frac{\varphi_q}{\varphi_2} \log F_2 \tag{3}$$

and often $c_q \sim 0$ was found. This relation has been shown[4] to be valid in a large class of cascade models including

(a) the 2D α-model in the 1D projection which is a generalization of the 1D α-model[1]

(b) a scale invariant cascade model which describes the degradation of a large initial mass W until a final hadron mass is reached[2,5]

(c) the QCD parton cascade in e^+e^- annihilation[6]

(d) the same as (c) but including cluster hadronisation.

Examples of such results are shown in fig.1.

The relation (3) has been derived analytically for the 2D α-model in the limit of large cascade length. The relation also implies that the ratio of local slopes φ_q/φ_2 where $\varphi_q = \partial \log F_q / \partial \log \delta y$ is independent of δy and this ratio can be reliably determined from a plot of $\log F_q$ vs. $\log F_2$.

The same relation (3) has also been found if the cascade models (b) - (c) above are analysed in two and three dimensional phase space, i.e. in the space of variables $y, \phi, \log p_T$, where ϕ is the azimuthal angle around the event axis and p_T is the transverse momentum; for the simple α-models this is not necessarily the case.

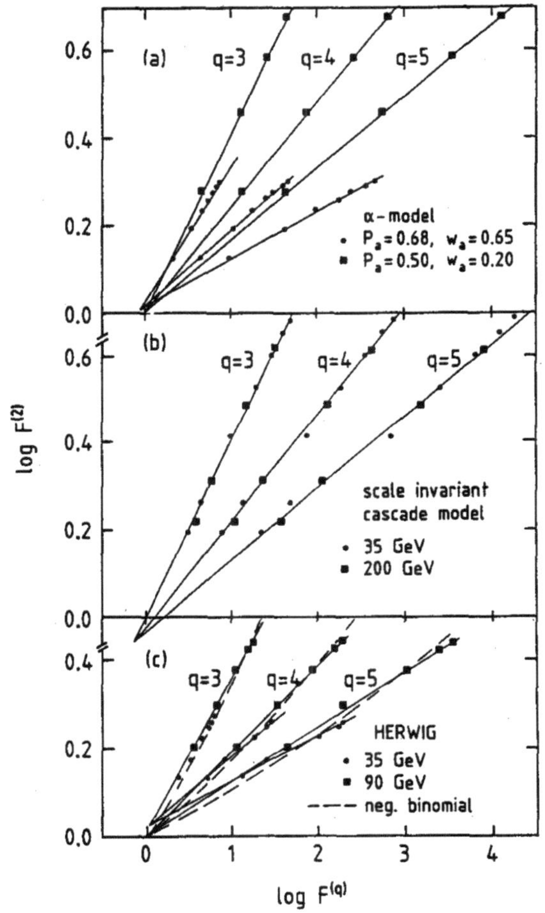

Fig. 1: Normalised factorial moments testing the linear relation (3) *for (a) 2D α-model for two sets of parameters*[4] *(b) scale invariant cascade model*[2,5] *and (c) QCD parton cascade model with hadronisation HERWIG*[6] *for* $|y| < 2$ *(from ref. 4).*

2.2. Observations on the data.

The experimental findings can be summarized as follows:

1. The data are in general consistent with a power law. A recent example for e^+e^- annihilation with data from DELPHI[7] and OPAL[8] is shown in fig. 2.
2. The relation (3) for normalised moments is largely independent of the average multiplicity $\bar{n}$ (at different energies the same F_2 occurs for different $\bar{n}$).
3. The slope ratio φ_q/φ_2 is independent of the dimension ($d = 1, 2, 3$). This

evidence has been given for πp and μp collisions by the NA22[9] and EMC[10] data presented at this workshop (for a recent compilation see also[11]).

4. The slope ratios φ_q/φ_2 are largely independent of the reaction type[12]. In particular, e^+e^- and nuclear data follow the same trend. An exception are the *hh* collisions at high energies (UA1 and UA5) which have slightly smaller φ_q/φ_2 values whereas the new data at lower energies by NA22 also follow the trend of the other processes.

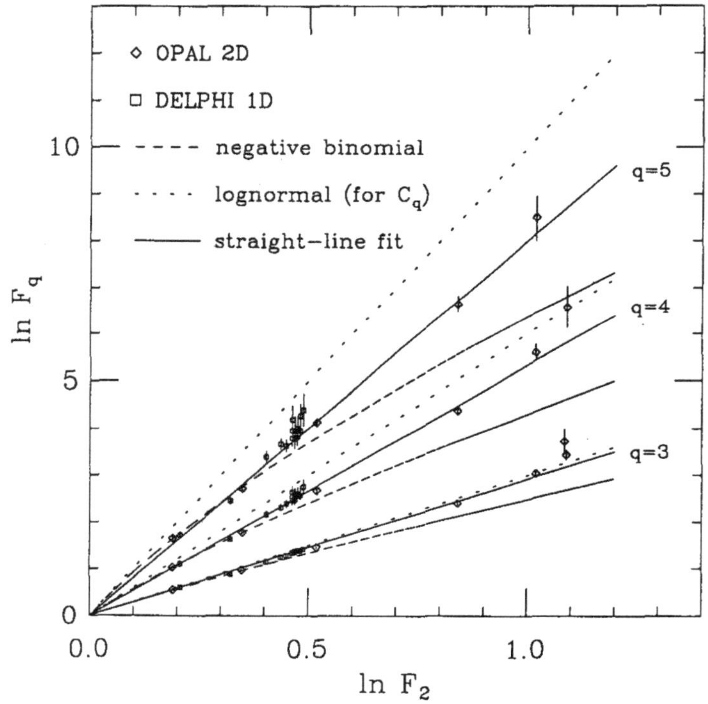

Fig. 2: Test of relation (3) *(straight lines) for* e^+e^- *data from LEP experiments*[7,8]. *We also compare with predictions to the negative binomial distribution and the log normal distribution. In the latter case a linear behaviour is predicted for the* C_q *moments, the corresponding* F_q *moments would become different for small multiplicities but still deviate from the data (from ref. 15).*

Two results are surprising and particularly noteworthy.

1. Nuclear collision processes are similar to others. A difference could have been expected if there had been a formation of a quark gluon plasma. This was a suggestion by Białas and Hwa[13] who argued that for a system at the critical point of a second order phase transition the intermittency slopes should behave monofractal, i.e. $\varphi_q/\varphi_2 \sim q-1$ in analogy to findings in the Ising model for a magnetised lattice. Present data on nuclei do not yet give evidence for such a behaviour, rather they are similar to the data of other processes. The 1D and 2D F_2 moments for nuclear collisions are typically rather small. In this case the only nontrivial correlations are the second cumulant correlations[14], and the higher factorial moments are related to the second moment by

$$F_q \approx 1 + \frac{q(q-1)}{2}(F_2 - 1) \tag{4}$$

In this limit (F_2 near 1) also the negative binomial and the log normal distributions approach the behaviour (4) and the same is also true for the data from other collision processes, for example e^+e^- annihilation as can be seen in fig. 2.

It would be interesting to study nuclear collisions further with higher statistics to see whether for larger values of F_2 (for example in 3D analysis) a linear behaviour for the higher moments as in fig. 2 emerges in deviation from (4).

2. Multiplicity distributions are approximately universal and depend on only 2 parameters. The approximately universal pattern for all collision processes in the log F_q vs. log F_2 plot implies, that for given $\bar{n}$ and F_2 the higher moments are determined (this has actually been shown only so far for the moments up to $q \approx 5$), so all multiplicity distributions, once the ratios φ_q/φ_2 are known, depend essentially only on two parameters for any phase space domain under consideration in arbitrary dimension. This is a remarkably simple structure for the rather complex multiparticle final states. The approximate universality seems to be a fact whereas the validity of a power behaviour (3) may be restricted to not too small values of F_2 and some curvature in the log F_q vs. F_2 plot may eventually appear for small F_2 as in eq. (4). The following questions remain: what determines the slope ratios φ_q/φ_2 and what is the corresponding shape of the multiplicity distribution. Next we turn to a discussion of these questions.

3. Description of multiplicity distributions in case of large density fluctuations*

First we discuss two popular forms for the multiplicity distribution which describe well the data for large phase space domains and small fluctuations, but fail in case of large fluctuations. Then we consider two forms which can also be applied in the latter case. All distributions can be derived from scale invariant cascade models.

3.1. negative binomial distribution (NBD)

This distribution occurs in a variety of models[16], in particular in the simplified parton shower model by Giovannini and Van Hove[17] as the compound of a Poisson

*based on work with K. Fiałkowski and I. Sarcevic[15].

and a logarithmic distribution. The higher order factorial moments can be derived from F_2 by the recursion relation

$$F_q = F_{q-1}(1 + (q-1)(F_2 - 1)) \tag{5}$$

independent of the multiplicity $\bar{n}$. The parameter free predictions are also shown as the dashed curves in fig. 2. They approach the data for small log F_2 but deviate more and more for increasing log F_2.

3.2. lognormal distribution (LND)

This distribution occurs if the particle density X is given by a product of random numbers

$$X_N = w_0 \cdot w_1 \dots w_N \tag{6}$$

with the same distribution $r(w)$ (self-similarity). Then the distribution of log X_N approaches a normal distribution, provided that the distribution of log w has a finite width and the central limit theorem can be applied. Then one obtains the LND

$$f(X_N) = \frac{1}{\sqrt{2\pi}\sigma X_N} \exp\left(-\frac{(\log X_N - \langle \log X_N \rangle)^2}{2\sigma^2} \right) \tag{7}$$

which depends on 2 parameters. It has been fitted to the multiplicity distribution of non-diffractive $p\bar{p}$ collisions at collider energies[18]. Both $p\bar{p}$ and e^+e^- data have been fitted by the Warsaw group[19] after a third parameter c has been added in (7) by $X_N \to X_N + C$. Unfortunately one looses the interpretation in terms of a cascade model in this case. From (7) one obtains for the moments $C_q = \langle n^q \rangle / \langle n \rangle^q$

$$\begin{aligned} C_q &= \exp(\sigma^2 \frac{q(q-1)}{2}) \\ C_q &= (C_2)^{\frac{q(q-1)}{2}} \end{aligned} \tag{8}$$

This relation is also shown in fig. 2. For small multiplicities the F_q moments would deviate from the C_q moments, but still the deviation of the LND from the data would remain.

3.3. logbinomial distribution

There are two reasons why the LND could be inappropriate: a) for finite cascade length N there are deviations from the Gaussian approximation, b) the central limit theorem does not apply.

To illustrate the first case we consider again the model (6) with a two valued w-distribution $r(w) = p_a\delta(w - w_a) + p_b\delta(w - w_b)$ as in the α-model[1]. Then the distribution of X_N assumes the values

$$X_N^{(k)} = w_a^k w_b^{N-k}, k = 0, 1, \dots N \tag{9}$$

with binomial probability. The distribution of particle multiplicity is obtained by convoluting the distribution of X_N with a Poisson distribution, which leads finally to the logbinomial distribution

$$P(n) = \sum_{k=0}^{N} \frac{(X_N^{(k)})^n}{n!} e^{-X_N^{(k)}} \binom{N}{k} p_a^k (1-p_a)^{N-k} \tag{10}$$

Note that from a distribution with standard moments C_q one obtains after convolution with a Poisson distribution the factorial moments F_q. The distribution (10) has 4 parameters: w_a, w_b, p_a and N ($p_a + p_b = 1$). It turns out that N is essentially a redundant parameter if it is not too small. The finite cascade length leads to the binomial distribution which is asymmetric for $p_a \neq \frac{1}{2}$ and approaches the symmetric Gaussian for $N \to \infty$.

Whereas two parameters can be related to $\bar{n}$ and F_2 a third parameter can be used to adjust the ratio F_3/F_2, so also the slope ratio φ_3/φ_2 can be adjusted to the data, contrary to the LND.

3.4. The log-Levy distribution

Another possible reason for the failure of the LND is a distribution $r(\log w)$ for the random number w in (6) which decreases slower than $|\log w|^{-3}$ for large $|\log w|$. Then no finite dispersion exists and the central limit theorem cannot be applied. In this case $\log X_N$ approaches an asymmetric stable Levy law $L_\mu(\log X_N)$ if $r(\log w) \sim |\log w|^{-1-\mu}$ for $|\log w| \to +\infty$. The function L_μ has to be evaluated numerically[15]. After convoluting with a Posson distribution we obtain the log-Levy distribution

$$P(n) = \int_{-\infty}^{+\infty} dX \frac{X^n}{n!} e^{-X} \frac{1}{X} L_\mu \Big(-\frac{\log X - \langle \log X \rangle}{\sigma} \Big) \frac{1}{\sigma}. \tag{11}$$

The moments from this model have already been calculated by Brax and Peschanski[20] to be given by

$$\frac{\varphi_q}{\varphi_2} = \frac{q^\mu - q}{2^\mu - 2} \tag{12}$$

which interpolates smoothly between $(q-1)$ and $q(q-1)/2$ for $0 < \mu < 2$, the cases of NBD for $F_2 \to \infty$ (5) and of LND (8). In this case the Levy index μ allows to adjust the slopes in fig. 2.

3.5. Comparison with MC data.

As no experimental data on multiplicity distributions in small phase space intervals are available we compared these distributions to the results from the HERWIG-Monte Carlo which is based on the parton cascade evolution and subsequent cluster hadronisation[6] and describes many experimental data well.

The results are shown in fig. 3. From the χ^2 values of the fits one can conclude that the logbinomial distribution fits best, followed by log-Levy. The LND fits are performed with 3 parameters as suggested in ref. 19; the model with 2 parameters as in (7) which can be related to a cascade model would fit worse.

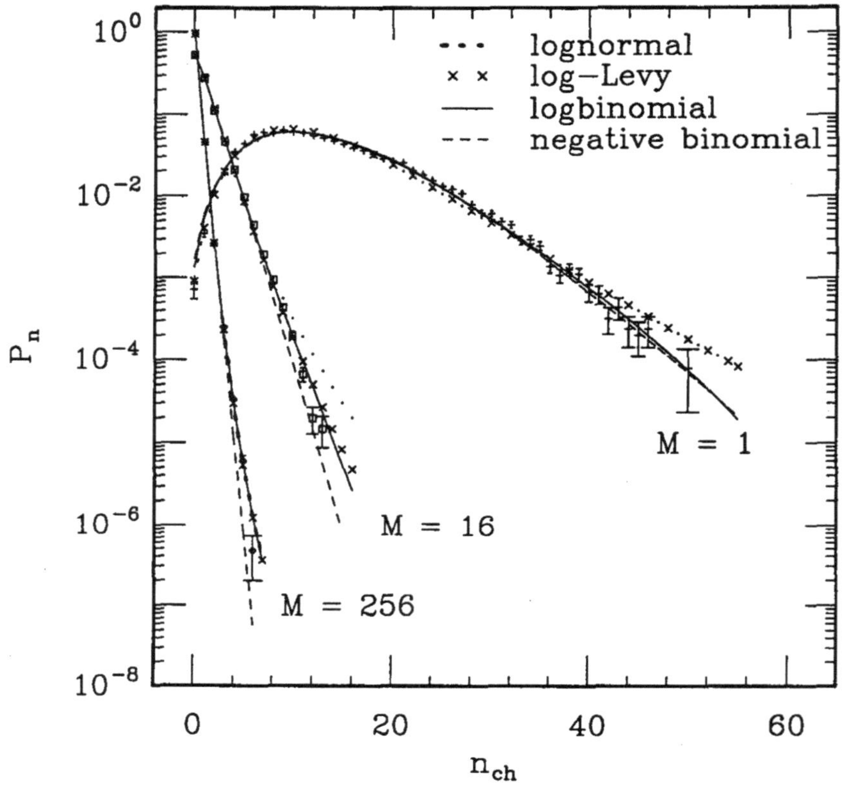

Fig. 3: Multiplicity distributions for different numbers M of 2D intervals in rapidity ($-2 \leq y \leq 2$) and azimuthal angle ($0 \leq \phi \leq 2\pi$), as obtained from the cluster model (HERWIG[6]), in comparison with several simple forms suggested by cascade models (from ref. 15).

We conclude that the essential features of multiplicity distributions can be fitted by 3-parameter models of the cascade type where two of them correspond to $\bar{n}$ and σ (or F_2) whereas a third parameter can be used to adjust the slope ratio φ_q/φ_2.

5. Extended parton - hadron duality?

The study of particle density fluctuations in small phase space intervals also allows for new tests of the hadronisation models. The data on e^+e^- - annihilation in general are well reproduced by parton cascade models with subsequent hadronisation, in particular the Lund-string model[21] and the cluster model[6], whereas a satisfactory description of the other processes in terms of theoretical models has not been achieved so far.

Here we want to discuss another approach towards the description of hadronic final states, called 'local parton hadron duality'(LPHD)[3]. It has been found that the inclusive momentum spectra of hadrons of mass m_h can be well described up to a normalisation constant by the inclusive momentum spectra of partons as calculated within the modified leading log approximation (MLLA) of QCD with m_h as the mass cutoff in the parton cascade; for a review, see e.g. ref. 22. This is a description of great simplicity, as only the Λ_{QCD} parameter can be adjusted but there is not yet a real theoretical understanding of this phenomenological success.

In this situation one may wonder whether the idea can be applied also to other observables beyond the single inclusive spectra. One proposal is to require LPHD for multi-inclusive distributions (correlation functions)[23], which has been motivated by the LUND hadronisation model an the negative binomial regularity. Equivalently one can consider the duality between factorial moments of hadrons and partons which are obtained from integrals over the correlation functions. Such a similarity has been shown to be satified by the experimental data at the Z resonance by Kittel and Botterweck[24], although a cutoff mass of $Q_0^2 = 0.4$ GeV^2 was required which is larger than what was needed for the inclusive spectra. Another proposal is to extend the LPHD to clusters of partons or hadrons[25]. This is verified for large cluster masses $m^2 = ys$ with $y \gtrsim 0.01$ from the multijet analyses and it has been suggested that the duality may work down to masses in the 1 - 2 GeV region.

Recently we remarked that in low multiplicity events, contrary to the case of single- and multi-inclusive distributions, the usual string- or cluster-hadronisation models give results different from LPHD if the application of the latter scheme is extended to such selected events[30]. A low multiplicity parton event (in the extreme case with only two partons) will decay according to the string breaking mechanism into many hadrons with limited p_T. A low multiplicity hadron event than is expected to show weaker intermittency effects then a low multiplicity parton event in which the partons are strongly collimated in angle. This expectation is confirmed by an explicit calculation using the HERWIG - Monte Carlo at the parton and hadron level. (In the parton cascade we included only light quarks and chose $Q_o = m_q = m_g = 0.32$ GeV and $\Lambda_{QCD} = 150$ MeV).

One can see in fig. 4a that the intermittency slope φ_2 increase at the hadron level and decrease at the parton level as function of the multiplicity. The question then arises whether the experimental data follow the hadronisation model, i.e. the 'hadron' data points in fig. 4a or whether they are closer to the 'parton' data what would be expected if LPHD could be extended to these selected multiplicities.

It has been found by the UA1 collaboration[26] that the experimental slopes φ_i indeed decrease with multiplicity whereas the slopes predicted from various models which also apply the string fragmentation scheme stay constant or increase (fig. 4b). This could be interpreted as an indication for an extended LPHD. It is therefore of

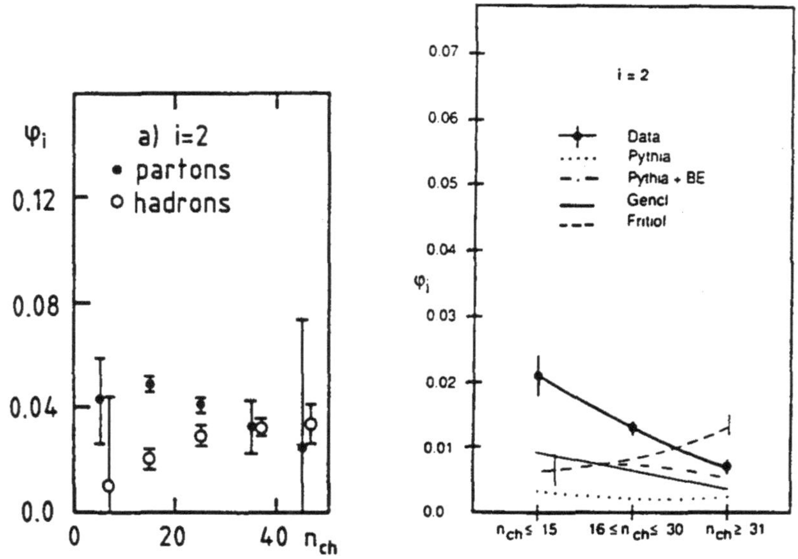

Fig. 4: Slopes φ_2 of factorial moments vs. particle (parton) multiplicity (a) as derived from the HERWIG MC for $e^+e^- \to$ hadrons at the parton and hadron level (calculated from M=6 and M=20 subdivisions of the rapidity interval $|y| \leq 2$ at $\sqrt{s}$=91 GeV); (b) $p\bar{p}$ collisions at $\sqrt{s}$=630 GeV from UA1 experiment[26].

interest to study the same problem in e^+e^- annihilation where the models are more reliable.

Such results have been presented at this workshop and are shown in fig 5 for the moments F_2 which have the highest precision. It turns out that the Lund JETSET model[21] describes the data very well for the large multiplicities of CELLO and OPAL (not shown) experiments, some small deviation is visible for DELPHI. On the other hand all experiments for the lowest multiplicity interval show a stronger rise of the moments as expected from the model. However, all experiments also mention experimental difficulties for low multiplicities. In particular, CELLO finds a reduction of the discrepancy if events with low multiplicities (≤ 4) in both hemispheres, which are interpreted as τ -lepton events, are removed. This effect is presumably less important at LEP because of the higher multiplicities. It would therefore be interesting to clarify the situation at the higher LEP energy. If we determine the slope φ_i from the moments F_2 for $M = 6$ and $M = 20$ as in fig. 4a we find for the 1D Delphi data a discrepency of $\sim 30\%$ in the smallest multiplicity bin, CELLO estimates $\sim 20\%$ whereas for $p\bar{p}$ at $\sqrt{s} = 630$ GeV (UA1) the discrepancy is $\gtrsim 100\%$. It was also reported[29] that the different models (JETSET, HERWIG ...) lead to quite different predictions for the moments in the low multiplicity region. This confirms the expectation that the predictions on low multiplicity events are sensitiv to the hadronisation mecha-

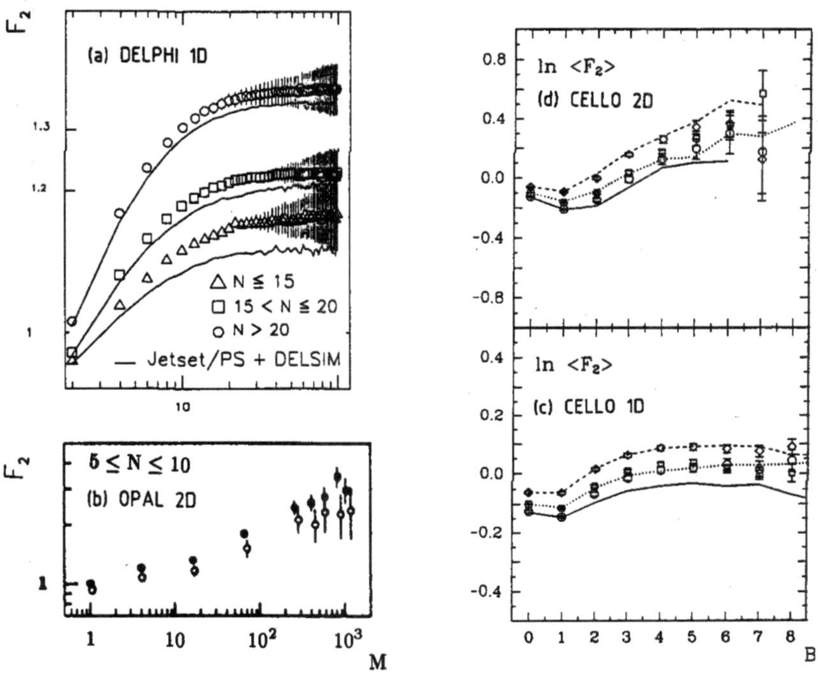

Fig. 5: Factorial moments (uncorrected) for different intervals of charged particle multiplicity N from a 1D (y) or 2D (y-ϕ) analysis vs. the number of subdivisions M or bisections $B = \log M/(d \log 2)$, compared with the JETSET MC: a) DELPHI[27]*, b) OPAL*[28]*, experimental data (full points) and model (open points), c) and d) CELLO*[29] *for multiplicities $\bar{N}$ = 6.8, 9.9 and 13.5 (moving upwards).*

nism. There is some hint in favor of an extended LPHD but the effect is appparently not as pronounced as in fig 4a.

5. Conclusion

Our theoretical understanding is different for the various collision processes. In e^+e^- annihilation the parton cascade model based on QCD with subsequent hadronisation has met much success and also gives a good overall description of multiplicity fluctuations. In the μp and hadron hadron collisions the role of perturbative QCD is less clear not to speak about nuclear collisions where collective phenomena like quark gluon plasma formation could occur. In this situation a comparison of the phenomenology of the various processes becomes important in order to find differences which could be a hint for new phenomena.

The results on factorial moments presented at this workshop confirm a rather universal pattern of correlations, eq.(3), with similar slope ratios φ_q/φ_2 in all processes (with the exception of the high energy hh data). These findings imply that multiplicity distributions in any domain of phase space depend essentially on two parameters $(\bar{n}, F_2)$. Specific forms of multiplicity distributions (logbinomial and log-Levy) have been derived from simple 3-parametric cascade models which give in general a reasonable description of Monte Carlo data, especially also for small phase space intervals. The third parameter in these models is used to fit the slope ratios φ_q/φ_2. It would be interesting to study nuclear collisions with higher statistics to see whether this pattern also occurs in the case of large fluctuations not yet accessible so far as in the case of other processes.

A possible problem for the usual hadronisation models occurs for low multiplicities. There is a large discrepency between models and data in $p\bar{p}$ collisions, a smaller one in e^+e^- annihilation, and it is not yet clear whether in the latter case the effect could be due to instrumental or background problems. If it were real there could be an interpretation in terms of an extended form of the parton hadron duality hypothesis. It will therefore be interesting to clarify the experimental situation.

6. References

1. A. Białas, R. Peschanski, *Nucl. Phys.* **B 273** (1986) 703, **B308** (1988) 857.
2. W. Ochs, J. Wosiek, *Phys. Lett.* **B 214** (1988) 617.
3. Ya.I. Azimov, Yu. L. Dokshitzer, V.A. Khoze, S.I. Troyan, *Z. Phys.* **C 27** (1985) 65; *Z. Phys.* **C 31** (1986) 231.
4. W. Ochs, *Phys. Lett.* **B 247** (1990) 101.
5. W. Ochs, *Z. Phys.* **C 23** (1984) 131.
6. G. Marchesini, B.R. Webber, *Nucl. Phys.* **B 238** (1984) 1; **B310** (1988) 461.
7. P. Abreu et al. (DELPHI), *Phys. Lett.* **B 247** (1990) 137.
8. M.Z. Akrawy et al. (OPAL), *Phys. Lett.* **B 262** (1991) 351.
9. F. Botterweck, W. Kittel, F.K. Rizatdinova, L.N. Smirnova (NA22), presented by W. Kittel at this workshop and Nijmegen preprint HEN-348, Sept. 1991.
10. I. Derado et al., presented by N. Schmitz at this workshop; MPI Munich preprint MPI-PhE/91-08.
11. A. De Angelis, P. Lipa, W. Ochs, LP-HEP conference, Geneva, July 1991, to be published in the proceedings, preprint MPI-Ph/91-93.
12. W. Ochs, *Z. Phys.* **C50** (1991) 50.
13. A. Białas and R. C. Hwa, *Phys. Lett.* **B 253** (1991) 436.

14. H.C. Eggers, *Phys. Rev.* **D 44** (1991) 1975; H.C. Eggers, University of Arizona Thesis, Tuecson 1991.

15. K. Fiałkowski, W. Ochs, I. Sarcevic, MPI Munich preprint MPI-Ph/91-100, in preparation.

16. P. Carruthers, C.C. Shih, Int.J.Mod.Phys. **A2** (1987) 1447.

17. A. Giovannini, L. Van Hove, *Acta Phys. Pol.* **B19** (1988) 931.

18. S. Carius, G. Ingelman, *Phys. Lett.* **B 252** (1990) 647.

19. M. Gaździcki, R. Szwed, G. Wrochna, A.K. Wróblewski, *Mod. Phys. Lett.* **A6** (1991) 981.

20. Ph. Brax and R. Peschanski, *Phys. Lett.* **B 253** (1991) 225.

21. T. Sjöstrand, M. Bengtsson, *Comp. Phys. Comm.* **43**, 367 (1987); B. Andersson, G. Gustafson, G. Ingelman, T. Sjöstrand, *Phys. Rep.* **97** (1983) 31.

22. V.A. Khoze, Yu.L. Dokshitzer, S.I. Troyan, XX Int. Symp. on Multiparticle Dynamics 1990, Dortmund, eds. R. Baier, D. Wegener, World Scientific, Singapore, 1991, p. 288.

23. A. Giovannini, L. Van Hove, *Acta Phys. Pol.* **B19** (1988) 917; M. Garetto, A. Giovannini, T. Sjöstrand, L. van Hove, Proc. of "Perugia Workshop on multiparticle production", loc. cit., p. 181.

24. W. Kittel, XX Int. Symp. on Multip. Dyn., loc. cit., p. 401.

25. G. Cohen Tannoudji, W. Ochs, *Z. Phys.* **C 39** (1988) 513.

26. C. Albajar et al. (UA1), *Nucl. Phys.* **B 345** (1990) 1.

27. A. De Angelis, N. Demaria (DELPHI), this workshop and Udine preprint UDPHIR 91/05/AA (1991).

28. N. Geddes (OPAL), this workshop.

29. O. Podobrin (CELLO), this workshop.

30. W. Ochs, XX Int. Symp. on Multip. Dyn., loc. cit., p. 434.

THE SEARCH FOR THE ORIGIN OF NB REGULARITY THE SIMPLIFIED PARTON SHOWER MODEL

A. GIOVANNINI and R. UGOCCIONI
Dipartimento di Fisica Teorica, Università di Torino
and INFN, Sezione di Torino, Italy

ABSTRACT
The new findings on NB behavior at LEP by the DELPHI Collaboration on separated 2-, 3- and 4-jet contributions and our previous work on $q\bar{q}$ and gg systems with the JETSET 6.3 Lund Parton Shower strongly suggest to study single parton shower properties, where a higher degree of simplicity is expected. We propose a Simplified Parton Shower (SPS) model based on essentials of QCD in a correct kinematical framework. NB regularity is found to occur in the SPS for single parton shower both in full phase space and in symmetric rapidity windows.

1. Introduction

Motivated by the need to find the original simplicity of negative binomial (NB) regularity, we decided to study single parton shower properties in the framework of a Simplified Parton Shower (SPS) model, based on essentials of QCD. In Section 2 we examine critically the new experimental results on multiplicity distributions (MD's) which led to the discovery of the NB regularity at a more fundamental level than previously expected. Then we review our previous work on $q\bar{q}$ and gg systems and discuss the attempts to explain theoretically the NB behavior at perturbative level. The need for the study of single parton shower properties emerges. This study is carried on in Section 3 by introducing the SPS model; we find that simplicity is revealed at final parton level by NB behavior in different rapidity intervals and in full phase space. Conclusions are drawn at the end.

2. The search for the origin of the NB regularity

2.1. New experimental facts on multiplicity distributions

The fact that three classes of multihadron producing reactions (e^+e^- annihilation, deep inelastic lepton-hadron collisions and hadron-hadron collisions) show

---

Work supported in part by M.U.R.S.T. (Italy) under Grant 1990

approximate NB behavior in their MD's in full phase space and in symmetric rapidity windows is still controversial. Experimental data by UA5 Collaboration[1] at $Sp\bar{p}S$ Collider at c.m. energy $\sqrt{s} = 900$ GeV in large rapidity intervals and more recently by DELPHI Collaboration[2] at LEP at c.m. energy $\sqrt{s} = 91$ GeV in symmetric rapidity intervals $|y| < y_c$, have shown indeed that NB regularity is probably violated at such top energies; in both cases, it should be noticed, fitted data are referred to the *full sample of events*.

Encouraged by these findings theorists started to build new phenomenological models for interpreting in other frameworks[3] the very large amount of data described by NBMD's and by related clan structure, and experimentalists to produce alternative fits incorporating the new shoulder structure observed in the data. The best fits turn out to be given both in $p\bar{p}$ collision[4] and in e^+e^- annihilation[5] by the sum of two NBMD's

$$P_n(\alpha, \bar{n}_1, k_1, \bar{n}_2, k_2) = (1-\alpha)P^{(\mathrm{NB})}(n; \bar{n}_1, k_1) + \alpha P^{(\mathrm{NB})}(n; \bar{n}_2, k_2) \qquad (1)$$

The real problem is the physical interpretation of this result. Of course, as it stands, eq. (1) has not much information in it. It is clear that these new discoveries, if confirmed, would definitely weaken the *simplicity* and the *universality* of NB behavior, which is based on a single, two-parameter MD. It was a consequence of the two mentioned properties of the NBMD that a unified description of a complex phenomenon like multihadron production in *different* high energy collisions became possible in terms of the *same elementary subprocesses* (QCD parton showers). Furthermore on the same basis we claim that *essentials of QCD* can be reduced to *gluon self-interaction*[6], the dominant multiplicative mechanism of the showers described by the Altarelli-Parisi equations[7] which, incorporated in the JETSET 6.3 model with Lund string fragmentation prescription[8], led to NB behavior at final parton level.

The hadronization process goes, in this scheme, via Generalized Local Parton Hadron Duality (GLPHD)[9], which is a much stronger statement than other standard formulations of LPHD originated from the idea of preconfinement[10]. GLPHD says that inclusive rapidity distributions of any order n at partonic level are proportional to the corresponding ones at hadronic level, the proportionality factor being $\rho \simeq 2Q_0/1$ GeV to power n, where Q_0 is the parton virtuality cut-off at which the shower development is stopped and hadronization takes place. This fact is far rich in consequences: if NB regularity is verified at one of the two levels (hadronic or partonic), GLPHD implies that the regularity is necessarily valid at the other level, as it has been shown in ref. [9], page 923. The observed NB regularity at hadronic level calls therefore via GLPHD for a similar behavior at partonic level. In addition, Monte Carlo experiments performed with JETSET 6.3 Lund Parton Shower model show a higher degree of simplicity at parton than at hadron level[9,11]. Finally the formulation of GLPHD in terms of inclusive rapidity distributions strongly suggests an integrated description of multiparticle correlations and MD's according to the new scenario disclosed by intermittency[12].

In this fascinating perspective we read also a tentative answer to the wish of theorists in the 40's, when high multiplicities were first discovered in cosmic ray collisions, to understand them in terms of non-linear strong interaction field theories[13]. It is interesting in this respect to remark that a new element of consistency is added to our speculation by the discovery by MacKeown and Wolfendale in 1966 of NB behavior in the charged pions MD generated by a primary nucleon in cosmic rays[14].

In spite of the beauty and simplicity of this scheme (which at the moment cannot be deduced from first principles, but only induced from experimental facts), we were ready to abandon it, with great satisfaction of at least a part of our community, if data so demanded.

It came as a surprise the very recent discovery by DELPHI Collaboration that NB regularity is valid again in different rapidity intervals and in full phase space when the *full sample of events* is separated into *2- 3- and 4-jet contributions*[15]. NB regularity, which was found to be violated at LEP energy in the whole sample of events (thus weakening our views on the unifying mechanism of multihadron production), turns out to be verified at a more fundamental level, when the 2- 3- and 4-jet contributions are separated using the JADE algorithm. Moreover the mentioned best fits in terms of two NBMD's proposed for multihadron production in e^+e^- annihilation could find in this framework a natural interpretation in terms of the relative 2-jet and 3-jet contributions. Maybe the same argument could be applied to the shoulder problem in $p\bar{p}$ collision at c.m. energy $\sqrt{s} = 900$ GeV. The occurrence of two NBMD[4] —i.e., the explanation of the shoulder effect— might be also in this case an indication of multi-jet contributions. Unfortunately, the UA5 Collaboration did not measure particles' momenta and therefore it is not possible to test the above speculation[16]. Of course the result of the DELPHI Collaboration opens a new problem on the identification of jets: the JADE mechanism may be insufficient, or the fact that the quality of the NB fits on the separated samples of 2- 3- and 4-jet contribution improves for smaller $y_{\min}$ might have some physical meaning. On top of that the identification in the data of *single* jet properties and in particular of its MD structure is now becoming a challenging problem for experimentalists.

It should also be remembered that previous findings by HRS Collaboration in e^+e^- annihilation at c.m. energy $\sqrt{s} = 29$ GeV agree with those by DELPHI Collaboration in that NB behavior was shown to be much better (it gave better χ^2/NDF) for the 2-jet sample of events than for the whole sample[17]. In this respect it is striking to compare the behavior of the average number of clans at parton, $\bar{N}_p$, and hadron, $\bar{N}_h$, level in the 2-jet contribution in e^+e^- annihilation at c.m. energies $\sqrt{s} = 29$ GeV and $\sqrt{s} = 91$ GeV[18]. We found that $\bar{N}_h$ is *approximately* constant with energy in rapidity windows $|y| < y_c$ for $y_c = 0.5, 1, 2$ and that $\bar{N}_p$ is constant with energy *within errors*. Furthermore the density of partonic clans per unit of rapidity turns out to be again constant and equal to 1 throughout the allowed rapidity range. Both facts are consistent with the view that a *higher degree of simplicity* is visible in the *2-jet final parton* MD than in the corresponding final charged hadron MD, and support the idea that an even greater simplicity should be present at the *single* parton shower level.

In summary, NB regularity is valid at a more fundamental level than previously expected; accordingly, a higher degree of simplicity should be sought at the single parton shower level.

2.2. Single parton shower and JETSET 6.3 Monte Carlo "experiment"

The relevance of the study of the single parton shower for understanding multiparticle production emerged from previous work on the QCD Parton Shower Model for e^+e^- annihilation as implemented in the program JETSET 6.3. This study started when it was discovered that the behavior of final partons —although related— was simpler than that of final charged hadrons. This was indeed the result of the paper on charged hadron and parton MD's in symmetric center-of-mass rapidity windows produced in $q\bar{q}$ and gg systems at c.m. energy $\sqrt{s} = 29$ GeV, 200 GeV and 2000 GeV[9,11]. Following the pioneering work by W. Kittel[19], it was shown that MD's are well described both at parton and hadron level by NBMD's and in addition that GLPHD —as already mentioned— is contained in the model for c.m. energies $\sqrt{s} \geq 200$ GeV.

In a subsequent investigation it was decided to strengthen the previous conclusions by studying the dependence of the above regularities on the parton virtuality cut-off Q_0 ($Q_0 = 1 \div 2$ GeV) and by examining the relevance of the angular ordering option incorporated in the Lund Parton Shower Model on MD's[20]. It was found that Lund hadronization prescription leads to infrared stable results and that angular ordering does not change the main results on MD's: it affects the average multiplicity $\bar{n}$ of NBMD only but not the parameter k (which is related to the dispersion D by $D^2/\bar{n}^2 = 1/\bar{n} + 1/k$).

In the search for the original simplicity of multiparticle production we moved next to the analysis of single g-jet (or g shower) and q-jet (or q shower) properties at different c.m. energies ($\sqrt{s} = 200$ GeV and 2000 GeV) in the $q\bar{q}$ and gg systems and in the $q(\bar{q})$ and g ancestor frames[21]. Single jets were found to have remarkably simple properties, simpler than the compound systems, and NB behavior in symmetric rapidity windows was found also at the single parton shower level. The same trend was confirmed by the analysis on the channel $e^+e^- \to b\bar{b}$. The evolution of the $b(\bar{b})$-ancestor has been followed from the initial to the final state at parton level, and then to B-hadrons at hadron level. $b(\bar{b})$-showers are showing again simple properties and NB behavior in symmetric rapidity windows.

In conclusion it was found that approximate NB regularity and related clan structure are the way simplicity shows up in single parton showers. Furthermore simplicity is higher at parton than at hadron level and in single parton shower than in the compound parton-parton system.

2.3. Single parton shower MD's in the KUV equations

One should mention the well known result that an approximate solution of the Konishi-Ukawa-Veneziano (KUV) equations[22] obtained within the simplified exclusive approach to single QCD jet evolution in terms of Markov branching processes in the variable $\mathcal{Y}$ shows NB behavior. (Here $\mathcal{Y} = \frac{6}{11\mathcal{N}_c - 2\mathcal{N}_f} \ln \left[\frac{\ln(Q^2/\Lambda^2)}{\ln(Q_0^2/\Lambda^2)}\right]$, where $\mathcal{N}_c$ is the number of colors, $\mathcal{N}_f$ is the number of flavors, Q^2 is the initial parton virtuality, Q_0^2 is the infrared cut-off and Λ is the renormalization scale.)[23] Since KUV equations can be regarded in a certain framework as the differential counterpart of the Altarelli-Parisi (AP) equations[7], the importance of this fact is evident. It establishes a direct link between an experimental fact observed in the soft region and an approximate QCD result. The connection of this property of parton MD's to final parton MD's and then to observable charged particles MD's is not trivial at all. Interesting attempts in this direction are limited to full phase space only; the behavior of MD's in rapidity windows is not considered.

For instance, in ref. [24], approximate asymptotic NB behavior for final particle MD's in e^+e^- annihilation has been predicted in full phase space by including soft gluon contributions to all orders, within leading order perturbative QCD. This prediction concerns asymptotic energy only ($\alpha_s \leq 0.05$) contrary to the experimental findings which show NB behavior at much lower energies like those of PEP, PETRA and LEP. This fact is evident from Figure 1 in ref. [24]. It should also be noticed that the same calculation carried to next-to-leading order does not show NB behavior[25]. The above theoretical results do not add any new information on MD's to the well known fact that, being the evolution of the shower mainly controlled by gluon self-interaction according to AP or KUV eqs., *qualitative* NB behavior should be there as discussed in refs. [22,23]. The reason of dissatisfaction lies in that analytical calculation do not meet *quantitatively* the experimental results.

Other authors, by assuming NB behavior in full phase space in the same framework and by using the calculation of the dependence of the parameter k on the running coupling constant α_s directly from the dispersion D, predicted final particle MD's at the corresponding energy[26]. A careful comparison of theoretical expectations on the parameter $k_{\rm th}$ from the next-to-leading order formula (see eq. (7.4) in ref. [26])

$$1/k_{\rm th} = 0.375 - 0.756\sqrt{\alpha_s} \tag{2}$$

and experimental findings of the same quantity $k_{\rm exp}$ is again unsatisfactory. The estimates of α_s in eq. (2) come from jet multiplicities ($\alpha_s = 0.115 \pm 0.017$) and from hadronic width ($\alpha_s = 0.141 \pm 0.017$) at LEP energy[27]. One obtains from eq. (2) $k_{\rm th}(\alpha_s = 0.115) = 8.4 \pm 1.3$ and $k_{\rm th}(\alpha_s = 0.141) = 11 \pm 2$. Both values are quite far from $k_{\rm exp} = 23.20 \pm 0.97$ measured in ref. [2]. It should be noticed that the found $k_{\rm th}$ values are expected to occur, according to JETSET 6.3, at c.m. energy $\sqrt{s} \geq 2000$ GeV in the $q\bar{q}$ system; on the other hand, by inserting the value of $k_{\rm exp}$ measured at $\sqrt{s} = 91$ GeV in eq. (2) one determines $\alpha_s = 0.192 \pm 0.002$, which should be compared to the quoted experimental value. This fact suggests to calculate eq. (2) to the next order, i.e., $(\sqrt{\alpha_s})^2$. This can be considered a challenging test of perturbative QCD on MD's.

The real question at this point is the following: can one connect the nice result on parton MD's for a single jet (parton shower) obtained in ref. [23] to what we know up to now only from JETSET 6.3 for final parton MD's (partons are in the final stage of their evolution and ready to hadronize)? Furthermore, and even more interestingly, can this study be extended to symmetric rapidity intervals both at parton and hadron level? In our opinion, on the answer to these questions depends our understanding of the origin of NB regularity.

Stimulated by the mentioned results by the DELPHI Collaboration, which show via GLPHD an higher degree of simplicity at parton level in the 2-jet sample of events than in the whole sample, convinced by previous work on JETSET 6.3 of the relevance of *single parton showers* as *elementary substructures* of multiparticle production in different reactions, in order to attempt an answer to the raised questions, we decided to carry on the study, initiated with Léon Van Hove, of a Simplified Parton Shower (SPS) model, which we propose to discuss in the next section.

3. The SPS model

Guided by the criterion of maximum simplicity extensively discussed in the previous section, we intend to describe a single shower initiated by a parton of given virtuality and to follow its evolution along its branchings in a correct kinematical framework to final parton level. The rapidity structure is assumed to be inspired by the essentials of QCD which we believe to be in the gluon self-interaction.

In the following we describe first the virtuality evolution of SPS, then its rapidity structure.

3.1. Virtuality evolution of the SPS

An initial parton of virtuality W is degrading to virtuality Q where it splits into two partons of virtuality Q_0 and Q_1. We require $Q \geq Q_0 + Q_1$ and that $Q =$ 1 GeV is the minimum allowed virtuality for a parton; this implies an automatic cut-off, i.e., no splitting occurs for $Q < 2$.

We denote the probability for a parton of virtuality W to split at Q, $p(Q|W)$; the probability for a parton (ancestor) of virtuality W to generate n final partons, $P_n(W)$; the probability for a parton which splits at virtuality Q to generate n final partons, $R_n(Q)$ (notice that for $Q < 2$ GeV, $R_n(Q) = \delta_{n1}$). All these probabilities are normalized to 1; the situation is illustrated in figure 1. A simple relation exists between the just defined probabilities:

$$P_n(W) = \int_1^W p(Q|W)R_n(Q)dQ \tag{3}$$

We assume that $p(Q|W)$ can be factorized

$$p(Q|W) = C(W)p_0(Q) \tag{4}$$

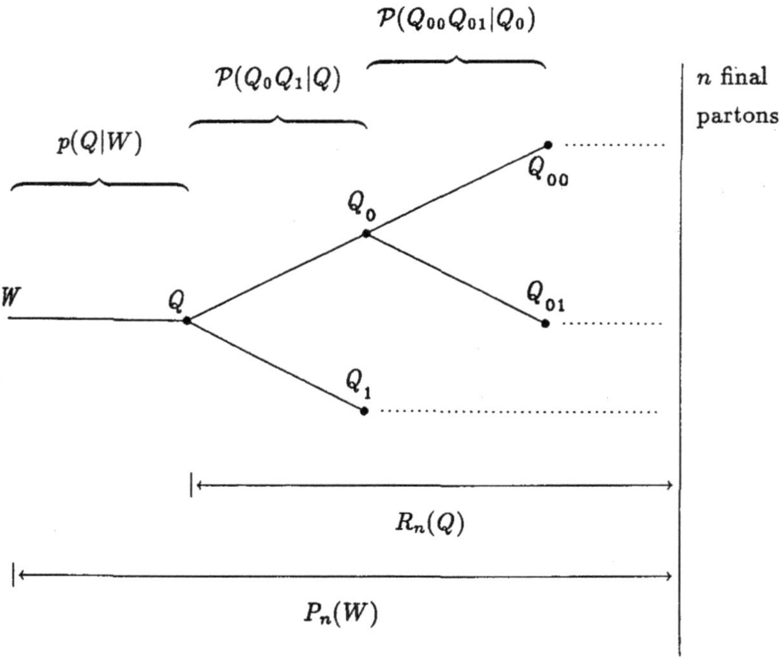

Fig. 1. Schematic representation of the virtuality evolution of the SPS, together with the relevant splitting probability distributions, as described in the text.

By differentiating eq. (3) with respect to W, by taking into account eq. (4) in the integral (3) and by using the normalization condition on (4) one gets:

$$\frac{\partial P_n(W)}{\partial W} = p(W|W)\left[R_n(W) - P_n(W)\right] \tag{5}$$

We define the joint probability of degrading from virtuality Q to virtuality Q_0 and Q_1, $\mathcal{P}(Q_0Q_1|Q)$ as follows

$$\mathcal{P}(Q_0Q_1|Q) = p(Q_0|Q)p(Q_1|Q)N(Q) \tag{6}$$

where $N(Q)$ is a normalization factor. We express then $R_n(Q)$ in terms of $\mathcal{P}(Q_0Q_1|Q)$:

$$R_n(Q) = \sum_{n'=1}^{n-1} \int_1^\infty dQ_0 \int_1^\infty dQ_1 \mathcal{P}(Q_0Q_1|Q) R_{n-n'}(Q_0) R_{n'}(Q_1) \theta(Q - Q_0 - Q_1) \tag{7}$$

In conclusion, eq. (5) says how the probability for an ancestor of virtuality W to generate n final partons varies in terms of the probability for a parton, which splits at W to Q_0 and Q_1, to generate n partons. $R_n(W)$ in (5) is calculated via eq. (7).

Define now the generating functions associated to $P_n(W)$ and $R_n(W)$:

$$f(z,W) = \sum_{n=1}^{\infty} z^{n-1} P_n(W) \tag{8}$$

$$g(z,Q) = \sum_{n=2}^{\infty} z^{n-2} R_n(Q) \qquad \text{for } Q \geq 2 \text{ GeV} \tag{9}$$

$$R_n(Q) = 0 \qquad \text{for } Q < 2 \text{ GeV}, \; n \geq 2 \tag{10}$$

It is easy to express the connection between $R_n(W)$ and $P_n(W)$ in terms of the generating functions (8) and (9). In order to be more explicit we divide the domain of integration in eq. (7) into three subdomains (see fig. 2):

(I) $Q_0 < 2$ GeV and $Q_1 < 2$ GeV: no-one of the two generated partons will split;

(II) $(1 \text{ GeV} < Q_0 < 2 \text{ GeV},\; Q_1 > 2 \text{ GeV}) \cup (1 \text{ GeV} < Q_1 < 2 \text{ GeV},\; Q_0 > 2 \text{ GeV})$: only one of the two generated partons will split;

(III) $Q_0 > 2$ GeV and $Q_1 > 2$ GeV: both generated partons will split.

It follows that

$$\begin{aligned} g(z,Q) = & \int_1^2 \int_1^2 \mathcal{P}(Q_0 Q_1|Q)\theta(Q - Q_0 - Q_1) dQ_0 dQ_1 \\ & + 2\int_1^2 dQ_0 \int_2^{\infty} dQ_1 \mathcal{P}(Q_0 Q_1|Q) z g(z,Q_1)\theta(Q - Q_0 - Q_1) \\ & + \int_2^{\infty}\int_2^{\infty} \mathcal{P}(Q_0 Q_1|Q) z g(z,Q_0) z g(z,Q_1) \theta(Q - Q_0 - Q_1) dQ_0 dQ_1 \end{aligned} \tag{11}$$

It results an integral non-linear Volterra equation; the non-linearity is of Lyapunov-Lichtenstein type. For $f(z,W)$ one obtains

$$f(z,W) = \int_1^2 p(Q|W) dQ + z\int_2^W g(z,Q) p(Q|W) dQ \tag{12}$$

or in differential form

$$\frac{\partial f(z,W)}{\partial W} = p(W|W)[z g(z,W) - f(z,W)] \tag{13}$$

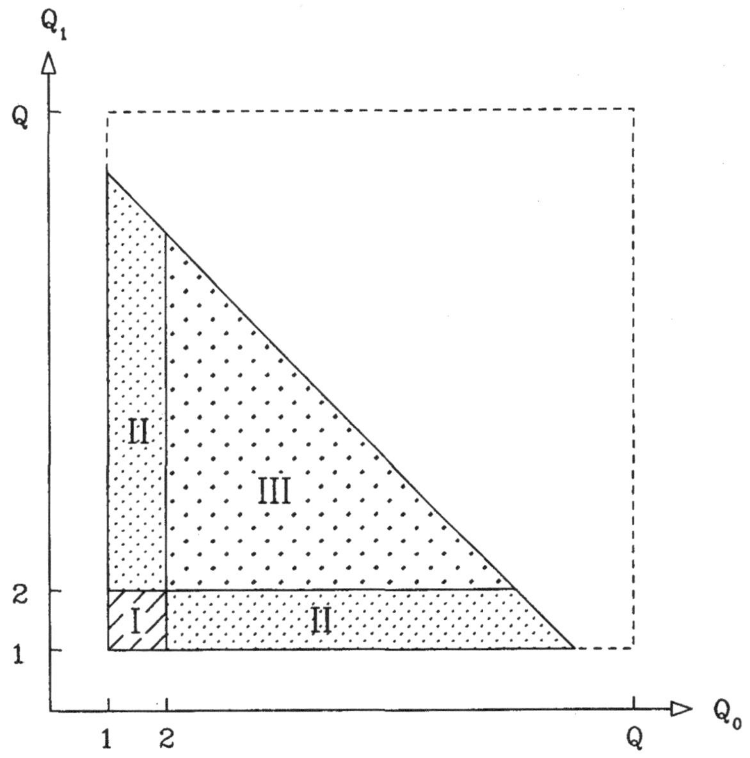

Fig. 2. The domain of integration in eq. (7) is divided into three regions, indicated by different shading and by the roman numerals.

Once $g(z, W)$ is known, $f(z, W)$ can be calculated from eq. (12).

The above equations can be solved in two relevant, although extreme, cases: (i) by limiting oneself to the domain where at most one of the two generated partons can split (regions I and II in fig. 2);
(ii) by extending the domain of integration in allowing $Q_0 + Q_1$ to be larger than Q, but requiring $Q_i \leq Q, \quad i = 0, 1$ (dashed box in fig. 2).

The corresponding solutions depend of the choice of the splitting function $p(Q|W)$ only through the variable u defined by $du = p(W|W)dW$; the solutions are:

$$\text{(i)} \qquad g(z, Q) = f(z, Q) \qquad f(z, W) = e^{u(z-1)}; \tag{14}$$

$$\text{(ii)} \qquad g(z, Q) = \{f(z, Q)\}^2 \qquad f(z, W) = [(e^u - 1)(1 - z) + 1]^{-1} \tag{15}$$

(i) is a shifted Poisson distribution with average number of final partons $\langle n \rangle = u+1$;
(ii) is a shifted Furry distribution with average number of final partons $\langle n \rangle = e^u$.

A third interesting, analytically solvable case is that in which the two generated partons are forced to have the same virtuality $Q_0 = Q_1$, but requiring $Q_0 \leq Q$. This corresponds to replacing the θ function in eqs. (7) and (11) by $\delta(Q_0 - Q_1)$. From eq. (11) we get directly the solution

$$g(z, Q) = [(e^{\tilde{u}} - 1)(1 - z^2) + 1]^{-1} \tag{16}$$

where $d\tilde{u} = \mathcal{P}(QQ|Q)dQ$ and the average number of final partons is $\langle n \rangle = 2e^{\tilde{u}}$. (16) is the generating function of a doubly shifted Furry distribution in which only the even multiplicities have a nonzero probability, i.e., only an even number of final partons is allowed: in fact the condition used here provides always either *two final* partons or *two splitting* partons. Use of eq. (12) allows in principle to calculate $f(z, W)$, but an explicit expression for it cannot be written down without specifying the splitting probability $p(Q|W)$; further discussion of the interesting features of this solution is therefore postponed to section 3.3.

3.2. Rapidity structure of SPS

We define, in the infinite momentum frame, the fraction $z_{0,1} = E_{0,1}/E$ of the splitting parton's energy E taken by the produced partons, where $E_{0,1}$ is the energy of parton 0 and 1 respectively. Energy conservation requires $z_0 + z_1 = 1$.

Enforcing the criterion of maximum simplicity (with an eye to QCD) we assume that the distribution in energy of the produced partons is regulated by

$$P(z_i)dz_i \propto \left(\frac{1}{z_i} + \frac{1}{1 - z_i}\right) dz_i \qquad i = 0, 1 \tag{17}$$

which is invariant for $z_0 \to z_1 = 1 - z_0$.

Now we calculate the limits of variation of z_0, which correspond, in the c.m. frame of the splitting to the two extreme configurations

(a) (b)

i.e., $z_0^{(a)} \leq z_0 \leq z_0^{(b)}$ or $z_1^{(b)} \leq z_1 \leq z_1^{(a)}$. In the infinite momentum frame we can express the variable z in terms of the rapidity variable:

$$z_i^{(a)} = \frac{Q_i}{Q} e^{\delta_i}, \quad z_i^{(b)} = \frac{Q_i}{Q} e^{-\delta_i} \quad (i = 0, 1) \tag{18}$$

where Q_i, Q are the virtualities of the produced and splitting partons respectively, and δ_i is the difference in rapidity between the generated and splitting partons. The explicit expression of $z_0^{(a),(b)}$ in terms of Q_i/Q are:

$$z_0^{(a),(b)} = B \mp \sqrt{B^2 - \frac{Q_0^2}{Q^2}}, \quad B = \frac{1}{2}\left(1 + \frac{Q_0^2}{Q^2} - \frac{Q_1^2}{Q^2}\right) \tag{19}$$

The situation in which z_0 lies between $z_0^{(a)}$ and $z_0^{(b)}$ corresponds to nonzero transverse momentum $p_{T,i}$, i.e., we replace Q_i^2/Q^2 by $Q_i^2/Q^2 + \frac{|p_{T,i}|^2}{Q^2}$ with

$$0 \leq \frac{|p_{T,i}|^2}{Q^2} \leq \left(\frac{|p_{T,i}|^2}{Q^2}\right)_{\max} \quad \text{and} \quad \sum_{i=0,1} \sqrt{\frac{Q_i^2}{Q^2} + \left(\frac{|p_{T,i}|^2}{Q^2}\right)_{\max}} = 1 \tag{20}$$

It follows

$$\left(\frac{|p_{T,i}|^2}{Q^2}\right)_{\max} = B^2 - \frac{Q_0^2}{Q^2} \tag{21}$$

We find for the $p_{T,i}$ dependence in terms of z_0

$$\frac{|p_{T,i}|^2}{Q^2} = B^2 - \frac{Q_0^2}{Q^2} - (z_0 - B)^2 \tag{22}$$

which gives the dependence of the rapidity change Δy_0 on z_0 as

$$\Delta y_0 = \frac{1}{2} \ln \frac{B + |z_0 - B|}{B - |z_0 - B|} \tag{23}$$

We are now ready to implement the virtuality evolution described in the previous section, with the rapidity structure just examined, in a Monte Carlo simulation, whose results are presented in the following section.

3.3. Numerical solutions of SPS

In order to solve numerically eqs. (11) and (12) we must make a choice about the probability $p(Q|W)$ for an initial parton of virtuality W to split at Q. We will use the decay-like form:

$$p(Q|W)dQ = e^{A(\tau - \tau_W)} A d\tau \tag{24}$$

where the evolution parameter $\tau = \ln(\ln Q)$; τ_W is its initial value and A a free parameter.

Our simplified parton shower has been written as Monte Carlo code on the basis of the following procedure:

– the initial ancestor of initial virtuality W splits at the virtuality Q, generated with probability $p(Q|W)$, eq. (24).
– if the virtuality Q is larger than or equal to 2 GeV (automatic cut-off) two daughter partons of virtuality Q_0 and Q_1 respectively are generated according to the joint probability distribution $\mathcal{P}(Q_0Q_1|Q)$, eq. (6), with $Q_0 + Q_1 \leq Q$. Energy fractions z_0 and z_1, with $z_0 + z_1 = 1$ are then assigned to the daughters with probability distribution $P(z_0)$, eq. (17). The partons' azimuthal angle is assigned randomly in the c.m. frame of the splitting.
– the previous step is iterated for all daughter partons until all produced partons have virtuality less than 2 GeV.

Full phase space parton MD's generated by an initial parton at all virtualities (it should be remembered that the MD in full phase space is independent of the rapidity evolution of the parton shower) and for any choice of the parameter A (the only parameter in the model) are well fitted by NBMD; other standard fits are usually quite bad. The general trend of the NB parameters in full phase space, shown in Fig. 3, is consistent with what we know from NB behavior at parton level from JETSET 6.3 simulations for $q\bar{q}$ and gg systems: the average number of final partons $\bar{n}$ grows with initial virtuality W, while the parameter k decreases with increasing virtuality. The average number of clans $\bar{N}$ also grows with initial virtuality, the rate of increase being independent of A, and the average number of partons per clan $\bar{n}_c$ increases (the larger the value of A, the faster the increase).

The rapidity structure discussed in the previous section has also led to very interesting results (see Table II in ref. [28]). MD's in rapidity intervals $|y| < y_c$ with $y_c = 0.5, 1\ldots$ up to 5, are well fitted by NBMD for $A \geq 1$ (after neglecting the first bin), but not for $A < 1$ (where cascading is limited, see eq. (24)). The parameter k becomes less than 1 for $y_c = 0.5$ favoring more aggregation in $|y| < 0.5$; the average number of clans $\bar{N}$ does not depend on the ancestor's initial virtuality W in fixed rapidity intervals up to $|y| < 2$. The average number of partons per clan $\bar{n}_c$ does not vary with y_c at fixed W: this is a new feature, showing separation of clans in rapidity.

We come back now to eq. (16). By using eq. (24) one finds that the free parameter A must be greater than 0.5 in order to obtain in this case a normalizable joint probability distribution $\mathcal{P}(Q_0Q_1|Q)$. We observe an unexpected property of the new MD: as the ancestor's initial virtuality W increases the MD (16) tends to a limiting shape characterized by the parameter

$$e^{\bar{u}} = \frac{\gamma(2A-1, \ln Q)}{\gamma(2A-1, \ln 2)} \underset{Q\to\infty}{\longrightarrow} \frac{\Gamma(2A-1)}{\gamma(2A-1, \ln 2)} \tag{25}$$

where γ is the incomplete gamma function. It should be noticed that for larger values of A, the limiting shape is reached for larger values of W as can be seen from eq. (25). This interesting finding deserves further investigation, which we decided to postpone to future work.

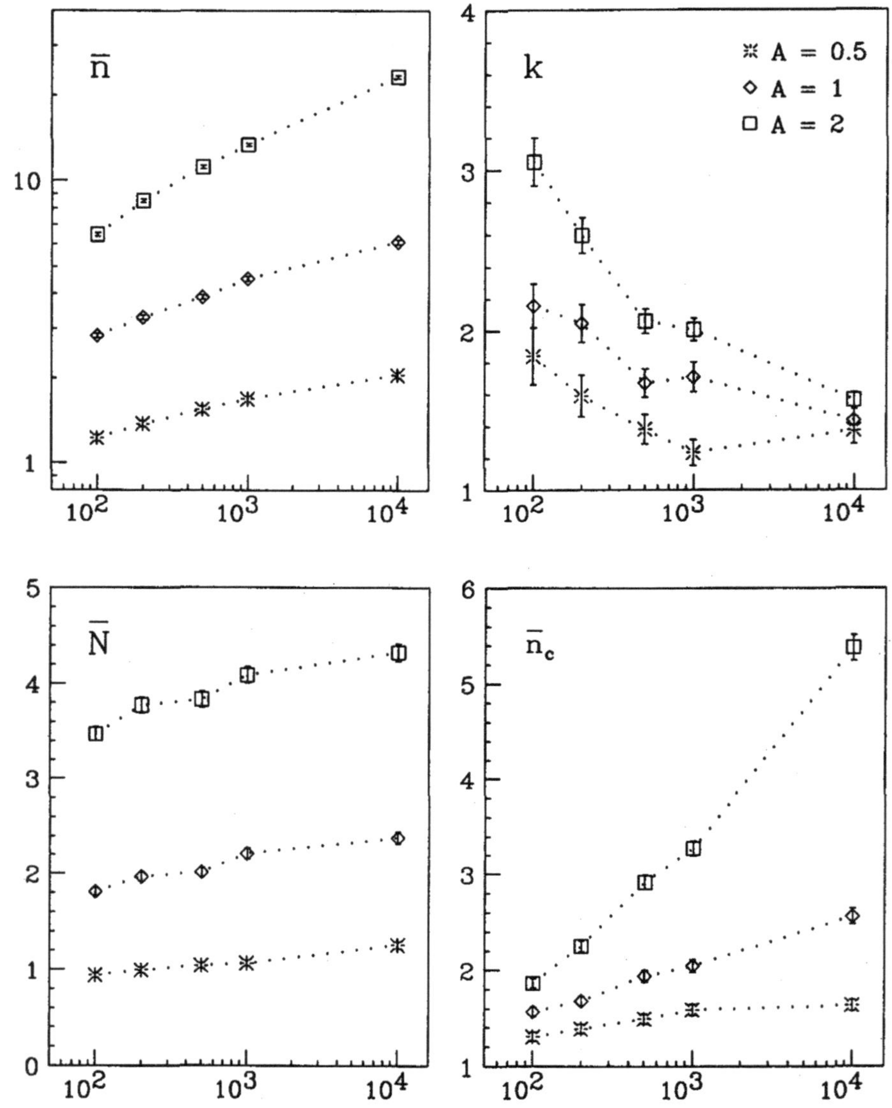

Fig. 3. Standard NB parameters ($\bar{n}$, k), corresponding average number of clans ($\bar{N}$) and average clan multiplicity ($\bar{n}_c$) of final parton MD's in full phase space for $A = 0.5$ (stars), $A = 1$ (diamonds) and $A = 2$ (squares) are plotted vs the initial parton virtuality W. The lines are drawn to guide the eye.

4. Conclusions

Continuing our search for the origin of NB regularity in high energy collisions, it was shown that new experimental data by the DELPHI Collaboration and previous theoretical work strongly suggest to study *single parton shower* evolution in a correct kinematical framework inspired by essentials of QCD.

Guided by the criterion of maximum simplicity we followed a single parton shower, initiated by a parton of given virtuality, along its branchings to the final parton level. The rapidity structure has been assumed to be controlled by gluon self-interaction. This procedure led us to an integral non-linear Volterra equation for the generating function associated to the probability for a parton which splits at virtuality Q to generate n final partons. The equation was solved analytically in extreme cases which resulted of particular interest.

After studying the rapidity structure under the above mentioned assumption we decided to implement it, with the virtuality evolution, in a Monte Carlo simulation. The Simplified Parton Shower (SPS) model which we propose depends on one parameter only, A, defined in eq. (24). The parameter A regulates the length of the parton cascade. NB regularity is found to occur for any choice of the parameter A in full phase space. Other fits are definitely quite bad. In symmetric rapidity windows the regularity occurs for $A \geq 1$ only. The cascades are too short for $A < 1$ and consequently they do not show NB behavior.

Two peculiar features of the SPS model are that the average number of clans, $\bar{N}$, does not depend on the ancestor's initial virtuality W in fixed rapidity intervals up to $y_c < 2$ and that clans are naturally separated in rapidity.

These first results suggest to focus our future work on the following points:
(i) study the composition of two parton cascades in order to meet, via GLPHD, experimental findings at charged hadron level.
(ii) give an integrated description of correlations and multiplicity distributions in the SPS model.
(iii) analyse intermittency properties of factorial moments in the SPS model by choosing symmetric rapidity windows below $y_c = 0.5$. It is interesting in this respect the fact that aggregation is higher in the interval $|y| < 0.5$ than in larger intervals.

We are of course aware of the many limitations of our approach to the SPS model. Essentially, more QCD should be included. For the moment it should be considered an encouraging finding that a *complex* and *puzzling experimental fact* like NB regularity can be justified in the SPS model in terms of *elementary parton structures* with *simple properties*.

Acknowledgments

One of us (A.G.) would like to express his gratitude to N. Schmitz and to the Organizing Committee for the stimulating and friendly atmosphere that was created at this workshop.

References

1. R.E. Ansorge et al., UA5 Collaboration, *Z. Phys.* **C43** (1989) 357.
2. P. Abreu et al., DELPHI Collaboration, "Charged Particle Multiplicity Distribution in Restricted Rapidity Intervals in Z^0 Hadronic Decays", CERN–PPE/91–78.
3. G.J.H. Burgers, C. Fuglesang, R. Hagedorn and V. Kuvshinov, *Z. Phys.* **C46** (1990) 465;
   S. Barshay, *Mod. Phys. Lett.* **A6** (1991) 55, and references therein.
4. C. Fuglesang, in *Multiparticle Dynamics (Festschrift for Léon Van Hove)*, La Thuile, Italy, eds. A. Giovannini and W. Kittel, World Scientific, Singapore 1990, p. 193.
5. V. Uvarov, private communication.
6. L. Van Hove and A. Giovannini, "Multihadron Production and QCD Parton Showers" CERN-TH.5885/90, to appear in *Proceedings of XXVth International Conference on High Energy Physics*, Singapore, 1990.
7. G. Altarelli and G. Parisi, *Nucl. Phys.* **B126** (1977) 298.
8. M. Bengstsson and T. Sjöstrand, *Nucl. Phys.* **B289** (1987) 810;
   T. Sjöstrand and M. Bengstsson, *Computer Physics Commun.* **43** (1987) 367.
9. L. Van Hove and A. Giovannini, *Acta Phys. Pol.* **B19** (1988) 917.
10. Ya.I. Azimov, Yu.L. Dokshitzer, V.A. Khoze and S.I. Troyan, *Z. Phys.* **C27** (1985) 65; *Z. Phys.* **C31** (1986) 213;
    D. Amati and G. Veneziano, *Phys. Lett.* **B83** (1979) 87;
    G. Marchesini, L. Trentadue and G. Veneziano, *Nucl. Phys.* **B181** (1981) 335.
11. A. Giovannini and L. Van Hove, *Acta Phys. Pol.* **B19** (1988) 931.
12. B. Buschbeck, P. Lipa and R. Peschanski, *Phys. Lett.* **B215** (1988) 788;
    E.A. De Wolf, *Acta Phys. Pol.* **B21** (1990) 509.
13. W. Heisenberg, *Z. Phys.* **113** (1939) 61;
    W. Heisenberg and coll., in *Vorträge über Kosmische Strahlung*, Springer Verlag, Berlin 1943, p. 57 and 115; *Z. Phys.* **126** (1949) 569; *Z. Phys.* **133** (1952) 65;
    J.R. Oppenheimer, H.W. Lewis and S.H. Wouthuysen, *Phys. Rev.* **73** (1948) 127.
14. P.K. MacKeown and A.W. Wolfendale, *Proc. Phys. Soc.* **89** (1966) 553.
15. DELPHI Collaboration, "Charged Particle Multiplicity of quarks and gluon Jets in Z^0 Hadronic Decays", contributed paper n. 392 to the Geneva EPS-LP/91 Conference, Geneva, 24 July – 1 Aug 1991.
16. We thank G. Ekspong for interesting and useful comments on this point.

17. M. Derrick et al., HRS Collaboration, *Phys. Lett.* **B168** (1986) 299; M. Derrick et al., HRS Collaboration, *Phys. Rev.* **D34** (1986) 3304.
18. A. Giovannini, S. Lupia and R. Ugoccioni, "Multiplicity Distributions in High Energy Collisions", DFTT 33/91, to appear in the *Proceedings of the 4th "Blois" Workshop on Elastic and Diffractive Scattering*, Isola d'Elba, 22-25 May 1991.
19. W. Kittel, in *Proceedings of Workshop on Physics with Future Accelerators*, La Thuile 1987, ed. J. Mulvay, CERN Yellow Rep. 87-7, Vol. II, p. 424.
20. M. Garetto, A. Giovannini, T. Sjöstrand and L. Van Hove, in *Proceedings of the Perugia Workshop on Multiparticle Production*, eds. R.C. Hwa, G. Pancheri and Y. Srivastava, World Scientific, Singapore 1989, p. 181.
21. A. Giovannini, R. Ugoccioni and L. Van Hove, in *Proceedings of the XXVth Rencontres de Moriond* at Les Arc, ed. J. Trân Thanh Vân, Edition Frontières, 1990, p. 299; A. Giovannini, R. Ugoccioni and L. Van Hove, in *Proceedings of the International Workshop on Correlations and Multiparticle Production* (CAMP), Marburg, eds. M. Plümer, S. Raha and R.M. Weiner, World Scientific, Singapore 1991, p. 219
22. K. Konishi, A. Ukawa and G. Veneziano, *Nucl. Phys.* **B157** (1979) 45.
23. A. Giovannini, *Nucl. Phys.* **B161** (1979) 429.
24. E. D. Malaza and B. R. Webber, *Nucl. Phys.* **B267** (1986) 702.
25. E. D. Malaza, *Z. Phys.* **C31** (1986) 143.
26. Z. Kunszt and P. Nason, in *Z Physics at LEP 1*, Geneva 1989, eds. G. Altarelli, R. Kleiss and C. Verzegnassi, CERN Yellow Rep. 89-08, p. 373.
27. T. Hebbeker, "QCD Studies at LEP", review talk given at the Geneva EPS-LP/91 Conference, Geneva, 24 July - 1 Aug 1991.
28. R. Ugoccioni and A. Giovannini, "Virtuality Evolution and Rapidity Structure of a Simplified Parton Shower", DFTT 14/91, to be published in *Z. Phys. C - Particles and Fields*.

DISCUSSION OF SOME PROPERTIES OF G MOMENTS

W. Florkowski

Gesellschaft für Schwerionenforschung GSI

Postfach 110541, D-W-6100 Darmstadt, Germany

and

H. Niewodniczański Institute of Nuclear Physics

ul. Radzikowskiego 152, 31-342 Kraków, Poland.

ABSTRACT

We discuss the general features of the G_q moments in the case where they are applied in the analysis of the fluctuations of the rapidity distributions. In particular their dependence on the statistical background is studied.

1. G and F moments

For simplicity we consider the events with fixed total multiplicity N. If the initial rapidity window is divided into M bins and N_j $(j = 1, ..., M)$ denotes the multiplicity in the jth bin then the G_q moments for one event are [1]

$$G_q(M,N) = \sum_{j=1}^{M}{}'(N_j/N)^q, \tag{1}$$

where the prime signifies that the sum is over non-empty bins only. Consequently, q can be an arbitrary real number, positive or negative. Knowing the *multiplicity distribution* $Q(N; N_1, ..., N_M)$ one can calculate the average value of G_q according to the expression

$$< G_q(M,N) > = \sum_{N_1,...,N_M} Q(N; N_1, ..., N_M) \sum_{j=1}^{M}{}'(N_j/N)^q. \tag{2}$$

Here the sum runs over the all possible choices of $N_j's$ which satisfy the condition $N = N_1 + ... + N_M$. Denoting by f_j the fraction of particles in the jth bin we can schematically rewrite Eq. 2 in the following form.

$$< G_q(M,N) > = \int df_1...df_M \tilde{Q}(N; f_1, ..., f_M) \sum_{j=1}^{M}{}' f_j^q, \tag{3}$$

where $\tilde{Q}(N; f_1, ..., f_M)$ is the probability that in the jth bin we have f_jN particles.

It is interesting to write down the corresponding definitions of the *factorial moments* [2]. We again consider the events with fixed total multiplicity N and for one event we use the formula

$$F_q(M,N) = M^{q-1} \sum_{j=1}^{M} \frac{N_j(N_j-1)...(N_j-1+q)}{N(N-1)...(N-1+q)}, \tag{4}$$

where q is a positive integer. For the sample of events the average value of the moments is given by

$$< F_q(M,N) > = \sum_{N_1,...,N_M} Q(N; N_1, ..., N_M) M^{q-1} \sum_{j=1}^{M} \frac{N_j(N_j-1)...(N_j-1+q)}{N(N-1)...(N-1+q)}, \tag{5}$$

with $Q(N; N_1, ..., N_M)$ being the same multiplicity distribution as that introduced above in Eq. 2. If one assumes that the latter can be represented by the following convolution

$$Q(N; N_1, ..., N_M) = \int dp_1...dp_M P(N; p_1, ..., p_M) B(p_1, ..., p_M; N_1, ..., N_M), \tag{6}$$

where $P(N; p_j)$ describes the dynamical part of fluctuations and $B(p_j; N_j)$ is the multinomial distribution, then the average factorial moments are given by [2]

$$< F_q(M,N) > = \int dp_1...dp_M P(N; p_1, ..., p_M) \sum_{j=1}^{M} p_j^q. \tag{7}$$

One can observe that Eqs. 3 and 7 have the same mathematical structure but their physical interpretation is different. Firstly, $f_j's$ in Eq. 3 denote fractions of particles in the jth bin whereas $p_j's$ in Eq. 7 are the probabilities that a particle will fall into the jth bin. Secondly, the functions in the integrands of Eqs. 3 and 7 are different. The measurements based on Eqs. 3 and 7 would be equivalent in the limit of infinite number of particles but this limit is not a realistic one, especially when we study moments as functions of decreasing rapidity interval. Consequently, the results concerning the fractal structure, which are obtained by using the G_q moments, will be in general different from those obtained by the measurements of the factorial moments.

2. Splitting functions

Let us now discuss the form of $P(N; p_1, ..., p_M)$ which is scale invariant and leads to the power law behavior of the factorial moments; we follow here the example of the α model [2]. The probabilities p_j are constrained by the normalization condition so it is convenient to introduce the new variables which are independent of each other. For the first partition, $M = 2$, we define the variable x_1 in such a way that $p_1^{(1)} = x_1$ and $p_2^{(1)} = 1 - x_1$. For the second partition, $M = 4$, we define the variables x_{11} and x_{22} through the relations: $p_1^{(2)} = x_1 x_{11}, p_2^{(2)} = x_1(1 - x_{11}), p_3^{(2)} = (1 - x_1)x_{22}$ and $p_4^{(2)} = (1-x_1)(1-x_{22})$, see Fig. 1. For the next partitions we repeat the same scheme. Assuming that the variables x can be treated as independent random numbers having the same distribution $\psi(x)$ we get from Eq. 7 the following results:

$$< F_q(2, N) >= \int dx_1 \psi(x_1)[x_1^q + (1 - x_1)^q] \tag{8}$$

and

$$\begin{aligned} < F_q(4, N) > &= \int dx_1 dx_{11} dx_{22} \psi(x_1)\psi(x_{11})\psi(x_{22}) \left[x_1^q x_{11}^q + x_1^q(1 - x_{11})^q \right. \\ &\left. + (1 - x_1)^q x_{22}^q + (1 - x_1)^q (1 - x_{22})^q\right]. \end{aligned} \tag{9}$$

One can see that the above integral factorizes and using the normalization condition for $\psi(x)$ we get the result $< F_q(4, N) >=< F_q(2, N) >^2$. For an arbitrary number of partitions we obtain the *power law behavior* $< F_q(2^\mu, N) >=< F_q(2, N) >^\mu$.

If we considered the splitting functions describing the distribution of the fractions of particles then the analogous procedure to that outlined above could be applied for the G_q moments. It would work, however, only in the case of infinite number of particles because for finite and different multiplicities it is impossible to use the same splitting function. This argument and the observation of certain trends in the behavior of the measured G_q moments, *i.e.* saturation of the moments for small rapidity bins in the $\ln < G_q(M,N) >$ vs $\ln M$ plots, led us [3] to the consideration of the *multiplicity dependent splitting functions.*

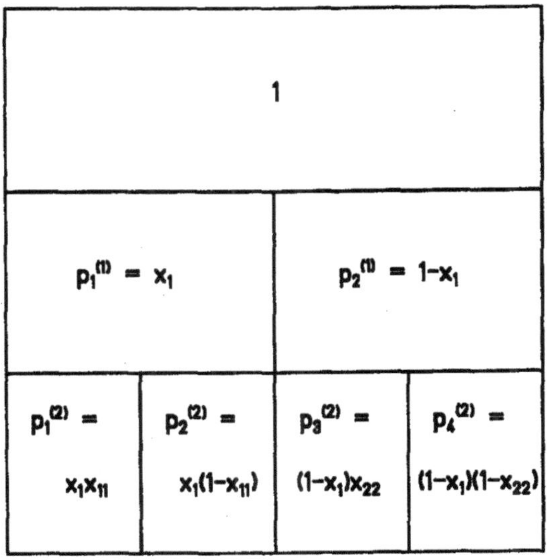

FIGURE 1 Parametrization of the probabilities that a particle will fall into a given bin; the same parametrization can be used to describe the fractions of particles found in the bins.

In this approach the production of particles can be simulated by a stochastic process in which a given number of them, corresponding to the final multiplicity, is randomly distributed into smaller and smaller bins. At the beginning we separate particles into two bins using the multiplicity dependent splitting function $\psi(N,x)$. As the result we obtain x_1N particles in the left bin and $(1-x_1)N$ particles in the right one. For the next partition we use the same function but with different multiplicity arguments, *i.e.* $\psi(x_1N,x)$ and $\psi((1-x_1)N,x)$ for the left and right bin respectively. In this case the average values of the G_q moments are the following

$$< G_q(2,N) > = \int dx_1 \psi(N,x_1)[x_1^q + (1-x_1)^q] \tag{10}$$

and

$$< G_q \quad (4,N) >= \int dx_1 dx_{11} dx_{22} \psi(N,x_1)\psi(x_1 N, x_{11})\psi((1-x_1)N, x_{22})$$
$$\times[x_1^q x_{11}^q + x_1^q(1-x_{11})^q + (1-x_1)^q x_{22}^q + (1-x_1)^q(1-x_{22})^q]. \qquad (11)$$

Because the splitting function depends now on the multiplicity the last integral does not factorize and we do not recover the strict power law behavior. Nevertheless $\psi(N,x)$ is independent of the bin size and, consequently, applicable at all steps of splittings. This indicates that a characteristic *scale-invariance* is present in our approach.

3. Universality relation for the G moments

Let us now consider the case when $\psi(N,x)$ is the function peaked at $x = 1/2$. In such a situation we expect that $< G_q(4,N) >$ can be approximated by the following expression

$$< G_q(4,N) >= \int dx_1 \psi(N,x_1)[x_1^q + (1-x_1)^q] \int dx \psi(N/2,x)[x^q + (1-x)^q]. \qquad (12)$$

Similarly, for an arbitrary number of partitions we obtain

$$< G_q(\mu+1,\nu) >=< G_q(\mu,\nu) > A_q(\mu-\nu), \qquad (13)$$

where we have introduced the notation:

$$N = 2^\nu, \quad M = 2^\mu \qquad (14)$$

and

$$A_q(\mu-\nu) = \int dx \psi(2^{\nu-\mu},x)[x^q + (1-x)^q]. \qquad (15)$$

One can check by simple substitution that the solution of Eq. 13 can be written as the following series

$$\ln < G_q(\mu,\nu) >= \sum_{\mu'=\mu_0'}^{\mu-1} \ln A_q(\mu'-\nu) - \sum_{\mu'=\mu_0'}^{-1} \ln A_q(\mu'-\nu), \qquad (16)$$

where the second term on the right hand side comes from the requirement of normalization of the G_q moments: $\ln < G_q(0,\nu) >= 0$. One of the consequences of the above result is that for fixed q the quantity

$$\Gamma_q(\mu-\nu) \equiv \ln < G_q(\mu,\nu) > - \ln < G_q(\nu,\nu) > \quad (17)$$

is a function of the difference $\mu - \nu$ only, so we find that

$$\ln < G_q(\mu,\nu) >= \Gamma_q(\mu-\nu) - \Gamma_q(-\nu). \quad (18)$$

This *universality relation* [3] expresses the scaling behavior of the G_q moments. Their dependence on two arguments, μ and ν, can be described in terms of a scaling function of one variable only. Knowing $\Gamma_q(\xi)$ we can determine the G_q moments as functions of μ for all values of ν. As shown in Fig. 2 one can obtain the graph of $\ln < G_q(\mu,\nu) >$ vs μ by placing the origin of the $\ln < G_q(\mu,\nu) >$ coordinate system on the $\Gamma_q(\xi)$ curve at the point $\xi = -\nu$.

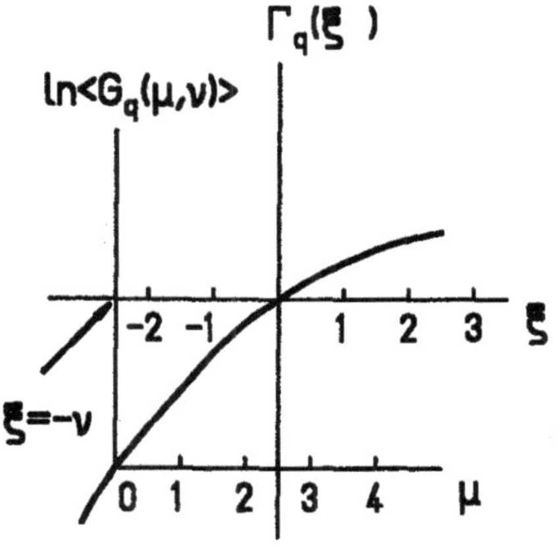

FIGURE 2 Illustration of how $\ln < G_q(\mu,\nu) >$ can be determined from $\Gamma_q(\xi)$. Here, for example, $\nu = 2.5$.

4. G moments for the statistical background

Let us consider the case when the multiplicity splitting function corresponds to the *binomial distribution*, *i.e.*, $\psi(N, x_1) = NQ_{st}(N; N_1, N_2)$ where

$$Q_{st}(N; N_1, N_2) = \frac{N!}{N_1!N_2!}\delta_{N,N_1+N_2}2^{-N} \tag{19}$$

and the fractions of particles in a bin have discrete values $x_j = N_j/N$ $(N_j = 0, ..., N)$. The binomial distribution describes the behavior of particles which are not correlated and, therefore, it can be regarded as a purely *statistical distribution.* Consequently, using Eq. 2 with the distribution (19) one calculates the average value of the G_q moments corresponding to the statistical background

$$< G_q(2, N) >_{st} = \sum_{N_1,N_2=0}^{N} Q_{st}(N; N_1, N_2)[N_1^q + N_2^q]N^{-q}. \tag{20}$$

For larger numbers of partitions we represent the statistical distributions by the products of the binomial distributions. For example, in the case of two partitions we have $Q_{st}(N; N_1, N_2, N_3, N_4) = Q_{st}(N; N_1 + N_2, N_3 + N_4)Q_{st}(N_1 + N_2; N_1, N_2)$ $\times Q_{st}(N_3 + N_4; N_3, N_4)$ and

$$< G_q(4, N) >_{st} = \sum_{N_1,N_2,N_3,N_4=0}^{N} Q_{st}(N; N_1, N_2, N_3, N_4)[N_1^q + N_2^q + N_3^q + N_4^q]N^{-q}. \tag{21}$$

In this way we retain the structure typical for Eqs. (9) and (11). One can also check that Q_{st} in Eq. 21 is the multinomial distribution.

Let us now discuss the behavior of the G_q moments calculated for the statistical background. In our approach we will use the mathematics and the therminology borrowed from the fractal theory [4]. The basic quantity is the slope

$$\tau(q, M) = \frac{\Delta(\ln < G_q(M, N) >)}{\Delta \ln M}, \tag{22}$$

from which α_q can be calculated as

$$\alpha_q = \frac{\partial}{\partial q}\tau(q, M) \tag{23}$$

and the spectral function $f(\alpha)$ is obtained by the Legendre transform

$$f(\alpha_q) = q\alpha_q - \tau(q, M). \qquad (24)$$

In contrast to usual fractal theory treatment we have an M-dependence in Eqs. 22 - 24, *i.e.* we define the slope (22) locally. In the analysis of the G_q moments one has to consider such a local slope because the moments saturate for large M's and the plot $\ln < G_q(M,N) >$ vs $\ln M$ is, in general, not a straight line. Consequently, we omit the limit $\delta = 1/M \rightarrow 0$ and investigate the behavior of the moments in finite bins. Although such an approach is in this point different from that used in the fractal theory, one can see that in several cases the physical properties of the multiplicity distributions can be well interpreted in terms of fractal expressions.

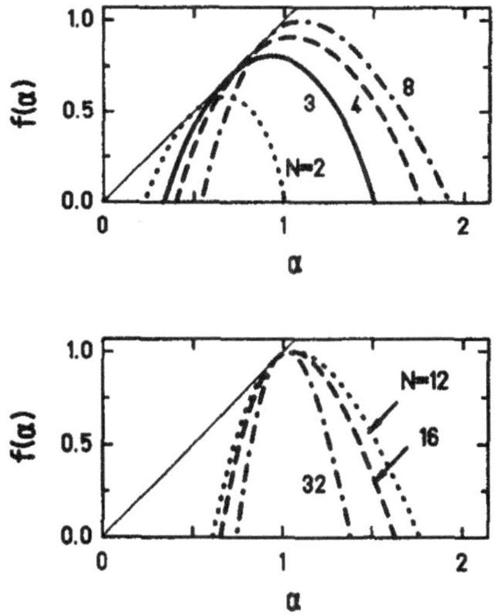

FIGURE 3 The spectral functions $f(\alpha)$ calculated for the binomial distribution (19) for N ranging from 2 to 32 .

As an example we calculated the fractal dimension D_0 of the distribution (19). We first found the slope for the first partition, $\tau(q,2) = \ln < G_q(2,N) > / \ln 2$, and then from Eqs. 22 - 24 obtained the following result

$$D_0 \equiv f(\alpha_0) = 1 + \ln[1 - Q_{st}(N; 0, N)] / \ln 2. \qquad (25)$$

For small N the probability that one bin is empty is large and it makes D_0 much smaller than 1. For larger values of N the dimension D_0 increases and saturates at unity. We can see this behavior in Fig. 3 where the full spectral function $f(\alpha)$ are shown for various values of N. The fractal dimension D_0 corresponds in each case to the maximum of the $f(\alpha)$ curve.

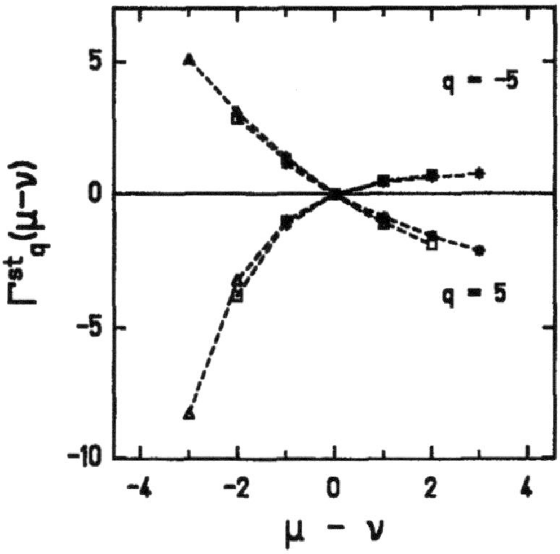

FIGURE 4 The values of the moments of the fifth order calculated for the statistical background and plotted in the form which exhibits their universal behavior: triangels, squares and stars correspond to the cases $\nu = 4,3$ and 2.

It is also interesting to check the validity of the universality relation (19) in the case of pure statistical fluctuations. We therefore calculate the average G_q moments for different values of μ and ν and plot them in the $\Gamma_q(\xi)$ coordinate system, placing the $\ln < G_q(\mu = \nu, \nu) >$ points at its origin. The results of our calculation, in the special case of $q = \pm 5$, are shown in Fig. 4. The points lie with a good approximation on the same curves confirming the idea of universal behavior.

We compared our $\Gamma^{st}_q(\xi)$ curve to that extracted from the UA1 data (Fig. 6 of Ref. 5) and found that they are practically the same. This fact indicates that the experimentally observed universality can be explained by the presence of pure statistical fluctuations in the system. We also found that the difference between the values of the moments for the data and statistics becomes very small for large μ because the G_q moments have a well defined asymptotics which leaves little freedom for dynamical fluctuations to show up. Contrary, in large bins the data differ from statistics substantially. In this region the analysis of the non-trivial correlations can be performed by simple comparing the data to the corresponding statistical values.

5. Summary

The aim of this note was to discuss several aspects of the application of the G_q moments in the measurements of fluctuations in high energy physics. At first we pointed out the differences between the approach based on the G_q moments and that based on the factorial ones. Afterwards using the concept of splitting functions we discussed the cases of strict self-similarity and of universality. Finally, using both the therminology taken from the theory of fractals and the idea of universality we studied the dependence of the G_q moments on the statistical background. We realized that on the level of $\Gamma_q(\xi)$ function one cannot filter out the dynamical content of the fluctuations.

6. Acknowledgments

The main results presented in this note where obtained in collaboration with R. C. Hwa. I would like to thank him for numerous discussions, inspirations and encouragement. I also thank H.-P. Pavel for critical remarks and A. Tucholski for interesting discussions.

References

1. R. C. Hwa, *Phys. Rev.* **D41** (1990) 1456
2. A. Bialas and R. Peschanski, *Nucl. Phys.* **B273** (1986) 703; **B308** (1988) 857
3. W. Florkowski and R. C. Hwa, *Phys. Rev.* **D43** (1991) 1548
4. T. Tél, *Z. Naturforsch.* **43a** (1988) 1154
5. H. Dibon and M. Markytan, *preprint* HEPHY-PUB 541/90

Intermittency and Multifractality in Monte Carlo e^+e^- Annihilation

Charles B. Chiu

Center for Particle Physics and Department of Physics
University of Texas, Austin, Texas 78712

ABSTRACT

Properties of intermittency and multifractality in e^+e^- annihilation are investigated based on the JETSET Monte Carlo code of the Lund parton shower model. Dependence on energy, rapidity window width and multiplicity is found to be complicated. Universal property of the intermittency indices is explored. With the application of a certain remedial cut procedures, an approximate universal curve in multifractal analysis is found.

1 Introduction

I would like to talk about a recent work[1] which I did in collaboration with Hwa. The study of self-similarity in multiplicity fluctuations is an active area of recent research.[2] It began with the proposal[3] of the intermittency phenomena, i.e. a power law behavior in the normalized factorial moments as functions of the rapidity bin width. Since then, this phenomenon has been observed in various collision processes.[2] An extension of that description to multifractal analysis has also been considered subsequently.[4,5,6]

The main objective of our investigation is to look for regularities and, more ambitiously, "universal" properties in the intermittency and the multifractal phenomena. As a first step in this direction, in the present work we confine our attention to the commonly believed simplest process, the e^+e^- annihilation process, and study the multiplicity fluctuation as a function of

S : the square of the *cm* energy,
Y : the width of the rapidity window, and
N : the multiplicity in the window.

The JETSET Monte Carlo program,[7] which has successfully described the factorial moment data in e^+e^- annihilation process, provides a natural tool for us to carry out this study. Of course we realize that Monte Carlo prediction is not yet the data. Also we do not necessary wish to adhere to the detail dynamics of the Monte Carlo program. However, we take the point of view that Monte Carlo predictions give a reasonable interpolation of the data and they provide an intelligent guess when extrapolated to the unexplored territory. Anyhow, we believe that it is worthwhile to see what we can learn from JETSET on the SYN dependence.

For the intermittency phenomena, we look for observables which are relatively insensitive to the SYN dependence. Presumably they would serve as the common observables for comparing intermittency phenomena in different processes. In the realm of multifractal analysis, our earlier investigation was based on different versions of the branching models.[6] Now with more realistic predictions of the JETSET model, we are able to take a closer look at the scaling behavior. We suggest a remedial cut-procedure to accompany the multifractal analysis. This enhances the scaling behavior signal. Based on the selected samples, we identify a certain universal property in the multifractal phenomena.

2 Intermittency analysis

Let me begin with the definition of "horizontal analysis" and the "vertical analysis". Consider a rapidity window centered about $y = 0$, with width Y. The window is divided evenly into M bins labelled by $j = 1, 2, \ldots M$, with the bin width $\delta = Y/M$. The individual Monte Carlo events are labelled by: $k = 1, 2 \ldots K$, where K is the total number of events. The multiplicity in the jth bin of the kth event is designated by $n_j(k)$.

For the kth event, the "horizontally analyzed" normalized factorial moment of order q is given by

$$F_q^h = \frac{\frac{1}{M} \sum_j n_j(n_j - 1) \cdots (n_j - q + 1)}{[N(k)/M]^q} . \tag{1}$$

We consider two ways in averaging the logarithm of the F-moment. The first is to take the logarithm in the last step, i.e. after averaging over all events. The second is to take the logarithm of F for each event, and then average over all events. They are respectively designated by: $\ell n F^{h,v}$ and $(\ell n F^h)^v$. Correspondingly the "vertically analyzed" normalized factorial moment of order q is given by:

$$F_q^v = \frac{\frac{1}{K} \sum_k [n_j(n_j - 1) \cdots (n_j - q + 1)]_k}{(N_j/K)^q} . \tag{2}$$

The logarithm of F based on these two different averaging procedures are:

$$\ell n F^{v,h} \text{ and } (\ell n F^v)^h$$

Intermittency occurs when the factorial moment exhibits a power behavior as a function of the bin width, i.e.

$$F_q(\delta) \propto \delta^{-a_q} \propto M^{a_q} , \tag{3}$$

where the power a_q is the "intermittency-index", or "IM-index". In practice a_q is determined by the slope in the "$\ell n F_q$ vs $\ell n M$ (or $\mu = \ell n M/\ell n 2$) plot", or the "log-log plot".

For the SY-dependence, we consider following 4 cases:

$$(\sqrt{s} \text{ in TeV}, Y): \quad (1,2),\ (1,4),\ (10,2) \text{ and } (10,4)\ .$$

For the N-dependence, we collect those events, each with its multiplicity fall within a narrow width about a central value of $N = 2^\nu$ and perform averages over that sample. We then compare results obtained with different values of ν.

We present final results in terms of the anomalous fractal dimension:

$$d_q = \frac{a_q}{q-1}, \tag{4}$$

where for each case the IM-index, a_q is determined in the $0 \le \mu \le 4$ interval.

In Fig. 1, the q-dependence of anomalous fractal dimension d_q, determined from the average quantity: $\ell n F_q^{h,v}$, for the four SY-cases is displayed as the solid curves for a range of ν-values. For $\nu > 3$, the shapes of the curves are approximately independent of ν, although their magnitudes do vary with ν. The dashed curves are designated as the "all" curves, where the d-values are determined based on all events, i.e., events with all multiplicities.

The corresponding situation determined from the average quantity $(\ell n F^h)^v$ is shown in Fig. 2. Notice that the variations of d-curves as a function of ν are again not very significant. For the "all" curves, they are almost independent of q. This is to be in contrast with the corresponding monotonically increasing behavior shown in Fig. 1.

The d_q curves determined from the two averages of vertical analysis: $\ell n F_q^{v,h}$ and $(\ell n F_q^v)^h$ turned out to be essentially the same. The d_q curves from $\ell n F_q^{v,h}$ are shown in Fig. 3. Note that the curves are of the same approximate shape and their magnitudes differ somewhat for the four SY cases. The dashed curves, or the "all" curves are all above the solid curves. This can be traced to be due to the fact that for the present case, the quotient of the average of numerator functions to the average of the denominator functions is larger than the average of the corresponding quotients.

We now compare the anomalous fractal dimension of "all cases", $\bar{d}_q$'s. As mentioned earlier, they are the dashed curves shown in Fig. 1, 2 and 3. These dashed curves together with the corresponding curves, obtained from the quantities: $(\ell n F_q^v)^h$, are shown in Fig. 4. Again, notice the general trend. Within each type of averaging procedure, the curve shapes for all four SY-cases are similar. Their magnitude do vary with S and Y.

We proceed to show that the SY dependence shown in Fig. 4 can be further consolidated into the dependence of one variable. Fig. 5 displays $\bar{d}_q$ vs dN/dy. Notice the approximate universality of a common positive slope. The positivity of the slope given here is oppositive to the trend found by Buschbeck and Lipa[8] in other processes at lower energies. It is important to follow up the experimental verification of this Monte Carlo prediction.

We turn to the quantities $(\ell n F_q^h)^v$, which have very mild q-dependence. For this case, the S-dependence and the Y-depedence may be combined into one single average multiplicity dependence. Fig 6 shows that the $\bar{d}$-function is approximately linear in $\ell n\langle N\rangle$.

In summary we have studied factorial moments of the Monte Carlo data based on both the horizontal and the vertical analyses. For each analysis, two types of averaging procedures were used. We find in general the IM-index have complicated dependence on SYN. One lesson from the present analysis is that, apparently it is not meaningful to compare, *a priori*, the IM-index of e^+e^- annhilation process determined at particular values of SYN, with the IM-index obtained from another process at different values of SYN. A possible universal feature we have found is that, the plot of $\bar{d}_q$ vs dN/dy has approximately an equal slope, insensitive to the SYN dependence. If we regard this feature as universal characteristics of e^+e^- annihilation process, it would be natural to compare the same quantity in different processes.

3 Multifractal analysis

In our present work, the multifractal analysis is confined to the horizontal analysis. The formalism for the horizontal analysis[6] is similar to that of the vertical analysis.[5] We recall that the horizontally analyzed G-moment of a given event is defined by

$$G_q = \sum_{j=1}^{M}{}' p_j^q, \text{ where } p_j = \frac{n_j}{N}. \tag{5}$$

where superscript prime designates summing over nonempty bins only. The scaling behavior of the G-moment follows from, among other things, two basic assumptions.

1. Particles in the inidividual bins are assumed to have the power law scaling behavior i.e.
$$p_j \sim \delta^{\alpha_j}, \text{ or } p_j^q \sim (\delta^{\alpha_j})^q. \tag{6}$$
2. The G-moment has the integral representation
$$G_q = \int d\alpha' W(\alpha', \delta)\delta^{\alpha' q} \tag{7}$$

If the weight function has a multifractal property, the weight function would take on the form:

$$W(\alpha, \delta) \sim \frac{\rho(\alpha)}{\delta^{f(\alpha)}}, \tag{8}$$

i.e. in the denominator the bin-width δ is raised to a power f, which is the multifractal dimension. To see the meaning of f, for a moment just suppose that the weight function did not have a fractal property. The usual geometric measure in one dimension gives a factor of δ^{-1}, or $f = 1$. Monofractality would correspond to the case, where f is a fractional constant, between 0 and 1. Multifractality states that the fractal dimension of the weight function, $f \leq 1$, and it varies with the power: α.

Substituting Eq. (8) into Eq. (7), we get

$$G_q(\delta) \sim \int d\alpha' \rho(\alpha') \exp\left\{-\left[\alpha' q - f(\alpha')\right] \ell n \frac{1}{\delta}\right\} . \tag{9}$$

Since we are interested in the small δ, or the large $\ell n \frac{1}{\delta}$ behavior of G, assume there is a stable stationary point in the integrand (the location of the stationary point may in general depend on q), the α' integral can be readily carried out by the method of steepest descent. Let the stable stationary point be at $\alpha' = \alpha$, expanding the exponent about this point gives

$$\alpha' q - f(\alpha') = \left[\alpha q - f(\alpha)\right] + \left[q - \left.\frac{df(\alpha')}{d\alpha'}\right|_{\alpha}\right](\alpha' - \alpha) - \frac{1}{2} \left.\frac{d^2 f(\alpha')}{d\alpha'^2}\right|_{\alpha} (\alpha' - \alpha)^2 + \cdots , \tag{10}$$

with

$$q = \frac{df(\alpha)}{d\alpha} , \tag{11}$$

and

$$\frac{d^2 f}{d\alpha^2} < 0 . \tag{12}$$

For large $\ell n \frac{1}{\delta}$, integrating over α' leads to:

$$G_q(\delta) \sim \rho(\alpha) \sqrt{\frac{2\pi}{\frac{d^2 f}{d\alpha^2} \ell n \frac{1}{\delta}}} \, \delta^{\alpha q - f(\alpha)} \sim \delta^{\alpha q - f(\alpha)} \equiv \delta^{\tau(q)} . \tag{13}$$

In the second to the last step of Eq. (13), to conform to the usual practice, the $\ell n \frac{1}{\delta}$ variation is ignored. This then gives the fundamental relation in the multifractal analysis, where the scaling index of the G-moment,

$$\tau(q) = \alpha q - f(\alpha), \text{ or } f(\alpha) = \alpha q - \tau(q) . \tag{14}$$

In other words, the multifractal dimension f and the scaling index τ are Legendre transforms of each other.

As with the case of intermittency analysis, we may consider taking two different averages for the logarithm of the G-moments.

$$\ell n \, G^{h,v} \text{ and } (\ell n \, G^h)^v .$$

A typical log-log plot for the multiplicity-bin $\nu = 5$ case is shown in Fig. 7. For now, let us focus our attention on the general trend of the dotted curves. When $q = 1$, $\ell n G_q = 0$. This is a dividing line. For $q < 1$, $\ell n G_q$ is positive, while for the $q > 1$, it is negative. As q decreases from 0, the fixed-q curve starts to develop a negative curvature. The more negative is q, the more pronounced is this negative curvature. The curves appear to bend toward the horizontal direction, suggesting the presence of "saturation" effect.

We look for a criterion to select a subset of events[9] which would give approximately a linear behavior in the log-log plot and would lead to a certain universal property in the G-analysis. From the very definition of the horizontally analyzed G-moment, it is clear that such selections must be applied to individual events, not to different bins within a given event.

We first carry out a *preliminary multifractal* analysis on $\ell n(G_q^h)^v$ and on $\ell n G_q^{h,v}$ based on the initial slope in the μ-region: from 1 to 2. We find that the analysis on the first quantities occassionally lead to an $f(\alpha)$ curve which violates the negative curvature property of Eq. (12), inàppropriate to multifractal description. So we concentrate on the $(\ell n G_q^h)^v$ quantities.

Analogous to the application of the p_T cut in selecting jet events, we find it is useful to apply cuts to exclude events where the "mean magnitude of ℓnG" is "large". Qualitatively speaking, cut on large positive ℓnG tends to, among other things, suppress the "deep" dip events and cuts on large negative ℓnG tends to, among other things, select events with prominent peaks.

In particular, based on the preliminary G-analysis, we assign a "typical mangitude" of the positive ℓnG value to be $C_+ = (\ell n G_+^h)^v$, where "+" labels a typical point, say at $\mu = 3$, $q = 3$. The corresponding "typical magnitude" of the negative ℓnG value is $C_- = (\ell n G_-^h)^v$, where "−" labels a typical point, say at $\mu = 3$, $q = -2$.

Our selection criteria is then to accept events only with:

$$\ell n\, G_+ \geq ZC_+ \quad \text{and} \quad \ell n G_- \leq ZC_- \tag{15}$$

Here we have introduced a multiplicative factor Z to adjust the amount of the cut involved. For $Z \gg 1$, the above inequalities do not impose any effect of cut, since by definition, the inequalitites will be satisfied for all events. As Z decreases down to the neighborhood of unity, the filtering effect becomes more pronounced. At the same time, one finds that the saturation effect in the log-log plot, especially for large negative q becomes less pronounced.

A typical situation in the log-log plot, where cut is applied with $Z = 0.8$, for $\nu = 5$, is shown as the solid curves in Fig. 7. More details are illustrated in Fig. 8. A glance at Fig. 7 and Fig. 8 reveals that there is an approximate linearity in the range $\mu = 0$ to 4 throughout the negative q region shown. In the positive q region the linearity is not as good. For uniformity, we take the range: $\mu = 0$ to 4 for the G-analysis for all q values.

The corresponding $f(\alpha)$ function for this case with cut is shown in Fig. 9. To sharpen the relationship among the various ν curves, we rescale those curves shown in Fig. 9 in the following way. From Eq. (14), it can be shown that $f(\alpha) = \alpha$ is always true at $q = 1$. Since at this point, α_1 is the "information dimension", i.e.

$$\alpha_1 = \lim_{\delta \to 0} \left({\sum_{j=1}^{M}}' p_j \ell n p_j \right) \Big/ \ell n \delta \,. \tag{16}$$

We refer this point on the $f(\alpha)$ curve as the "information point". Now define the rescaled quantities

$$\tilde{\alpha} = \frac{\alpha}{\alpha_1}, \; \tilde{f} = \frac{f}{f(\alpha_1)} = \frac{f}{\alpha_1}. \tag{17}$$

If we do this for all ν curves, then the rescaled curves: $\tilde{f}(\tilde{\alpha})$ will coincide at $\tilde{\alpha} = \tilde{f}(\tilde{\alpha}) = 1$. Fig. 10 is the superposition of rescaled curves of (1, 2) case together the corresponding curves for the other three SY-cases. Here we have included those curves only with $\nu > 3$. Notice that there is approximate coincidence in the region where $\alpha \leq 1$, or $q \geq 1$. We interprete it to be a form of approximate universality that unifies the multifractal structure in this region.

Above the information point, where $\alpha > 1$ (or $q < 1$), the universality fails to hold true. This is not totally unexpected, since the cut sample is designed to select on peak events and events which do not involve deep dips, so we would be too optimistic to expect simplicity to arise also for $q < 1$ moments, which characterizes the dips. To reiterate, if we focus on the peak situation as the intermittency case does, we have found the universality in the $\tilde{f}$ curves among the large multiplicity events, i.e. for $\nu > 3$.

We now come back to the "all" events, i.e. the events which include all multiplicities. Fig. 11 is the log-log plot based on the multiplicities of the (1, 2) case before the application of cuts. We apply the same cut procedure, i.e. first find the corresponding average magnitudes of C_+ and C_- and then apply the cut based on the inequality of Eq. (15). We use the same value $Z = 0.8$. The scaling behavior for the "all" case here turned out to be not as good as that for the fixed multiplicity analysis. One could in principle vary Z value to obtain better linear behavior. However, it is unclear whether it is justified to consider different Z values between the fixed multiplicity analysis and the "all" analysis. Anyhow for $Z = 0.8$, the corresponding rescaled curves for the (1, 2) case together with those for the other 3 SY-cases are shown in Fig. 12. The approximate universality for $q > 1$ is again obtained. In fact the shapes of the rescaled curves here are comparable to those obtained from the fixed-ν analysis.

Let us sum up the situation on the G-analysis. We have outlined a procedure for obtaining the approximate linearity in the log-log plot with one adjustable parameter Z. We find this procedure works particularly well for negative q and for fixed ν analysis. It leads certain regularities for the f-curves in question and an approximate universality in the rescaled curves in the region where $q > 1$.

One of the concerns in the G-moment analysis is the presence of statistical background contribution in the $f(\alpha)$ curve. It is our hope that the application of the cut procedure, which leads to the approximate linearity in the log-log plot, suppresses the statistical background, since the statistical background is not expected to have the scaling properties. Of course this point is needed to be studied more carefully. If it turns out to be affirmative, then the cut procedure will serve as a mean to enhance the scaling signal. Also different choices of the "typical points" leading to the determination of $C_\pm$, which appear in Eq. (15) are needed to be explored.

Experimental demonstration of the approximate scaling behavior for some selected sample of events together with a certain universal features in the rescaled $\tilde{f}(\tilde{\alpha})$ curves are of great interest, since this would provide first experimental evidence that in multiparticle fluctuation phenomena there exists a quantifiable universal behavior.

4 Acknowledgements

I am grateful to Rudy Hwa for his inspiring and untiring efforts in our collaboration. We are grateful to T. Sjostrand for providing us the Monte Carlo code and the manual for JETSET 7.3, without which this work could not have been done. Thanks to Alessandro de Angelis and H. Dibon for informative discussions. This work was supported in part by the U. S. Department of Energy under Grant No. DE-FG06-65ER-4022.

5 References

1. C. B. Chiu and R. C. Hwa, *Intermittency and multifractality in e + e− annihilation*, 1991, submitted for publication.

2. Proceedings of the Santa Fe Workshop, March 1990, edited by F. Cooper, R. C. Hwa and I. Sarcevic (World Scientific Singapore, 1991); W. Kittel, Proc. of the 20th Int. Symposium on Multiparticle Dynamics, Gut Holmecke, Germany, Sept. 1990.

3. A. Bialas and R. Peschanski, *Nucl. Phys.* **B273**, 703 (1986).

4. I. M. Dremin, *Mod. Phys. Lett.* **A3**, 1333 (1988); P. Lipa and B. Buschbeck, *Phys. Lett.* **B223**, 465(1988); P. Carruthers, *Int. J. Mod. Phys.* **A4**, 5587 (1989).

5. R. C. Hwa, *Phys. Rev.* **D41**, 1456 (1990).

6. C. B. Chiu and R. C. Hwa, *Phys. Rev.* **D43**, 100(1991); Santa Fe Workshop, ref. 2.

7. T. Sjostrand and M. Bengtsson, *Computer Phys. Comm.* **43**, 367 (1987);
T. Sjostrand, *Int. J. Mod. Phys.* **A3**, 751(1988);
T. Sjostrand, *JETSET* 7.3 (1990).

8. B. Buschbeck and P. Lipa, Santa Fe Workshop, ref. 2.

9. R. C. Hwa, OITS-459, *Phys. Rev. D* (to be published).

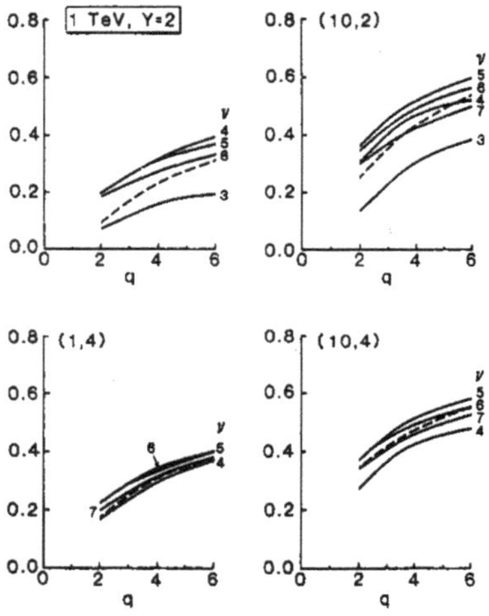

Fig. 1 q-dependence of analomous fractal dimenison d_q obtained from $\ell n F^{h,v}$. The four plots are for the four SY cases considered. Solid curves are based on events selected at various ν-bins. The dashed curves are obtained from events of all multiplicities.

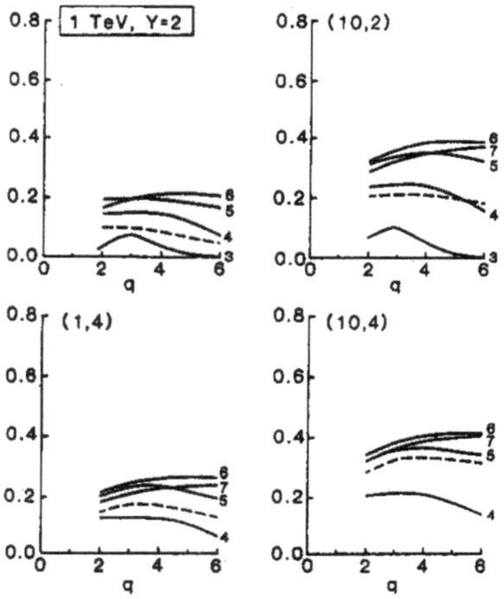

Fig. 2 q-dependence of analomous fractal dimenison d_q obtained from $(\ell n F^h)^v$. The four plots are for the four SY cases considered. Solid curves are based on events selected at various ν-bins. The dashed curves are obtained from events of all multiplicities.

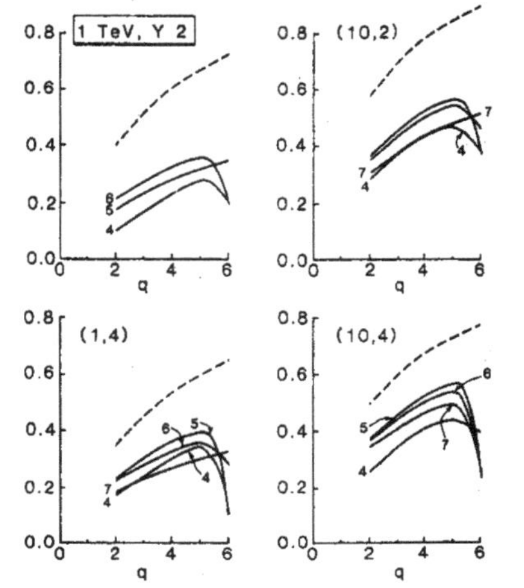

Fig. 3 q-dependence of analomous fractal dimenison d_q obtained from $\ell n F^{v,h}$.The four plots are for the four SY cases considered. Solid curves are based on events selected at various ν-bins. The dashed curves are obtained from events of all multiplicities

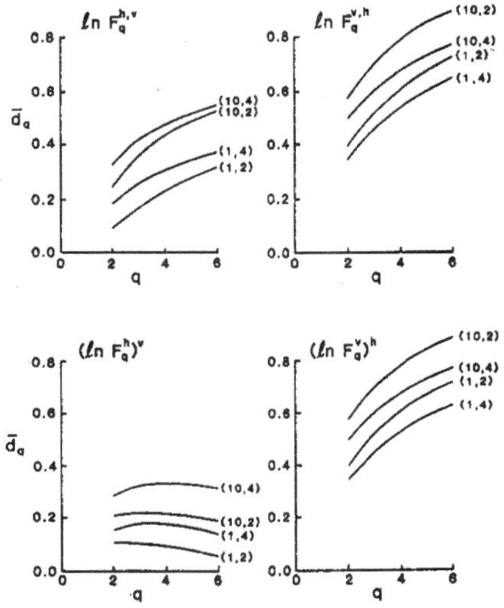

Fig. 4 q-dependence of analomous fractal dimension $\bar{d}_q$ obtained based on events of all multiplicities. The four plots are for two averages of the horizontal analyses and for two averages of the vertical analyses. In each plot, the curves are for the four SY cases.

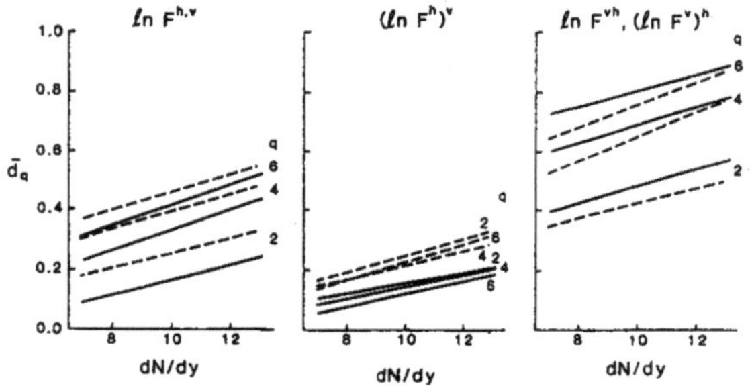

Fig. 5 Analomous fractal dimension $\bar{d}_q$ as a function of the mean ratio of multiplicity divided by the window width. First two plots from the left are for the two averages of the horizontal analyses the last plot is for two averages of the vertical analyses.

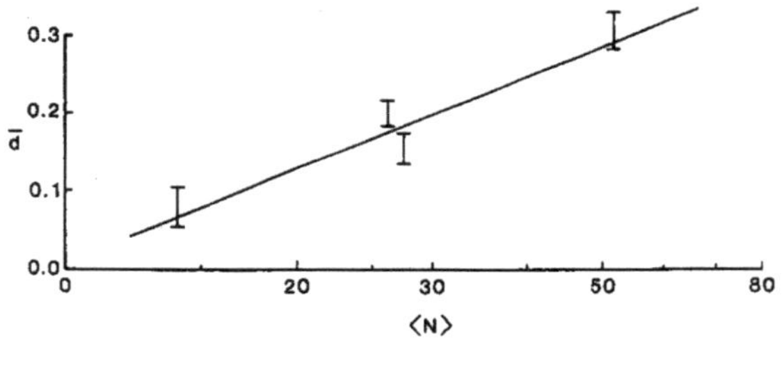

Fig. 6 Analomous fractal dimenison $\bar{d}$ obtained from $(\ell n F^h)^v$ as a function of the $\ell n\langle N\rangle$. Error bars indicate variations in q.

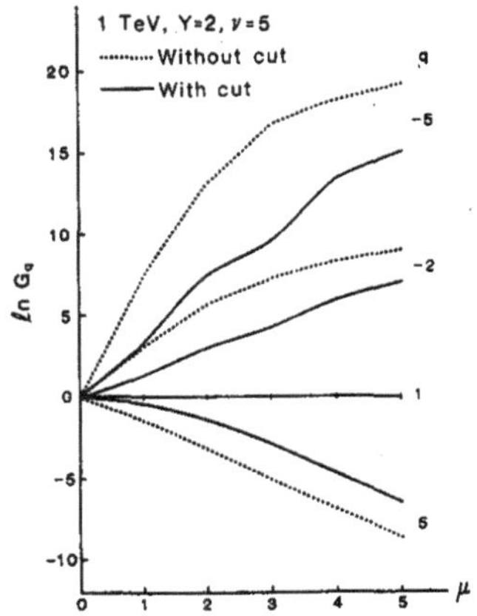

Fig. 7 The log-log plot of the G-moment for the multiplicity bin $\nu = 5$, for the (1, 2) case. The solid curves are with cut and the dotted curves without cut.

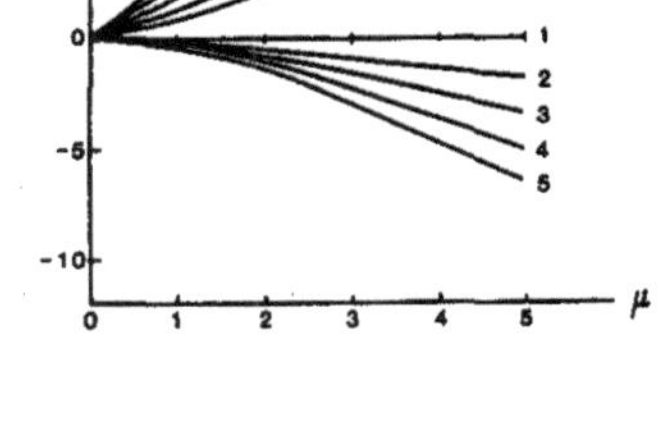

Fig. 8 More detail illustration of the log-log plot of the G-moment with the multiplicity bin $\nu =$ 5, for the (1, 2) case. Here the cut is applied.

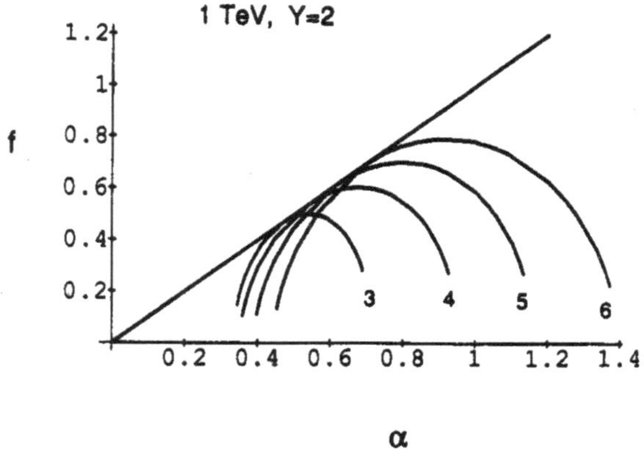

Fig. 9 α-dependence of the f-function for multiplicity bins $\nu = 3, 4, 5, 6$, for the (1, 2) case, after the cut is applied.

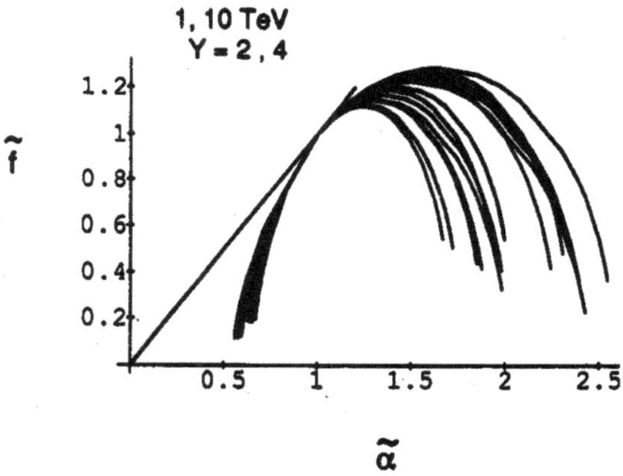

Fig. 10 The superposition of the $\tilde{f}$-function for multiplicity bins with $\nu > 3$, for all SY cases considered.

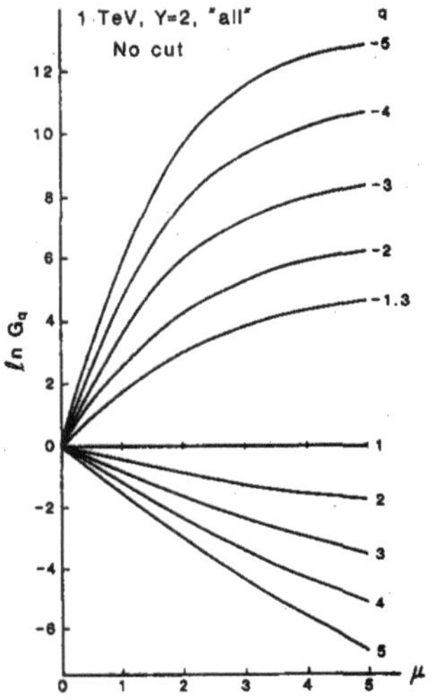

Fig. 11 The log-log plot of the G-moment based on all multiplcities for the (1, 2) case applied here before the application of cut.

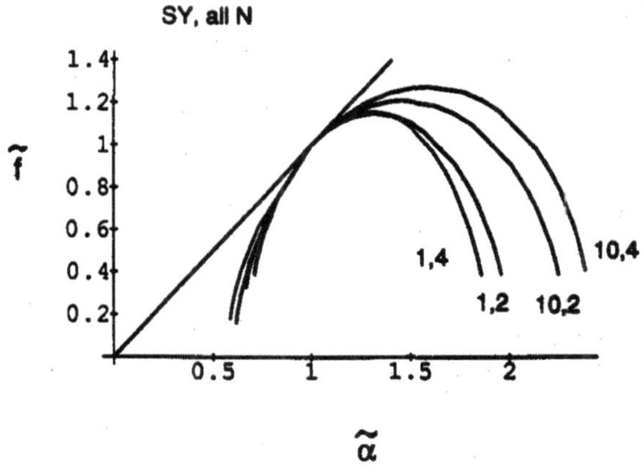

Fig. 12 The superposition of the $\tilde{f}$-function based on all multiplicity events for the four SY cases considered, where cut is applied.

INTERMITTENCY, MULTIFRACTALITY AND HADRONIC COLLISIONS

Rudolph C. Hwa
Institute of Theoretical Science and Department of Physics
University of Oregon, Eugene, OR 97403-5203

We discuss the properties of the G moments as a means of determining the multifractal features of the intermittency phenomena in multiparticle production. We then summarize ECCO, a Monte Carlo code for simulating hadronic collisions in the framework of the Geometrical Branching Model. It is shown that ECCO can reproduce the intermittency data.

There are two aspects to the study of intermittency in multiparticle production. One is to learn about the various properties of the phenomenon. The other is to learn from those properties the nature of hadronization. So far it is the former aspect that has received extensive investigation, while progress in the latter aspect has been meager. Of course, the ultimate objective for studying intermittency or any other property of multiparticle production is to understand the dynamics of leptonic, hadronic and nuclear collisions and to describe the many-parton problem in a way more feasible for computation than what QCD provides. In this paper I report on an effort to construct a model of hadronic collisions that can yield the observed intermittency. But before that I discuss the multifractal properties of intermittency.

When the factorial moment [1] of the variety usually analyzed, i.e.

$$\langle F_q \rangle = \left\langle \frac{1}{M} \sum_{j=1}^{M} \frac{n_j(n_j-1)\dots(n_j-q+1)}{(\langle N \rangle / M)^q} \right\rangle \tag{1}$$

in the usual notations, develops a power-law behavior

$$\langle F_q \rangle \propto \delta^{-\phi_q} \tag{2}$$

one usually says that there is self-similarity in the system, which has fractal properties. The fractal anomalous dimension is taken to be

$$d_q = \phi_q/(q-1) \quad . \tag{3}$$

But the precise sense in which this is to be understood as a measure of fractality has not been made clear. Since (1) has no meaning for $q < 2$, the usual dimension of a fractal, defined at $q = 0$, and the information dimension, defined at $q = 1$, are not accessible as descriptions of the fractal system probed by F_q. To have a proper multifractal description of the multiplicity fluctuations an alternative to the factorial moments would have to be used.

The G moments [2] have been suggested as a means to achieve that end. It is defined by

$$G_q = \sum_{j=1}^{M}{}' p_j^q \tag{4}$$

where $p_j = n_j/N$ and the summation is over nonempty bins only. The definition allows q to be any real number, positive or negative. The properties of G_q have been investigated for specific models [3] as well as in the data [4]. While the multifractal spectrum can be determined from the power-law behavior of G_q, the major obstacle to a true revelation of the fractal properties of the underlying dynamics is the presence of the statistical contribution [5].

Cuts on event selection to enhance the dynamical content have been suggested [6], and that procedure has been upgraded in an actual application to the e^+e^- annihilation problem in the framework of JETSET [7], as reported by Chiu [8] in this Workshop. When the appropriate cuts are made, the G moments exhibit linear behavior in the log-log plot in M for both positive and negative q. Such power-law behaviors provide the basis for inferring the multifractal properties of the collision problem.

Another approach that facilitates a closer comparison between the F_q and G_q moments have been developed more recently [9], [10]. In that approach the cut is introduced in the definition of G_q:

$$G_q = \sum_{j=1}^{M}{}' p_j^q \theta(n_j - q) \tag{5}$$

where q is taken, for the moment, to be a real integer, positive or negative. The step function thus selects contributions from only those bins that contain particles greater or equal in number to q. In this respect it is similar to the selection criterion of F_q. p_q is still the bin multiplicity n_j divided by the total multiplicity of the event in the rapidity interval Y. Thus G_q is still normalized to 1 at $q = 1$, and is equal to the total number of nonempty bins at $q = 0$.

With the G moments defined in (5) it is found in the EMC data [9] that G_q exhibits asymptotic power-law behaviors at large M with exponents τ_q that are essentially independent of the statistical contamination in the multiplicity fluctuation for $q \geq 4$. Comparison with F_q can be more readily made if the latter is defined with $\langle N \rangle$ in (1) replaced by the event multiplicity N in Y. Thus in some loose sense G_q provides a continuation of F_q to $q = 1$ and 0. The saturation problem that G_q encounters for negative q is still there; that is in the nature of the collision problem when $\langle N \rangle$ is low. We expect that when the G moment analysis is applied to nuclear collisions where $\langle N \rangle$ is high, the properties of dips that are probed by the negative-q moments will also convey interesting information, as the positive-q moments have.

We now turn our attention from the properties of intermittency to the dynamics of hadronic collisions that can give rise to those properties. Before so doing, it should be pointed out that whether or not one regards intermittency as containing new physics, the data on F_q for hadronic collisions have not been reproduced by the Monte Carlo simulation of any existing models [11]. That fact alone seems sufficient to indicate the existence of some dynamical mechanism of hadronization that has not been taken into account. On the other hand, random cascade models, such as the α model [12], have been shown to give rise to intermittency. But such toy models do not contain enough physical reality to predict other properties of hadronic collisions. What we discuss below is the latest version of the geometrical branching model (GBM) [13], which has the attributes of both being physical and having random cascading. It has been tuned to fit essentially all the data that the model is at this point capable of predicting, including the F_q moments. Thus we believe that it is currently the only model that cannot be ruled out by the data on intermittency.

The GBM has been under investigation for several years, confronting various aspects of the multiparticle production data, such as KNO scaling [14], minijet production [15], proton-nucleus [16] and nucleus-nucleus collisions [17]. However, it has not been used by experimental groups to compare with their data because it lacked a MC code. In collaboration with J.C. Pan we now have developed such a code, called ECCO, which stands for Eikonal Cascade COde [13]. At this stage the code is still rather simple, with cascades being restricted to the rapidity space only and resonance production being neglected. Nevertheless, we have been able to generate results on $\langle N\rangle$, C_q, P_n, dN/dy, $C_2(y_1, y_2)$, F_q and dN/dp_T^2 that all agree with data in the energy range $10 < \sqrt{s} < 100 GeV$. Since our initial aim is to get the soft production process correctly, we have not considered jet production and other hard processes that are important at higher energies.

The geometrical part of GBM is based on the conventional eikonal formalism. In terms of the eikonal function $\Omega(R)$ the inelasticity factor $g(R) = 1 - e^{-2\Omega(R)}$ may be written in a series [16]

$$g(R) = \sum_{\mu=1}^{\infty} \pi_\mu(R) \tag{6}$$

where

$$\pi_\mu(R) = \frac{[2\Omega(R)]^\mu}{\mu!} e^{-2\Omega(R)} \tag{7}$$

The multiplicity distribution P_n is then

$$P_n = \int dR^2 \sum_{\mu=1}^{\infty} \pi_\mu(R) B_n^\mu \tag{8}$$

which is one of the basic formulas of GBM. The integral is over all impact parameters, the sum is over the number of so-called cut-Pomerons, and B_n^μ is the probability of producing n particles in a μ-cut-Pomeron term. Each cut-Pomeron contribution may be regarded as a Born term in a series expansion of inelasticity; $\pi_\mu(R)$ is its weight and B_n^μ is specified by the branching part of the GBM.

In earlier formulation of the GBM B_n^μ was taken to be the Furry distribution [14]-[17], which is the solution of a stochastic equation for Furry branching [18]. It is analytically tidy but lacks information on where particles are produced in phase space. In ECCO [13] branching is done by MC simulation in accordance to some prescription which represents the essence of the dynamics of soft production that cannot be formulated analytically. We summarize here the central features of the branching procedure.

A. Initial mean energy. Our interest is primarily in the inelastic production of mesons in, say, a pp collision. If W denotes the total energy of the particles produced (excluding leading baryons), then the average $\bar{W}$ must depend on $\Omega(R)$. We parametrize that dependence by

$$\frac{\bar{W}}{\sqrt{s}} = \left[\frac{\Omega(R)}{\Omega(0)}\right]^\lambda \tag{9}$$

where λ is a free parameter that turns out to be 0.1, when the average multiplicity at $\sqrt{s} = 22$ GeV is fitted.

B. Cascade mass distribution. In the cascade chain each cluster is partitioned into two daughter clusters whose masses m_1 and m_2 are distributed according to

$$D(m_1, m_2) = \left(\frac{m_1 + m_2}{m}\right)^\gamma \tag{10}$$

where m is the mass of the mother cluster. These clusters are not on-shell massive states, but represent collections and subsets of virtual partons. These mass variables are used mainly to keep track of the branching history, and signal the ending of a branch when m_i goes below $2m_\pi$. γ is a parameter that turns out to 0.2 or 0.3 depending on whether the ISR data [19] or the NA22 data [20] on C_q are fitted.

C. Rapidity width. The first step in dealing with the rapidity distribution is to specify the range of rapidity space in which the clusters are to populate. Since the momenta of the leading baryons are unknown from event to event, energy-momentum conservation cannot be used to determine the rapidities of the clusters. That turns out to be a positive attribute in our model, for otherwise energy-momentum conservation would lead to too much long-range correlation. The rapidity range is set by the parameter ζ in

$$y_{max} = \zeta\, \ell n \left(\sqrt{s}/m_p\right) \tag{11}$$

which turns out to be $\zeta = 0.75$, when the plateau width of dN/dy is fitted.

D. Rapidity distribution of clusters. This is the most distinctive part of the model that characterizes the crucial mechanism for generating intermittency. In the rest frame of the mother cluster the distribution of the daughter clusters in rapidity is

$$p(y) = p_0 y^{-\beta} \quad . \tag{12}$$

This is a distribution that has no scale (in contradistinction to any formula that involves a correlation length) and has a singularity at $y = 0$. These properties are essential in generating self-similarity in the fluctuation of multiplicities in varying y-bin sizes. The NA22 data [20] on F_q can be fitted when $\beta = 0.6$. Thus the singularity at $y = 0$ is a weak one, meaning that with low but nonvanishing probability successive steps of the cascade can all involve small y daughters and descendants, thereby giving rise to spikes in the rapidity distribution, but only very rarely. The result on F_q is shown in Fig.1.

It should be emphasized that Fig.1 is not the result of a fit using some form of two-particle correlation as input. Indeed, the correlation function is an output of ECCO. Our input is eikonal formalism and successive cascading with four adjustable parameters λ, γ, ζ and β, plus one more for the p_T distribution not discussed here. While there are other models that can yield the observed $\langle N \rangle$, P_n, dN/dy, d^2N/dy_1dy_2, etc., as we can with ECCO, none of them can give the correct F_q [11], as achieved in Fig. 1. Thus we believe that the data on intermittency have captured a central feature of particle production that places a highly discriminating demand on dynamical models.

Regarding ECCO now as a good code to simulate particle production in hadronic collisions, we can use it to calculate the properties of intermittency in nuclear collisions. The formalism for nuclear collision in the framework of GBM has been worked out already in Refs. [16] and [17]. It is therefore straightforward to apply ECCO to simulate the production of low-p_T particles in heavy-ion collision experiments currently being carried out. Because of the many incoherent sources of production, the fluctuation in multiplicity gets smoothed out, so the degree of intermittency becomes significantly reduced. In Fig. 2 we show the result of ECCO applied to *O-Au*, *S-Au* and *Cu-Au* collisions at 20 and 200 GeV/n [21], where ϕ_q and d_q are as defined in (2) and (3). Note that the intermittency indices ϕ_q get smaller with larger beam nucleus and higher energy. But they seem to approach a nonvanishing limit for each q. The anomalous fractral dimensions are extremely small, perhaps too small for experimental determination. Nevertheless, the simulated result indicates that d_q does not become constant in q. That nonconstancy of d_q is relevant to the speculation that constancy may signal quark-hadron phase transition [22]. The latter can be a meaningful signal only if d_q is not a constant when there is no quark-gluon plasma formation. Our ECCO simulation, which does not assume plasma formation, lends support to the nonconstancy of d_q.

ACKNOWLEDGMENT

Most of the work that is reported here was done in collaboration separately with C.B. Chiu and J.C. Pan, to whom I am grateful for many invaluable discussions. The work was supported in part by the U. S. Department of Energy under Grant No. DE-FG-06-85ER-40224 and DE-FG-06-91ER40637.

References

[1] A. Bialas and R. Peschanski, *Nucl. Phys. B* **273**, 703 (1986).

[2] R.C. Hwa, *Phys. Rev. D***41**, 1456 (1990).

[3] C.B. Chiu and R.C. Hwa, *Phys. Rev. D***43**, 100 (1991).

[4] See papers by K. Sugano and by F. Rimondi in the *Proc. of Correlations and Multiparticle Production*, Marburg 1990, edited by M. Plumer, S. Raha and R.M. Weiner (World Scientific, Singapore, 1991); see also talks by H. Dibon, by N. Schmitz and by W. Kittel in this Workshop.

[5] W. Florkowski, talk given in this Workshop.

[6] R. C. Hwa, *Phys. Rev. D***44**, 915 (1991).

[7] C.B. Chiu and R.C. Hwa, OITS-466 (1991).

[8] C.B. Chiu, talk given in this Workshop.

[9] R. C. Hwa, G. Jansco, I. Derado, N. Schmitz, et al. (to be published).

[10] R.C. Hwa and J.C. Pan, OITS-478 (1991).

[11] W. Kittel, *Proc. of the 20th Int. Symposium on Multiparticle Dynamics*, Gut Holmecke, Germany 1990, edited by R. Baier and D. Wegener (World Scientific, 1991).

[12] A. Bialas and R. Peschanski, *Nucl. Phys. B***308**, 857 (1988).

[13] R.C. Hwa and J.C. Pan, OITS-463 (to be published in Phys. Rev. D).

[14] W.R. Chen and R.C. Hwa, *Phys. Rev. D***36**, 760 (1987); W.R. Chen, R. C. Hwa, and X.N. Wang, *Phys. Rev. D***43**, 2425 (1991).

[15] R. C. Hwa, *Phys. Rev. D***37**, 1830 (1988); W.R. Chen and R.C. Hwa, Phys. Rev D**39**, 179 (1989); X.N. Wang and R.C. Hwa, *Phys. Rev. D***39**, 187 (1989).

[16] R.C. Hwa and X.N. Wang, *Phys. Rev.* *D***39**, 2561 (1989).

[17] R.C. Hwa and X. N. Wang, *Phys. Rev.* *D***42**, 2459 (1990).

[18] R.C. Hwa, in *Hadronic Multiparticle Production*, edited by P. Carruthers (World Scientific, Singapore, 1988).

[19] G.J. Alner et al., *Phys. Lett.* *B***160**, 193 (1985); B**167**, 476 (1986).

[20] M. Adamus et al., *Z. Phys.* *C***37**, 215 (1988); I.V. Ajinenko et al., *Phys. Lett.* *B* **222**, 306 (1989);*B***235**, 373 (1990).

[21] R.C. Hwa and J.C. Pan, OITS preprint to appear.

[22] A. Bialas and R.C. Hwa, *Phys. Lett.* *B***253**, 436 (1991).

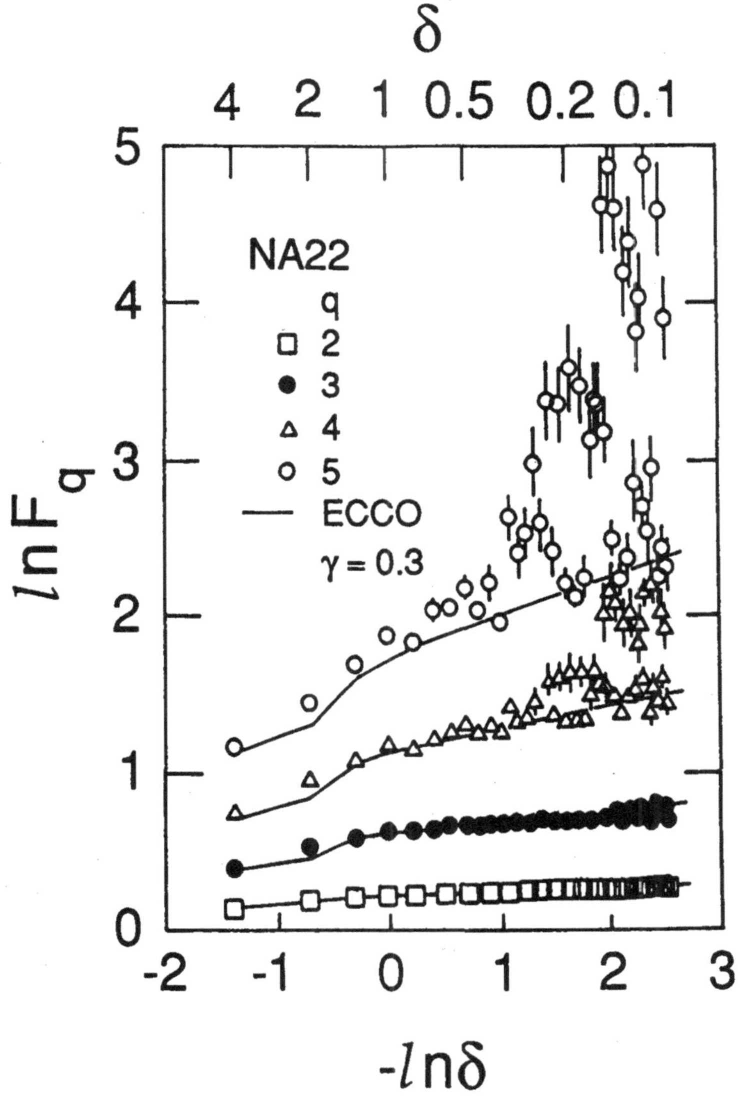

Fig. 1

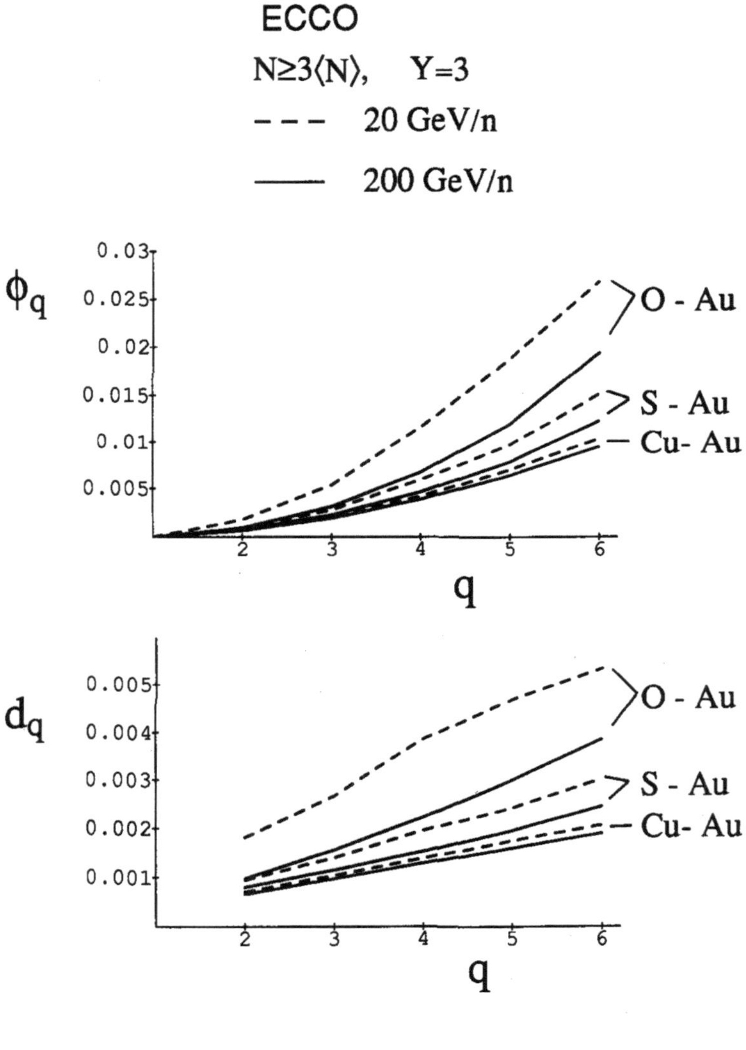
ECCO
N≥3⟨N⟩, Y=3
20 GeV/n
200 GeV/n
ϕ_q
0.03
0.025
0.02
0.015
0.01
0.005
O - Au
S - Au
Cu- Au
2
3
4
5
6
q
d_q
0.005
0.004
0.003
0.002
0.001
O - Au
S - Au
Cu- Au
2
3
4
5
6
q

Fig. 2

Status of Intermittency in the Two Component Dual-Parton-Model

F. W. BOPP
Fachbereich Physik, Universität Siegen
Siegen, Germany

about work done with
V. K. GUPTA[1] and J. RANFT[1]

using a Monte Carlo program written in collaboration with
P. AURENCHE[2], A. CAPELLA[3], K. HAHN[1], J. KWIECINSKI[4],
P. MAIRE[2], D. PERTERMANN[1] and TRAN THAN VAN[3]

ABSTRACT

The standard parametrisation does not completely reproduce the observed experimental "intermittency". It is, however, within a factor of two of the data. Considering the complex structure of short range correlations which depend on many details of the chain decay, we can not follow speculations about a completely new dynamical structure.

Intermittency and the Dual Parton Model

A central question is the dynamical significance of the "intermittency" seen in experimental data. For completeness, the main observation is the following: One divides the phase space in rapidity intervals of varying sizes and plots the logarithm of the k'th factorial moment in these intervals normalised to the k'th power of the corresponding particle density as function of the logarithm of the bin size. In the region roughly between rapidity intervals $\delta y = 0.1$ and $\delta y = 1$ one finds a linear dependence with a small but significantly non vanishing slope. In the context of branching or

[1]Sektion Physik, Universität Leipzig, Germany
[2]LAPP, Annecy, France
[3]L.P.T.H.E. Université de Paris XI, Orsay, France
[4]Institute of Nuclear Physics, Krakow, Poland

multi-step models such a dependence can be explained by self-similarity. Such a self-similarity is found in ("leading log") QCD for the cascading process of highly virtual partons in less virtual ones. A postulated "parton-hadron duality"[2] is thought to transfer the parton structure produced in this way to the observed hadrons.

It seems important to stress the following. "Intermittency" does not play a fundamental role in the Dual Parton Model as implemented in our Monte Carlo program. The model contains in the considered rapidity region dynamical significant correlations which invalidate the concept of parton-hadron duality. There can be no fundamental relation between the long-range processes and the not always hadronic decays of resonances responsible for short-range correlations.

The situation is not yet clear but there are two possible final conclusions. Either models with independent resonance decays are missing something important or fractal concepts are not applicable.

Let us make this point more vividly. The beautiful, but not truly old Ringberg castle somehow inspires to dream up a new variation of Andersen's old fairy tale, in which a little boy sneaks up the intermittency court and observes

there are beautiful clothes
and there is an emperor

but unfortunately they are not quite at the same place. There are new concepts of a new field of physics which have the beauty of being extremely generally applicable. But because of the resonance decays they are applicable only in a region above $\delta y > 1$ while the experimental data supporting the concept are in a region $\delta y < 1$.

Why Do Resonances Destroy Intermittency?

Let us begin with a simple relation between the generating function and factorial moments[3]. The generating function is defined as

$$\mathcal{G}_{\delta y}(x) = \sum_{n=0}^{\infty} x^n \sigma_n(\delta y) / \sum_{n=0}^{\infty} \sigma_n(\delta y)$$

where $\sigma_n(\delta y)$ is the probability to have n particles in the interval δy. The factorial moments can then be expressed in the following way:

$$< n(n-1) \ \dots \ (n-k+1) > = \left. \frac{d^n}{dx^n} \mathcal{G}_{\delta y}(x) \right|_{x=1} .$$

Very generally, for inclusive cross sections there has to be an intrinsic uncertainty, which determines the sensitivity of the cross sections against kinematic variations. For particles produced from decays of low-spin resonances without transverse momenta the intrinsic uncertainty in rapidity is about unity, i.e.: $\Delta y \approx 1$. It is narrower if the resonance has a transverse momentum which is not small against the momenta the secondaries obtain in the resonance decay.

Let us now consider two rapidity bins δy and $a\delta y$ both considerably smaller than the intrinsic uncertainty Δy. The probability that a particle from the interval δy also lies in the say smaller interval $a\delta y$ is then just a question of phase space, i.e. it is a times the original probability. The generating function for not loosing a particle during reduction is therefore

$$G_{\delta y \to a\delta y}(x) = ax + (1-a)\,.$$

As the generating function of the combined process is just the combined function:

$$\mathcal{G}_{a\delta y}(x) = \mathcal{G}_{\delta y}\left(G_{\delta y \to a\delta y}(x)\right),$$

the derivative formula yields immediately

$$< n(n-1)\ \ldots\ (n-k+1) >|_{a\delta y} = a^k < n(n-1)\ \ldots\ (n-k+1) >|_{\delta y}\ .$$

It means for intervals below the intrinsic uncertainty, i.e. for $\delta y < \Delta y$, the considered ratio

$$< n(n-1)\ \ldots\ (n-k+1) >_{\delta y} / < n >^k_{\delta y}$$

reaches a constant value independent of δy.

Resonances in the Dual Parton Model

The rôle of resonances in the Dual Parton Model is not simple. In dual models there exists an argument for **Parton Hadron Duality** below the kinematic region of perturbative QCD. This argument was utilised in a simple model [4, 5] which tried to obtain the needed correlation in a Mueller Regge description without explicitly including resonances.

The basic idea of the argument has its origin in low energy phenomenology. It is needed to explain the strong suppression of backward scattering. The relation between forward and backward scattering

$$\sigma^{\text{Regge contribution}}_{\pi^- p \to \pi^- p} >> \sigma^{\text{Regge contribution}}_{\pi^- p \to p \pi^-}$$

is on first sight surprising since Dual Model amplitudes can be expressed as a sum over s-channel resonances. For each resonance decay forward and backward scattering are obviously of the same magnitude. The backward scattering suppression is obtained as resonances appear in exchange degenerate pairs which contribute to backward scattering amplitudes with opposite sign.

The same argument can be used in multiparticle production for the string decay. Neighbouring particles in dual strings can be described as a sum over resonance amplitudes. As illustrated in figure 1 exchange degeneracy leads to a suppression of non planar amplitudes. The rank ordering of secondaries is not broken or in a parton language the influence of resonances on the final parton velocities is significantly reduced.

Fig. 1: Exchange degeneracy and parton hadron duality

Origin	$\sqrt{s}$	b_2	b_3	b_4	b_5
UA1	630	0.012±0.002	0.028±0.005	0.049±0.010	0.062±0.026
DPM	630	0.0065±0.0024	0.016±0.004	0.028±0.009	0.042±0.024
DPM $n \leq 15$	630	0.0050±0.0019	0.019±0.004	0.042±0.009	0.063±0.025
DPM $15 < n \leq 30$	630	0.0036±0.0018	0.015±0.002	0.035±0.005	0.061±0.009
DPM $n > 30$	630	0.0028±0.0025	0.009±0.004	0.017±0.005	0.039±0.009
DPM	22	0.017±0.002	0.040±0.004	0.076±0.011	0.102±0.034
DPM	200	0.007±0.002	0.018±0.004	0.032±0.008	0.049±0.026
DPM	1800	0.006±0.004	0.015±0.006	0.027±0.011	0.041±0.021

Table 1: Slopes b_i of the factorial moments as calculated from our Dual Parton Model compared to experimental data [8]. The energy is given in GeV. Only charged particles were considered.

The argument holds in our opinion for high mass resonances. For low mass resonances as they are typically considered in string fragmentation models the exchange degeneracy is known to be broken by higher order terms in a topological expansion [7] and changes in the phases of the amplitudes will appear, which destroy the cancellation. The inclusion of low mass resonances with an isotropic forward-backward symmetric decay is a method to parametrise this effect. It was consistent with a large number of observed short range order effects.

The Results of the Monte Carlo Calculation

In our paper on correlation [9] first results were given on the normalized factorial moments $F_i =< n(n-1) \ \ldots \ (n-k+1) > / < n >^k$ as function of rapidity bins which are reproduced in table 1. It contains the values of the slope b_i obtained in the fit $\ln F_i = a_i - b_i \ln \delta y$.

The first two lines compare the experimental data of the UA1 collaboration[8] for the interval $-1.5 < \eta < 1.5$ with the prediction for the interval $-3.0 < y < 3.0$. We obtain in the model positive slopes roughly half as large as the ones found experimentally. The values in table 1 result from a fit in the region $0 \leq \ln(1/\delta y) \leq 1.65$ using the

Origin	$\sqrt{s}$	b_2	b_3	b_4	b_5
DPM	22	0.0074±0.006	0.027±0.002	0.062±0.007	0.122±0.022
NA22	22	0.012±0.001	0.048±0.002	0.14±0.01	0.31±0.02

Table 2: Slopes b_i of the factorial moments in rapidity bins as calculated from the Dual Parton Model compared to experimental data of the NA22 collaboration[6] for the interval $-2.0 < y < 2.0$. The energy is given in GeV. Only charged particles were considered.

Origin	$\sqrt{s}$	b_2	b_3	b_4	b_5
DPM	22	0.0015±0.0011	0.0044±0.0044	0.035±0.015	0.055±0.046
NA22	22	0.007±0.002	0.028±0.005	0.09±0.02	0.22±0.06

Table 3 Slopes b_i of the factorial moments in azimuthal angle bins calculated from the Dual Parton Model compared to experimental data of the NA22 collaboration[6] for the interval $0 < \phi < 2\pi$. The energy is given in GeV. Only charged particles were considered.

usual not correlated statistical errors for the F_i moments only.

The next lines in table 1 concern the multiplicity dependence. We calculate the factorial moments for three regions in the charged multiplicity. In the low multiplicity region $n \leq 15$ we find like the UA1-Collaboration[8] the slopes b_i to be about twice as large as when using the full multiplicity region data. The model reproduces again about half of the observed slope.

The last lines investigate the energy dependence. As observed in the data we find a decrease of the slope with energy.

Lower energy data are considered in detail in a comparison with NA22 data[6] described in tables 2 and 3. For this comparison the cutoff parameter between hard and soft was reduced (2 GeV instead of 3 GeV at 630 GeV) and the difference between proton-antiproton- and meson-proton-scattering was ignored. Table 2 displays the results of the rapidity dependent analysis which show the usual discrepancy.

The NA22 collaboration also investigated the dependence on the azimuthal angle using the same method ($\delta\varphi$ replacing δy). Again we obtain positive slopes yielding somewhat more than half of the experimental value (Table 3).

The observed slopes get larger if the intervals are restrained both in δy and $\delta\varphi$. The observation fits nice in a fractal picture[1]. For the small intervals the information about the direction of the colliding particles can get lost and as there is no preferred direction the observation of the "true" slope requires a multidimensional analysis. The model does not support the optimism, the results follow the data within their usual discrepancy.

The estimate of the natural width of particles originating in resonance decays drastically depends on the size of transverse momentum of the parent resonance in relation

Origin	$\sqrt{s}$	b_2	b_3	b_4	b_5
DPM	22	0.015±0.0016	0.066±0.012	0.44±0.07	1.44±0.23
NA22	22	0.027±0.002	0.092±0.012		

Table 4 Slopes b_i of the factorial moments as calculated from the Dual Parton Model compared to experimental data of the NA22 collaboration[6] for grids of the size $\delta\phi = 2\pi/M$ and $\delta y = 4.0/M$ with varying M. The energy is given in GeV. Only charged particles were considered.

Origin	$\sqrt{s}$	b_2	b_3	b_4	b_5
DPM	22	0.0143±0.0012	0.0511±0.0049	0.094±0.0188	0.0473±0.015
NA22	22	0.007±0.001	0.029±0.004	0.07±0.01	-0.05±0.04

Table 5: Slopes b_i of the factorial moments in rapidity bins of particles with $p_\perp > 0.3$ as calculated from the Dual Parton Model compared to experimental data of the NA22 collaboration[6] for the interval $-2.0 < y < 2.0$ .The energy is given in GeV. Only charged particles were considered.

to the momenta obtained during the decay. To investigate the role of this effect the NA22 group considered secondaries within a restricted transverse momentum range.

The comparison with our model (Table 5) now changes and the slopes are within large errors comparable. The disagreement is therefore not simply a consequence of too small transverse momenta of primarily produced resonances as the discrepancy has its origin in the low $p_\perp$ region.

How to clarify the situation?

Something is not yet correct in the present implementation of the model. To find out where, is not easy and the intermittency data are in our opinion less then optimally helpful.

For the unknown distribution the true statistical error cannot be calculated. Even if on disregards crazy examples of numerically significant extremely rare events, the error to be reasonably expected for a finite sample is hard to control. Meaningful calculations are therefore quite demanding on computer time.

The highly correlated data points (for different bin sizes) tend to be misleading. To illustrate the point we copied part of a figure containing a GENCL Monte Carlo fit to the UA1 data points of the rapidity bin size dependence. We do not understand why the calculated moment (shaded area) turns down even though the Monte Carlo GENCL is known not to contain negative short range correlations.

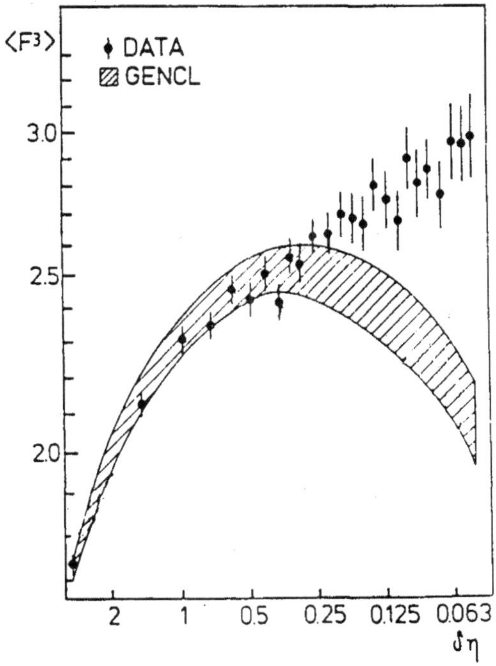

Fig. 2: The result of the GENCL Monte Carlo (shaded region) is compared with the UA1 data.

The best explanation we found - even the understanding of the Monte Carlo prediction is difficult - is the so called "truncation effect". This means that for an infinite sample the curve would turn flat and that the negative slope is just an artifact of the finite size of the sample.

To understand this effect one first has to realise that the fluctuation in F_3 arises from the numerator. Very often there will be no or almost no events in the sample having the necessary three particles within tiny bins considered ("empty bin effect") and eventually the moment will have to vanish. However, as the denominator is small for tiny bins there will be a significant contribution from the rare exceptions. In this way the average over samples can obtain the asymptotically required flat behaviour.

If this explanation is correct the standard way of analyses of the data drastically underestimates the statistical error. It would be correct to estimate the real statistical error from the deviation of the true flat distribution. The correct conclusion would be that the data are consistent with a flat curve within the statistical uncertainty determined from the discrepancy between the flat curve and the GENCL Monte Carlo.

For deciding where to adjust the model the usual divided two particle correlation

$$R_2(\delta y) = \frac{\int \rho_2(y_1, y_1 + \delta y)\, dy_1}{\int \rho_1(y_1)\rho_1(y_1 + \delta y)\, dy_1} - 1$$

seem more useful. As entries in neighbouring δy bins are uncorrelated the fluctuation allow to visualise the statistical uncertainty. If particle pairs are directly considered a logarithmic binning can be done in the δy range so far only reached in the intermittency data. The denominator can be calculated like the numerator just using consecutive events. An advantage of divided correlation is that the varying integration region presents no problem. Also there is the close connection to the intermittency ratio for which the discrepancy between model and experiment will have to be understood. It should be possible to ask additional questions – like wether the second particle belongs to the same or to the opposite hemisphere – to pin down exactly where and hopefully what additional correlations are needed.

References

[1] W. Ochs and J. Wosiek, Phys.Lett. B214 (1988) 617 and B232 (1989) 271.

[2] Y.I. Azimov et al., Z.Phys. C 27 (1985) Z.Phys. C 31 (1986) 213.

[3] see p.e. F. W. Bopp, Riv.Nouvo Cimento 1(1978) No.8.

[4] A.Capella, K.Fialkowski and A.Krzywicki, Orsay preprint LPTHE 89/21 (1989); Phys.Lett.B230 (1989) 149; discussions are thankfully acknowledged.

[5] P.Carruthers and I.Sarcevic, Phys.Rev.Lett.63,1562 (1989).

[6] I.V. Ajinenko et al. [EHS/NA22 Collaboration], Phys.Lett. B235 (1990) 373; W. Kittel, contribution to Santa Fee Workshop on intermittency in high–energy collisions, (1990).

[7] G.F. Chew and C. Rosenzweig, Phys.Rep. 41C (1978) No.5.

[8] C.Albajar et al. [UA1 Collaboration], Nucl.Phys. B345 (1990) 1.

[9] P. Aurenche et al., Z. Phys.C51 (1991) 99.

A Strategy for Multiparticle Dynamics *†

R.M. Weiner

Physics Department, University of Marburg, F.R. Germany
E.Mail: WEINER@DMRHRZ11

Abstract

The strategy consists in introducing a few characteristic quantities and express all observables, in particular fluctuations of multiplicities, in terms of these parameters. By comparing with experiment the phenomenological laws of multiparticle dynamics are derived. Quantum statistics is used to define these characteristic quantities. They are (besides the mean multiplicity) the chaoticity p and the correlation length ξ. It turns out that both p and ξ increase with energy.

1 Introduction

Multiparticle production is one of the most characteristic properties of strong interactions because it reflects the strong coupling nature of the underlying field theory which is assumed to be quantum chromodynamics (QCD). A theoretical approach to this chapter of physics must be phenomenological not only because QCD can be applied only in the perturbative regime but primarily because of the very nature of the problem which involves very many degrees of freedom. The situation is analogous to condensed matter physics where the underlying theory is well known (QED) but where microscopic concepts cannot replace macroscopic variables. Thus, far from being a deficiency and a provisional characteristic of multiparticle dynamics, the phenomenological nature of this theory is an intrinsic quality of it.

As in any phenomenology, the strategy is to find a few important physical quantities in terms of which most, if not all, observables can be expressed and to study these quantities - parameters - under various physical conditions, i.e. at various c.m. energies, for various particle reactions. The regularities which can be abstracted from such a study constitute in some sense the "laws" which govern this field and should in principle be derivable from a microscopic theory, much in the same way as certain "matter" constants and their dependence on state variables like temperature can be interpreted in terms of QED.

In the case of multiparticle dynamics the observables are of two types: average quantities like mean multiplicity, mean energy per particle etc. on the one hand and the fluctuations

*This lecture is dedicated to the 70th anniversary of my friend, collaborator and teacher G.N.Fowler in remembrance of our long standing collaboration. It is an attempt to summarize the newest developments in the realization of the program outlined in our joint paper [1].

†Invited talk at the Ringberg Workshop on Multiparticle Production, Fluctuation and Fractal Structure, June 1991

around these quantities on the other. These fluctuations are described by distributions like the multiplicity distribution $P(n)$ or energy (inelasticity) distribution $\chi(K)$, K being the inelasticity.
Fluctuations are, of course, a quite general phenomenon observed in many different chapters of physics and much experience and knowledge has been accumulated so far on this subject. It is then clear that any attempt to describe multiparticle dynamics has to use this knowledge, if it wants to avoid repetition.[1]
A particularly simple and powerful method for the description of fluctuations is represented by quantum statistics and is based on the density matrix approach. In this approach fluctuations are related to directly observable correlations. This method lies also at the basis of quantum optics including laser physics and it will be shown in the following that it can provide an economical and useful description also for phenomena observed in multiparticle dynamics.
As to the mean quantities, here, too, experience from other fields is available and has been used. In particular hydrodynamics and statistical thermodynamics may provide a framework for the description of these quantities in terms of temperature, which in its turn is related by an equation of state to the energy of the system. If hydrodynamics is not applicable, then either other more involved theoretical methods have to be applied or one just uses the experimental mean values. This last approach is used in the following.
The distributions which describe the fluctuations can be derived from the density matrix ρ. In particular

$$P_n = < n|\rho|n > . \tag{1}$$

It turns out that under certain circumstances the parametrization of the density matrix is easier than that of other fundamental quantities like the S matrix and this explains the wide applications the density matrix approach (i.e. the quantum statistical approach) has found among other things in optics. For identical particles this formalism can be taken over into hadron multiparticle production provided the energies are high enough so that the rest masses of particles and conservation laws can be neglected.
An important tool in this approach is the coherent state formalism which is particularly useful when coherence effects are present. Such coherence effects which, up to a certain point, correspond to classical fields and/or condensates exist in present theories of particle physics, although the experimental evidence for these effects is still disputed. A program for multiparticle dynamics based on the above ideas was outlined in ref. [1] where effects, characteristic for this type of approach, were predicted. Some of these effects were later found experimentally, others still await their confirmation, in particular a more direct evidence of coherence in Bose-Einstein correlations.

---

[1] Nevertheless, as we shall see in the following, sometimes such "rediscoveries" occured.

2 Quantum Statistical Approach to Correlations and Fluctuations of Coherent and Chaotic Fields

There are two extremes in the QS approach to fluctuations: "order" represented by coherent fields π_c and "disorder" represented by chaotic fields π_{ch}. In most cases known in optics, all fields $\pi(y)$ can be represented by a linear superposition of these two fields:

$$\pi(y) = \pi_c(y) + \pi_{ch}(y). \tag{2}$$

It appears reasonable to assume, as is done in most applications discussed below that the same superposition principle holds also in multiparticle production. An exception to this is discussed in section 4.5. y stands for several variables which characterize the field, either in momentum or configuration space. The difference between chaotic and coherent fields lies in the correlations of these fields. Thus for chaotic fields we have

$$< \pi^{\dagger}_{ch}(y_1)\pi_{ch}(y_2) >= Tr\left[\rho\pi^{\dagger}(y_1)\pi(y_2)\right] = \gamma(|y_1 - y_2|/\xi) \tag{3}$$

where ξ is the correlation length is in y space.
Eq. (3) expresses stationarity of the (chaotic) fields. If this condition is not satisfied in the variable y, then another variable, in terms of which (3) is satisfied has to be looked for. The simplest procedure in this case is to "average" out the non-stationarity or, equivalently, to consider narrow enough $|y_1 - y_2|$ intervals (cf. below).
For coherent fields, on the other hand, one has

$$< \pi^{\dagger}_c(y_1)\pi_c(y_2) >=< \pi^{\dagger}_c(y_1) >< \pi_c(y_2) >, \tag{4}$$

i.e. factorization.
The fields π are related to the multiplicities via

$$< |\pi|^2 >=< n > . \tag{5}$$

Then it is useful to characterize the chaoticity by the parameter

$$p =< n_{ch} > / < n >, \tag{6}$$

where

$$< n >=< n_c > + < n_{ch} >, \tag{7}$$

and the indices c and ch refer to coherence and chaos as defined above.
The distribution P_n or rather its moments can be expressed as integrals over the correlation function. Thus e.g. the second factorial moment is

$$\begin{aligned} < n(n-1) > &= \int \rho_2(k_1, k_2) dk, dk_2 = \\ &= \int C_2(k_1, k_2)\rho_1(k_1)\rho_1(k_2) dk_1 dk_2, \end{aligned} \tag{8}$$

where C_2 is the two particle correlation function:

$$C_2 = \frac{\rho_2(k_1 k_2)}{\rho_1(k_1)\rho_1(k_1)}. \tag{9}$$

ρ_1 and ρ_2 are one and two particle inclusive cross sections. Therefore it is reasonable to start our discussion with the correlation functions. In the following we shall review the developments in this field since 1989. As will be seen, some quite new and surprising facts emerged in this short period.

3 Correlations of identical bosons

Bose-Einstein Correlations (BEC)

3.1 Surprises due to the space-time approach in quantum statistics

The space-time approach has certain advantages as compared with the momentum (rapidity) approach because it does not assume stationarity in momentum space but rather in configuration space which is expected always to hold, as long as $|x_1 - x_2| < R$. Here x_i denote space-time coordinates and R is the radius (lifetime) of the source. We write thus the correlator for the fields

$$\begin{aligned} < \pi_m^\dagger(x_1)\pi_n(x_2) > &= \delta_{mn}\, C(x_1 - x_2), \qquad m,n = 1,2,3 \\ C(x_1 - x_2) &= C(0)\exp\left[-(x_1 - x_2)_0^2 L^2 - (\vec{x_1} - \vec{x_2})^2/L^2\right], \end{aligned} \tag{10}$$

where we have assumed a Gaussian parametrization and defined the configuration space correlation length L. It is clear that the correlation length L, as defined above, is different from, but related to, the correlation length ξ introduced in eq. (3) which refers ussually to rapidity space. The indices m, n refer to the isospin components 1, 2, 3 of the pion iso-triplet:

$$\pi_\pm = (\pi_1 \pm i\pi_2)/\sqrt{2} \qquad \pi_0 = \pi_3. \tag{11}$$

The space-time distribution of the source is for reasons of mathematical convenience also parametrized by a Gaussian:

$$\rho(x) \sim \exp(-x^2/R^2). \tag{12}$$

Define now the pion correlator in momentum space:

$$F(k_1, k_2) = (2\pi)^3\sqrt{4E_1E_2} < \pi_m^\dagger(k_1)\pi_m(k_2) >, \tag{13}$$

and

$$F(-k_1, k_2) = -(2\pi)^3\sqrt{4E_1E_2} < \pi_m(k_1)\pi_m(k_2) > . \tag{14}$$

In terms of the correlator (10) and the space-time distribution of the source (12) F can be written:

$$F(k_1,k_2) = \int \frac{d^4k}{(2\pi)^4} f^*(k_1+k)C(k)f(k+k_2) \quad k_r^0 = E_r, r = 1,2. \tag{15}$$

The advantage of the configuration space approach can now be seen immediately: With the standard assumption about f and C, it follows from (15) that not only (13) but also (14) does not vanish. This is at a first look surprising because in the assumption of stationarity in momentum space, (14) would vanish. This explicitly proves that this last assumption cannot hold in general and it leads to observable, although small, new effects (cf. ref. [2]).

Indeed from eqs. (13)-(15) follows:

$$\begin{aligned} C_2^{--} = C_2^{++}(\vec{k_1},\vec{k_2}) = 1 + |d_{12}|^2; \quad C_2^{+-}(\vec{k_1},\vec{k_2}) = 1 + |\tilde{d}_{12}|^2 \\ C_2^{-0} = C_2^{+0}(k_1,k_2) = 1; \quad C_2^{00}(\vec{k_1},\vec{k_2}) = 1 + |d_{12}|^2 + |\tilde{d}_{12}|^2. \end{aligned} \tag{16}$$

Here $+,-,0$ refer to positive, negative and neutral pions and d_{rs}, $\tilde{d}_{rs}$ are the normalized correlators

$$d_{rs} = \frac{F(k_r,k_s)}{[F(k_r,k_r)F(k_s,k_s)]^{\frac{1}{2}}} \qquad \tilde{d}_{rs} = \frac{F(k_r,-k_s)}{[F(k_r,k_r)F(k_s,k_s)]^{\frac{1}{2}}} \tag{17}$$

Eqs. (16) contain a few "surprises":

- The (BEC) for identical charged pions is different from those of neutral pions. In particular $MaxC_2^{00} = 3$ while $MaxC_2^{--} = 2$. [2]
- There exists a QS correlation C_2^{+-} between positive and negative pions.

With the parametrizations (10),(12) we have

$$\begin{aligned} d_{rs} &= \exp\left(-\frac{(\vec{k_1}-\vec{k_2})^2R^4}{8(R^2+L^2)}\right)\exp\left(-\frac{(E_1-E_2)^2R_0^4}{8(R_0^2+L^2)}\right) \\ \tilde{d}_{rs} &= \exp\left(-\frac{(\vec{k_1}+\vec{k_2})^2R^4}{8(R^2+L^2)}\right)\exp\left(-\frac{(E_1+E_2)^2R_0^4}{8(R_0^2+L^2)}\right). \end{aligned} \tag{18}$$

Here R_0 is the lifetime of the source of radius R.

It results that the "new" terms $\tilde{d}_{rs}$ are in general suppressed as compared with the "old" ones d_{rs} and that this suppression can be partially overcome for soft pions. The importance of the new terms $\tilde{d}$ lies in the fact that (i) they present two particle coherent states $< aa >$, i.e. squeezing [3] in the same way as terms of the form $< a >$ represent single particle coherent states; (ii) they prove explicitly that Bose-Einstein correlations are more complex than would be expected from the two-particle symmetrized wave function approach which does not lead to particle-antiparticle correlations; (iii) the normalization procedure in terms of $(+-)$ pairs which is used in experimental BEC studies has to be qualified for soft pions.

[2] It appears as if neutral pions would be "more equal" than charged ones.

Another consequence of the space-time approach (eq.(18)) is the fact that it provides a clear relation between the correlation length L and the geometrical radius R. Furthermore it appears that (BEC) do not measure R (as was thought so far) but rather a combination of R and L, i.e.

$$r_L^2 = \frac{R^4}{R^2 + L^2}. \tag{19}$$

The effects discussed above refer to chaotic fields. In the presence of coherence supplementary "surprises" emerge. Indeed it can be shown that the presence of coherence manifests itself not only through the modification of the intercept $C_2(0)$ but also in the dependence of C_2 on k_1, k_2.

While almost all experimental studies before 1989 considered an one exponential structure of C_2 and tried to estimate the chaoticity by extrapolating C_2 to the origin, QS in fact predicts in the presence of coherence a two-exponential structure, the two exponents being related in a very definite way. This result had in fact been pointed out already in 1978 by Biyajima, Miyamura, Nakai [4] but its relevance for the experimental determination of chaoticity was emphasized only much later [5]. Its practical importance lies among other things in the fact that the determination of the intercept $C_2(0)$ is hampered by the experimental difficulty of measuring very small values of $|\vec{k_1} - \vec{k_2}|$. The two-exponent structure shifts in part the burden of measurements to larger values of $|\vec{k_1} - \vec{k_2}|$, where these difficulties are present to a lesser degree. The first experimental attempt to detect the two-exponent structure has been reported in ref. [6]. Because of the limited statistics, however, no conclusive results could be obtained. Progress along these lines may be facilitated by the most recent development on this subject when it has been realized that the two-exponential structure may be observable only for soft pions [7],

Indeed, by using eqs. (3)-(4) we get

$$\begin{aligned} C_2(k_1,k_2) &= 1 + 2p(0)\,[1 - p(0)] \exp\left[-\frac{1}{4}(k_1^2 + k_2^2) r_L^2\right] \cdot \\ &\cdot \exp\left[-\frac{1}{8}(k_1 - k_2)^2 r_L^2\right] / \left[A(k_1)A(k_2)\right] \\ &+ p^2(0) \exp\left[-\frac{1}{4}(k_1 - k_2)^2 r_L^2\right] / \left[A(k_1)A(k_2)\right]. \end{aligned} \tag{20}$$

The metric is Euclidian, i.e. $k^2 = \vec{k}^2 + k_0^2$ and

$$A(k_i) = p(0) + [1 - p(0)] \exp\left[-\frac{1}{2} k_i^2 r_L^2\right]. \tag{21}$$

Here

$$p(0) = \frac{R_L^4 C(0)}{R_L^4 C(0) + R^4 |\pi_0|^2}, \quad R_L^2 = \frac{R^2 L^2}{R^2 + L^2}, \quad \pi_0 = \pi_c(0). \tag{22}$$

For simplicity in (20) we assumed $R = R_0$.

Eq. (20) contains the new elements mentioned above: There are two exponentials which differ by a factor of 2 and the weights of which are in the ratio $2p(0)[1-p(0)]/p^2(0)$. The radius and the correlation length appear again in the combination r_L^2 and, last but not least, the first exponential, which contains the interference between the coherent field $(1-p(0))$ and the chaotic field $p(0)$ is multiplied by a suppression factor $exp(-\frac{1}{4}(k_1^2+k_2^2)r_L^2)$.

3.2 Multiplicity dependence of BEC due to quantum statistical constraints

Another application of QS to BEC lies in the dependence of BEC on the multiplicity [8]. As is well known, the operators for the field (intensity) and number of particles do not commute. This means that measurements of "true" BEC can be performed only when no restriction on the multiplicity n, which fluctuates from event to event, is made. In practice, however, such restrictions are imposed, either simply because to measure a 2 identical particle correlation, one has to select events with at least 4 charged particles, or because of theoretical prejudices (centrality). This introduces a bias which has been estimated in [8] and we refer the reader to this reference for technical details. We quote here only some of the results (cf. Fig. 1):

1. C_2 decreases noticeably with the cut in n, n_{min}. This effect is more pronounced the larger n_{min} is.
2. C_2 at fixed n and at a given value of $|y_1-y_2|$ is always smaller than the corresponding unbiased C_2.

These results illustrate again the importance of a consistent QS treatment of BEC. The "old fashioned" two-particle wave function approach cannot yield similar results.

4 Multiplicity Distributions and Quantum Statistics

4.1 Rapidity dependence - "Beta" scaling

One of the straightforward consequences of QS applied to multiplicity distributions P_n is the prediction contained already in [1] of the broadening of P_n with the decrease of the width of the (symmetric) rapidity window δy. The surprise connected with the discovery of this effect in subsequent experiments at the CERN collider by the UA-5 collaboration was just due to the fact that the QS result was practically unknown to the high energy physics community and new attempts to interpret it were obviously condemned to repetition of old derivations. Indeed, the observed rapidity dependence follows from the dependence of the factorial cumulants μ_r on δy

$$\mu_r = \frac{<n>^r}{\nu^{r-1}}\left[(r-1)!p^r B_r + r!p^{r-1}(1-p)\tilde{B}_r\right]. \tag{23}$$

In this relation derived in quantum optics by Jaiswal and Mehta in 1970 and proposed for the description of multiparticle dynamics already in [1] ν is the number of "cells", $\nu = mf$, where m is the number of independent charge states and f the number of independent

sources, p is the chaoticity (the momentum dependence of p is not considered here) and $B, \tilde{B}$ are analytical functions of $\beta \simeq \delta y/\xi$. The form of $B, \tilde{B}$ determines the δy dependence of μ and the fact that μ does not depend on δy and ξ separately but rather on β was called "beta" scaling [9].
The functions B and $\tilde{B}$ can be calculated from the formulae:

$$B_r = \frac{1}{(\delta y)^r} \prod_{j=1}^{r} \int_0^{\delta y} dy_j \gamma(y_1, y_2) \cdots \gamma(y_{r-1}, y_r) \gamma(y_r, y_1) \tag{24}$$

$$\tilde{B}_r = \frac{1}{(\delta y)^r} \prod_{j=1}^{r} \int_0^{\delta y} dy_j \gamma(y_1, y_2) \cdots \gamma(y_{r-1}, y_r) \tag{25}$$

where the functions γ are defined by eq. (3).
Eq. (23) assumes stationarity in y and it is understood that the other variables (e.g. p_T) are averaged out. Eq. (23) reduces for $p = 1$ and $\beta = 0$ to
$\mu_r =< n >^r (r-1)!/\nu^{r-1}$ which is the negative binomial result, while for $p = 0$, $\mu_r = 0, r \geq 2$ we get the Poisson distribution. We see thus that the negative binomial (n.b.) distribution, which made headlines in the mid-eighties is a particular case of the distribution (23) put forward in 1978. Moreoever the QS formalism appears superior to the n.b. approach not only because it is more general but also because the theoretically not understood δy dependence of the n.b. parameter k has a very clear and definite meaning in QS being just a manifestation of the two-particle correlations γ which enter the functions B (linked-pair correlations) and ($\tilde{B}$) (ring-pair correlations).
A particular case of the rapidity dependence of μ_r is the forward-backward (F-B) correlation. It could be shown [9] that a consistent description of the experimental data of hadronic, semihadronic and e^+e^--reactions involving both the δy dependence and the F-B correlation is possible within this QS frame.

4.2 Quantum statistics and intermittency

The rapidity dependence of moments of multiplicity distributions has become in the last five years a testing ground for another interesting phenomenon, namely intermittency [10]. Intermittency leads to a power law dependence of the normalized factorial moments

$$\Phi_r = F_r / < n >^r \sim (\delta y)^{-\alpha}. \tag{26}$$

where α is an increasing function of r. These are related to the factorial cumulants μ_r by the relation

$$\exp\left[\sum_{n=1}^{\infty} \frac{\mu_n S^n}{n!}\right] = \sum_{k=0}^{\infty} \frac{F_k S^k}{k!} \tag{27}$$

which is to be understood as an identity in S. Given the fact that QS predicts qualitatively the same behaviour as that contained in (26), i.e. a decrease with δy which becomes more pronounced the higher the order r of the moment, it was interesting to check up to which point the predictions of QS and intermittency coincide. It is clear, however, that for very small values of δy eqs. (26) and (24) must differ because μ_r contains a finite scale in rapidity, ξ, while (26) assumes the absence of such a scale. The presence of an ultimate scale leads to a saturation of $ln\Phi_r$ as a function of $-ln\delta y$ (cf. Figs. 2,3). In ref. ([11]) a comparison of the prediction of QS represented by eq. (24) and data was performed for hadronic and nuclear reactions. Before discussing these results it is important to note that eq. (24) assumes among other things: (i) identical particles, (ii) stationarity in rapidity. On the other hand the data available in 1988, when this comparison was performed, referred to charged particles and the condition of stationarity was fulfilled only for the p-p data in the very central region. Nevertheless, as can be judged from Figs. 2,3, the agreement between eq. (23) and data is good, including the tendency of saturation. This was the first conventional explanation of the intermittency effect, and it was substantiated further in [12] where the importance of a three-dimensional analysis was emphasized. However, in [12] an apparently obvious "test" of the QS mechanism in intermittency was proposed and this test turned out to be negative. The test assumes that the QS mechanism is based entirely on Bose-Einstein correlations and therefore the intermittency slopes for identical particles α_{id} should be twice as large as those for charged ones α_{ch}. Experimental data on the other hand showed that $\alpha_{id} \simeq \alpha_{ch}$. This apparent failure of the QS interpretation of the intermittency data may have several explanations:

(i) it could mean that the intermittency phenomenon is real and that the hadronic scale ξ is, for reasons not yet understood, absent in the multiplicity data. However, before reaching this conclusion, other more conventional explanations have to be considered, within the QS framework. These are:

(ii) there exist correlations due to strong interactions between the charged particles (pions) which are similar to those between identical particles. This theoretical possibility was discussed already in ref. [9] in connection with the observation by the Na22 collaboration that the effective negative binomial parameter for charged data k_{ch} is related to the effective k of identical particle data k_{id} by the relation

$$k_{ch} = \frac{1}{2} k_{id} \tag{28}$$

rather than the naive expectation $k_{ch} = 2k_{id}$ which assumes independent charges. Support for this possibility comes from a recent investigation by Shih [13] which will be discussed in the next section.

(iii) the quantum statistical correlations between $\pi^{+'}s$ and $\pi^{-'}s$ discussed in section 3.1 (ref. [2]) contribute to the intermittency effect and thus the naive expectation

$$\alpha_{id} = 2\alpha_{ch} \tag{29}$$

does not hold. Indications in this direction come from the observation that the intermittency effect is strongly enhanced for soft pions [14] where also the QS $\pi^+\pi^-$ correlation is enhanced. A quantitative investigation of this possibility is under way [15].

4.3 Charge correlations and quantum statistics

The difference between quantum statistics as applied to optics and that specific for multiparticle dynamics lies in the nature of strong interactions. One of the consequences of strong interactions is the existence of charge correlations due e.g. to resonances. We mentioned in the preceding section the importance of this effect in the QS interpretation of multiplicity distributions. In this section we shall briefly outline an approach to this problem due to Shih [13] which could serve as an explanation why despite the existence of strong charge correlations, the multiplicity distributions of charged particles are nevertheless similar to those of identical particles.

The idea consists in trying to parametrize the charged field correlations in the same way as the identical field chaotic correlations, the difference being concentrated in the strength factor multiplying the corresponding correlations. One extends thus eq. (3) by writing

$$\Gamma(y_1, y_2; --) = \Gamma(y_1, Y_2; ++) = \frac{1}{A}\Gamma(y_1, y_2; +-) = \Gamma_0 \exp(-|y_1 - y_2|/\xi). \tag{30}$$

The chaoticity p and the correlation length ξ are assumed to be the same for charged and identical particles. This is one of the simplest possible generalizations of QS, demanding only one parameter more than in the case of identical particles, namely the specific charge-correlation strength A, and surprisingly enough it appears sufficient to describe the available intermittency data, both for charged and identical data. This is seen in Fig. 4. The fact that the parameter A turns out to be much larger than unity confirms the expectation that charge correlations are much stronger than identical particle correlations. Given the simplicity of the above ansatz, the phenomenological success of fitting both charged and identical data within the same QS formalism, with only 3 parameters (p, ξ, A) is quite remarkable and suggests that the quoted multiplicity data, at least in one dimension y, can be understood without invoking the intermittency conjecture.

4.4 Three dimensional analysis and non-stationary effects

The importance of a three-dimensional analysis of the multiplicity data was emphasized in ref. [12] and in [17] as well as in all subsequent experimental studies (cf. e.g. this volume). Furthermore the role of the stationarity assumption in rapidity in the QS approach had been realized from the very beginning [1]. The only support for this assumption lies in the presumed flatness of the rapidity distributions. As a matter of fact this flatness is restricted to the very central region. Moreover, it is only a necessary condition for stationarity as can be readily seen from eq. (3) which correlates two points y_1, y_2, rather than referring only to the particular case $y_1 = y_2$ which enters the rapidity distribution. A first investigation of the role of a three-dimensional analysis and of the stationarity assumption in the QS analysis of the multiplicity data has been performed in ref. [18].

One generalizes eq. (3) by the ansatz

$$< \pi^\dagger(y_1, p_{T1})\pi(y_2, p_{T2}) >= \gamma_y(y_1, y_2)\gamma_T(\vec{p}_{T1}, \vec{p}_{T2}). \tag{31}$$

This is still not quite general because it assumes factorization in (y, p_T). Partial support for this assumption comes from the observation that in a first approximation the single inclusive cross section factorizes.

Eq. (23) retains its form but the functions $B, \tilde{B}$ of eqs. (24),(25) are replaced by

$$B_r = B_{yr} B_{Tr}, \qquad \tilde{B}_r = \tilde{B}_{yr} \tilde{B}_{Tr} \tag{32}$$

where B_{yr} are the expressions (24),(25) and $B_{Tr}, \tilde{B}_{Tr}$ are the corresponding integrals in p_T space. γ_T is either parametrized by a Gaussian

$$\gamma_T \sim \exp(-|\vec{p}_{T1} - \vec{p}_{T2}|^2/\xi_T^2) \tag{33}$$

or an exponential

$$\gamma_T \sim \exp(-|\vec{p}_{T,1} - \vec{p}_{T,2}|/\xi_T). \tag{34}$$

The differences between these two forms appear to be small in applications. A further generalization of the QS treatment performed in [18] refers to non-stationary processes in rapidity. This problem has two aspects: (i) the change of multiplicity distribution with the change of the width δy of the symmetric rapidity window, (ii) the change of $P(n)$ with the shift of the rapidity window. Property (i) is already a consequence of the simplified formalism (23). To account for (ii) two approaches were used. The first interpreted the effect (ii) in terms of two independent sources, one completely coherent and another completely chaotic. A discussion of this mechanism will be given in the next section. The other approach attempts to describe this effect again by a linear superposition of a coherent and chaotic field (eq. (2)) as described below. This second approach applies as long as the deviation from stationarity is small. One generalizes the function $\gamma(y_1 - y_2)$ which enters eq. (3) by writing instead

$$\gamma(y_1, y_2) = \sqrt{p(y_1)p(y_2)\frac{dn}{dy_1}\frac{dn}{dy_2}} \exp(-|y_1 - y_2|/\xi). \tag{35}$$

The dependence of the chaoticity on rapidity reflects the fact mentioned above that multiplicity distributions change their shape when shifting the rapidity window. In particular it is known that in the center the distribution is broader than at large y. This is also reflected by the forward-backward correlation. As to the rapidity distribution dn/dy it is taken from experiment, considering the deviations from the plateau. After comparing this generalized QS formalism with data, one finds [18] that the generalized QS formalism leads to a "renormalization" of the parameter p and ξ. In particular an increase of p and a slight decrease of ξ is obtained, as compared with the one dimensional, stationary formalism. Agreement with the data for identical particles of the Na22 collaboration is satisfactory as can be seen from Fig. 5.

4.5 Two-component approach to multiplicity distribution

As mentioned in the previous section, an alternative interpretation of the change of $P(n)$ with the position of the rapidity window was given [19] in terms of two independent sources, one concentrated in the center and totally chaotic described by a negative binomial distribution P_{NB} and another concentrated at larger rapidities and completely coherent, described by a Poisson distribution P_P:

$$P(n) = \sum_{n_a,n_b} P_P(n_a)P_{NB}(n_b)\delta(n - n_a - n_b). \tag{36}$$

With this conjecture not only the multiplicity distribution in shifted rapidity windows could be explained but also the forward-backward asymmetry [20].
The remarkable outcome of the comparison of the consequences of this ansatz with data at various energies was that the mean multiplicity n_a emerged as a slow (logarithmic) function of s suggesting a bremsstrahlung-like mechanism while n_b appeared as a much stronger function ($s^{1/4}$) consistent with the expectations of the Landau hydrodynamical model. This consequence illustrates quite clearly the double-prong strategy outlined in the introduction for a description of fluctuations via QS on the one hand, and mean values via a hydrodynamical model on the other. Given the possible important implications of this result a more microscopic derivation of the two component model sketched above appeared desirable. Such a derivation was given in ref. [21] in terms of a stochastic branching equation:

$$\begin{aligned}\frac{\delta}{\delta t}P(n_a, n_b; t) &= \mu_0\,[P(n_a - 1, n_b; t) - P(n_a, n_b; t)] \\ &+ \lambda_0\,[P(n_a, n_b - 1; t) - P(n_a, n_b; t)] \\ &+ \lambda\,[(n_b - 1)P(n_a, n_b - 1; t) - n_b P(n_a, n_b; t)]\end{aligned} \tag{37}$$

where $\mu_o, \lambda_0, \lambda$ are production rates in the interval dt (cf. Fig. 6).

5 Conclusions

Energy dependence of multiplicity distributions in the quantum statistical approach

As stated in the introduction one of the main goals of the strategy outlined in this review is to determine from experiment the "laws" which govern multiparticle dynamics, i.e. the energy dependence of the main physical observables. The choice of these observables is dictated by very general considerations based on QS. In particular in our case these observables for identical particles are the mean multiplicity $< n >$, the chaoticity p and the correlation length L (ξ). The implementation of such a program is unfortunately hampered at present for hadronic reactions by the absence of data for identical particles at higher energies. The only available data for both charged and identical particles is due to the Na22 collaboration at $\sqrt{s} = 22$ GeV. On the other hand we have seen that (i) at this (rather low) energy both charged and identical data can be described by the QS formalism, the difference between charged and identical data being taken care of by the specific charge correlation parameter A (cf. eq. (32)); (ii) the charged data can be described at <u>all</u> energies available (up to $\sqrt{s} = 900$GeV) with the same QS formalism. This encouraged us to study the energy dependence of multiparticle production as reflected in the multiplicity distributions by the QS formalism and the results of this study will be presented below. Given the necessarily preliminary character of such a study, due to the absence of identical particle data, we have neglected, in a first approximation, the corrections for non-stationarity and three-dimentional kinematics. This also implies that the correlations are defined in rapidity space (eq.(3)) and not in configuration space. [3] Another aspect of the energy dependence is related to the fact that most multiplicity data refer to the central rapidity region. It is known that in this region the inelasticity K is also energy dependent. This means that the energy available for particle production

$$w = K\sqrt{s} \tag{38}$$

which determines the observables $< n >, p, \xi$ is a rather involved function of $\sqrt{s}$. This fact was taken into account in the analysis of the data.
The results of this analysis [22] are contained in the Figures (7-9) where the three characteristic quantities are represented as a function of s. In these figures predictions for Tevatron and SSC energies are also made. Besides the previously known increase of the mean multiplicity $< n >$ with s, the remarkable feature of these results is that both the rapidity correlation length ξ and the chaoticity p increase quite substantially with s. The absolute values of p as given in Fig. 7 are, of course, not realistic because the p_T dependence was not considered (cf. section 4.4). This omission amounts to the fact that the p values of Fig. 7 are strongly underestimated. They have to be "renormalized" by a factor of order 2. Nevertheless we believe that the overall s dependence is correct and reflects the broadening with s of the multiplicity distribution (deviation from KNO scaling). It appears that at low ($\sqrt{s}$ = 22GeV) energies, at least the chaoticity $p << 1$, i.e. there is an appreciable amount of coherence. This conclusion, which follows from the analysis of multiplicity data is consistent with the results of Bose-Einstein correlations obtained with the same Na22 data, although the interpretation of BEC still suffers from many ambiguities.One of the main tasks for future theoretical and experimental work is the elucidation of the role of coherence in Bose-Einstein correlations and a more detailed joint analysis of multiplicity and BEC data.
Since the chaoticity is an order parameter limited, by definition from above by 1, Fig. 7 gives an indication at what energy the distribution will become completely chaotic and beyond which the present trend will have to change. (The energy at which p reaches its maximum corresponds to a phase transition.)

This work was supported in part by the Federal Minister of Research and Technology under the contract number 06 MR 713 and Deutsche Forschungsgemeinschaft.

Figure Captions

1. Second order BEC as a function of $y_g = |y_1 - y_2|$ for a purely chaotic distribution ($p = 1$). The successive dashed curves represent the analytical result of the unbiased correlation function as given by QS [5]. In Fig. 1a the data points represent the

[3]The relation between the correlation length in momentum (rapidity) space and configuration space is the subject of present investigations. Preliminary results are presented in I.V. Andreev and R.M. Weiner, Multiparticle Dynamics 1990, eds. R. Baier, D. Wegener, World Scientific, p. 519

results of a Monte Carlo simulation without any bias on n. The other parts of Fig. 1 represent the results of the MC simulation with various fixed n_{min} (from ref. [8]).

2. Normalized factorial moments ϕ_q of order q in finite (pseudo) rapidity windows of width δy centered around $y_{CMS} = 0$, plotted against δy or $\delta\eta$ (pseudorapidity) for : (a) $p\bar{p}$ collisions at $\sqrt{s} = 540\text{GeV}$,(η) and (b) pp collisions at $\sqrt{s} = 22$ GeV(y). The order q is indicated next to the curves. These have been computed using eq. (23) with: (a) $p = 0.44$ and $\xi = 4.0$ and (b) $p = 0.32$ and $\xi = 1.0$ (from ref. [11]).

3. Averages of normalized factorial moments ϕ_q of order q in finite rapidity windows of width δy shifted across a wide rapidity range in each event, plotted against δy or $\delta\eta$ for: (a) pp collisions at $\sqrt{s} = 22\text{GeV}(y)$, and (b) $^{32}S + Au$ collisions at 200 A GeV (η). The order q is indicated next to the curves. These have been computed using eq. (24) with : (a) $p = 0.20$ and $\xi = 0.9$ and (b) $p = 0.015$ and $\xi = 0.2$ (from ref. [11]).

4. Factorial moments $f_{q,ch}$ and $f_{q,-}(q = 2-4)$ as a function of the rapidity window δy $(|y| < \delta y/2)$. (from ref. [13]). Dashed curves: $C = 2, p = 0.17, \xi = 0.70$. Solid curves: $C = 3.5, p = 0.11, \xi = 0.65$. Data points are experimental values obtained by the Na22 collaboration.

5. Comparison of experimental data on negative particles from Na22 experiment with values derived from Eqs. (31) through (35). Data points of the normalized factorial moments $f_q(q = 2, 3, 4)$ are derived from the experimental negative binomial fits of Na22 data [16] plotted against the rapidity window Y on a log-log plot, with $|y| < Y$ (from ref. [18]).

6. Typical evolution processes where two types of particles (partons) a and b are produced by a virtual source (s). The production vertices are characterized by couplings μ_0 and λ_0 respectively. The bottom process corresponds to the evolution (branching) of a single particle (partons) b. The corresponding production vertex is characterized by a coupling λ.

7. s-dependence of p. Errors are shown for values extrapolated to 2 TeV and 40 TeV from ref. [22]).

8. s-dependence of the correlation length ξ. Errors are shown for values extrapolated to 2 TeV and 40 TeV (from ref. [22]).

9. s-dependence of $< n >_{ch}$. Experimental data shown as diamonds; extrapolations are shown as asterisks; extrapolations according to the quadratic expression in lns are shown in squares (from ref. [22]).

10. s-dependence of normalized initial moments C_q of the charged multiplicity distribution for $q = 2, ..., 5$. Symbols are the same as in Fig. 9 (from ref. [22]).

References

[1] G.N. Fowler and R.M. Weiner, Phys. Rev. D17, 3118 (1978).

[2] I.V. Andreev, M. Plümer and R.M. Weiner, Univ. of Marburg preprint, August 1991, submitted for publication.

[3] A. Vourdas and R.M. Weiner, Phys.Rev. D38, 2209 (1988).

[4] M. Biyajima, O. Miyamura and T. Nakai, Proc. Conf. on Multiparticle Dynamics (Hakone, September 1978)(Research Institute for Fundamental Physics, Kyoto University, Kyoto) p. 139.

[5] R.M. Weiner, Phys. Lett. B232, 278 (1989); B242, 547 (1990).

[6] K. Kulka and B. Lörstad, Zs. f. Phys. C45, 581, (1990).

[7] I.V. Andreev and R.M. Weiner, Phys.Lett. B253, 416 (1991).

[8] G.N. Fowler, E.M. Friedlander, X.C. He, C.C. Shih and R.M. Weiner, Phys.Lett. B253, 421 (1991).

[9] G.N. Fowler, E.M. Friedlander, F.W. Pottag, R.M. Weiner, J. Wheeler and G. Wilk, Phys.Rev. D37, 3127 (1988); E.M. Friedlander, F.W. Pottag, R.M.Weiner, J. Phys. G15, 431 (1989).

[10] A. Bialas and R. Peschanski, Nucl.Phys. B273, 703 (1986); A. Bialas, these proceedings.

[11] P. Carruthers, E.M. Friedlander, C.C. Shih and R.M. Weiner, Phys. Lett. B222, 487 (1989).

[12] M. Gyulassy, in: Multiparticle Dynamics, L. Van Hove Festschrift, World Scientific 1990, p. 479.

[13] C.C. Shih, Phys.Lett. B259, 393 (1991).

[14] Na-22 Collaboration, N.M. Agababyan et al., Phys.Lett. B261, 165 (1991).

[15] I.V. Andreev, M. Plümer, R.M. Weiner, in preparation.

[16] EHS/Na-22 Collaboration, M. Adamus et al., Z.Phys. 37, 215 (1988); EHS/Na-22 Collaboration, I.V. Ajinenko et al., Phys.Lett. B222, 306 (1989).

[17] W. Ochs, Phys.Lett. B247, 101 (1990).

[18] E.M. Friedlander, X.C. He, C.C. Shih, R.M. Weiner, Phys.Rev. D44, 1396 (1991).

[19] G.N. Fowler, E.M. Friedlander, R.M. Weiner and G. Wilk, Phys. Lett. 57, 2119 (1986).

[20] M. Biyajima, G.N.Fowler, M. Suzuki, R.M. Weiner, and G. Wilk, Zs.f.Phys. C44, 199 (1989).

[21] M. Biyajima, G.N. Fowler, E.M. Friedlander, C.C. Shih, N. Suzuki, R.M. Weiner, and G. Wilk, Phys.Rev. D43, 1541 (1991).

[22] G.N. Fowler, E.M. Friedlander, F.S. Navarra, F.W. Pottag, R.M. Weiner, and G. Wilk, J.Phys. G16, 1439 (1990).

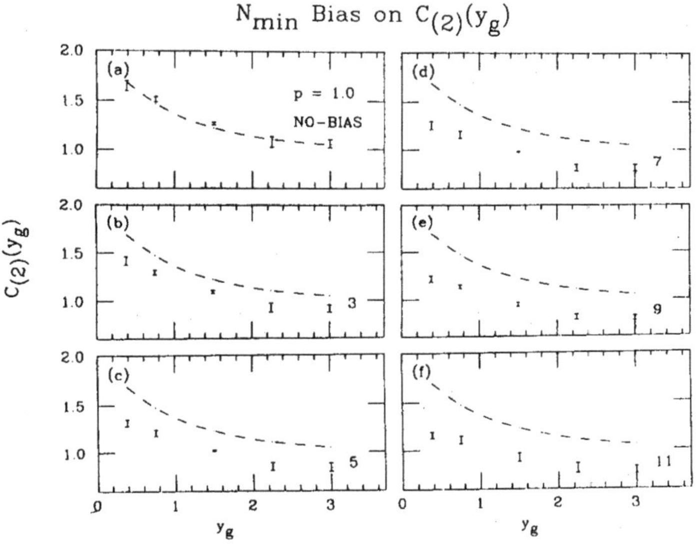

Fig. 1

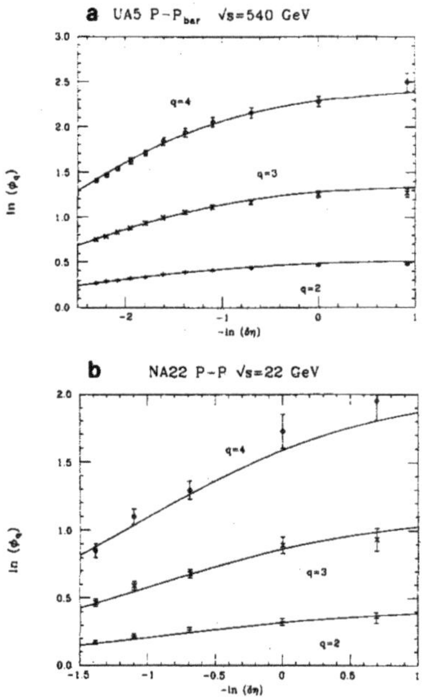

Fig 2

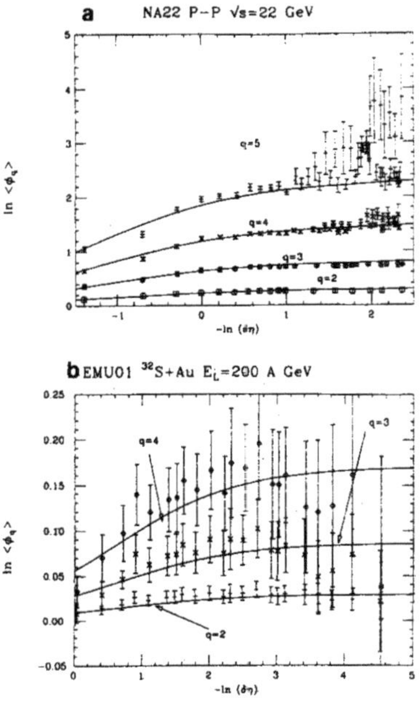

Fig. 3

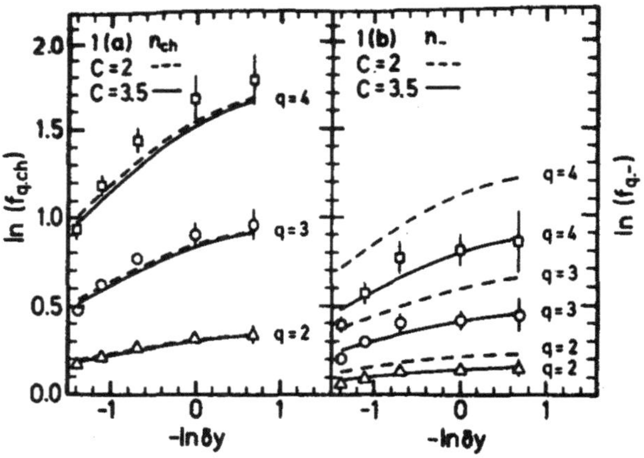

Fig. 4

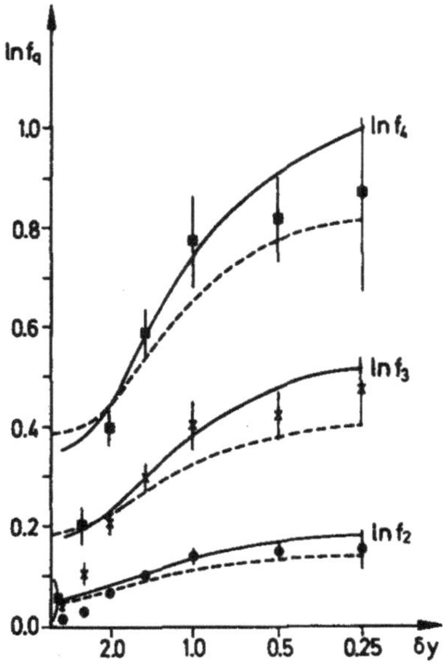

Fig. 5

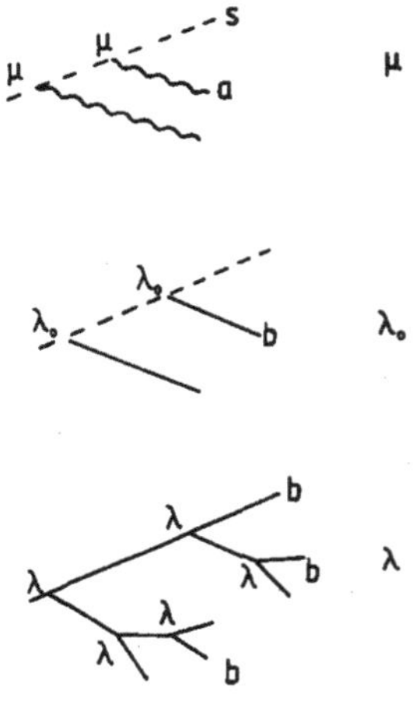

Fig. 6

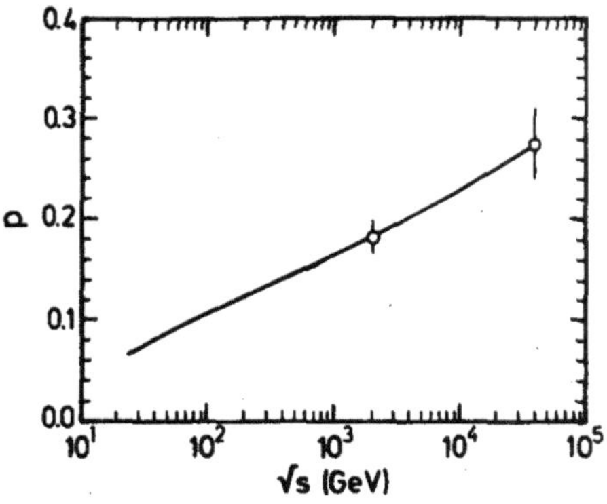

Fig. 7

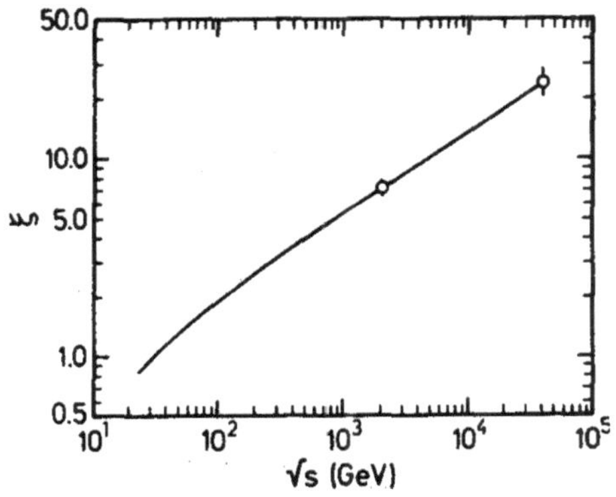

Fig. 8

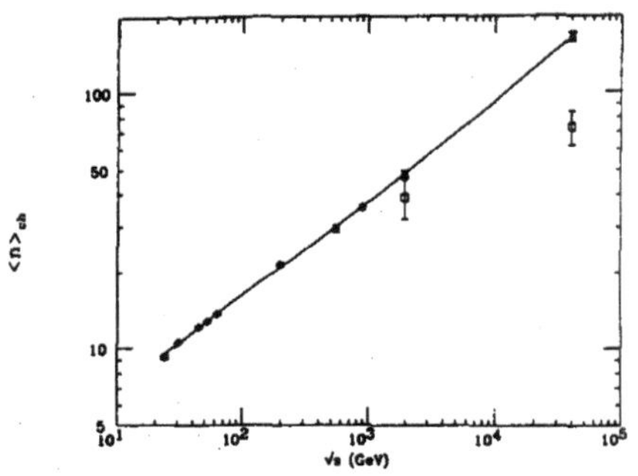

Fig. 9

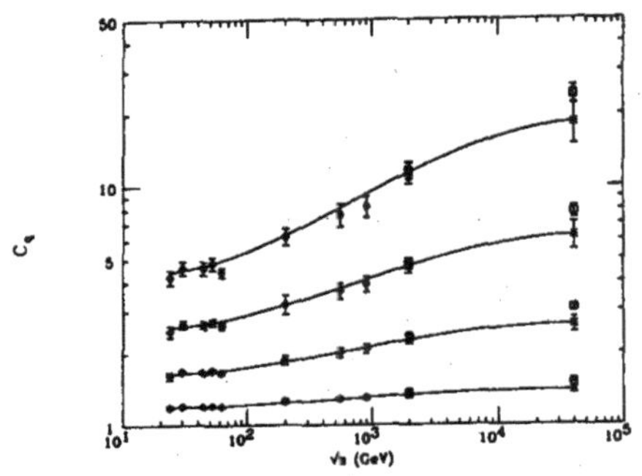

Fig. 10

PHASE TRANSITIONS AND SCALING IN MULTIPARTICLE PRODUCTION

I.M.Dremin
P.N.Lebedev Physics Institute, Moscow

M.T.Nazirov
Institute for High Energy Physics, Alma–Ata

Abstract

Some scaling laws experimentally observed in multiparticle production provoke the renormalization group approach for their treatment and appeal to the conformal symmetry. It could unite different experimental observations relating them to properties of phase transition from partons to hadrons and predict new features.

1. Introduction

Scaling behaviour of factorial moments which has been found recently in experiments with different particles and nuclei implies some general features of multiparticle processes. It demonstrates that the fluctuation of the number of particles belonging to a definite phase space bin does not saturate when the bin size decreases. These findings have been interpreted in terms of fractality and intermittency.

Deeper insight into the physical nature of the phenomenon can be gained when relating it to phase transitions and to their description in the framework of conformal theories. The purely mathematical relationship between these approaches is easily foreseen because the power–law behavior is common for factorial moments, critical phenomena and conformal theories. The way to it consists in demonstrating that Green functions of a conformal theory can be expressed as higher rank products of the order parameter describing a phase transition while the factorial moments are represented by some integrals from those Green functions.

The physical interpretation of this result relies on the general feature of all processes, namely, on the transition from partons to hadrons called hadronization. Therefore, the scaling laws should be a common trend of particle production for any colliding partners, even though the critical indices could differ. Moreover, the singularities appearing in correlation functions due to the phase transition can reveal itself in other properties of particles produced, in particular, in special behavior of their spectra at low transverse momenta or at low energies as well as in their correlations at close values of rapidities or transverse momenta.

Let us mention another interpretation of the power–law regularities observed experimentally. It is often attributed to the cascading nature of the process. Even though the cascades are not yet well developed at present energies and they hardly can be as general as phase transitions, it could happen that both mechanisms are at work. Further study of a diagram technique at the critical point is necessary to reveal the relative contributions of both pictures.

2. Main notations and relationships

As it is usually done in the fluctuation theory of phase transitions one should define the proper order parameter which is treated as a fluctuation field. As such, one could consider the function [1]

$$\varepsilon(\vec{x}) = \frac{\rho_e(\vec{x})}{\rho(\vec{x})} - 1 \tag{1}$$

where $\rho_e(\vec{x})$ and $\rho(\vec{x})$ are, correspondingly, the single–particle distribution in a given event (marked by e) and in a whole sample of events (an inclusive spectrum) at a phase space point $\vec{x} = (y,\vec{p}_T)$ ($y = 1/2\ \ln((E+p_\ell)/(E-p_\ell))$ is a rapidity and $\vec{p}_T$ is a transverse momentum). In what follows, we shall consider one–dimensional fluctuation $\varepsilon(y)$ only for the sake of simplicity since we shall not calculate critical indices directly but leave them yet as some undefined constants.

The function $\varepsilon(y)$ is a measure of fluctuations because it shows how strongly the single event distributions $\rho_e(y)$ decline from their averaged inclusive spectrum $\rho(y)$. It plays a role of an order parameter since its average values at parton and hadron phases differ. Namely, its average in the hadron phase is equal to zero:

$$<\varepsilon>_h = 0 \tag{2}$$

while in a parton phase it produces the ratio of inclusive parton to hadron spectra which provides in general:

$$<\varepsilon>_p \neq 0 \tag{3}$$

and according to the local parton–hadron duality hypothesis can be estimated about 0.2.

At the same time $\varepsilon(y)$ can be treated as a local fluctuation field and the whole field–theory machinery can be applied. It is well known that the partition function of any classical system can be written in the form analogous to S–matrix of a three–dimensional Euclidean field theory with a free Green function replacing the interaction potential. Namely, the probability of a given fluctuation $\varepsilon(y)$ has a general form:

$$w(\varepsilon) = Z^{-1} \exp[-F(\varepsilon)] \tag{4}$$

where $F(\varepsilon)$ is a Hamiltonian (more precisely, the free energy) and Z is a partition function. Adding to $F(\varepsilon)$ the term $J(\varepsilon)\cdot\varepsilon(y)$ with an external field J one traditionally gets the Green functions in a form

$$<\varepsilon_1 \ldots \varepsilon_q> = \frac{\delta^q \ln Z(J)}{\delta J(y_1)\ldots\delta J(y_q)}\Big|_{J=0} \tag{5}$$

According to (1) they are directly related to q–particle correlation functions $\rho(y_1,...,y_q)$ measured in experiment. For example,

$$<\varepsilon_1\varepsilon_2> = \frac{\rho(y_1,y_2)}{\rho(y_1)\rho(y_2)} - 1 \qquad (6)$$

where

$$\rho(y) = \frac{dn}{dy_1},\ \rho(y_1,y_2) = \frac{d^2n}{dy_1dy_2},\ \text{etc.} \qquad (7)$$

Finally, one can get the factorial moments as

$$F_q(\delta y) = (\delta y)^{1-q}\int_0^{\delta y} d\eta_1 ... \int_0^{\delta y} d\eta_{q-1} r_q(\eta_1,...,\eta_{q-1}), \qquad (8)$$

with $\eta_i = y_{i+1} - y_i$ and

$$r_q = \rho(y_1,...,y_q)/\rho(y_1)...\rho(y_q) \qquad (9)$$

Thus, formulae (4), (5), (8) show how the knowledge of the Hamiltonian of a fluctuation theory provides the Green function and, after integration, the factorial moments*). Looking from the other side, one can guess that the power–like behavior of factorial moments as functions of the bin width δy at $\delta y \to 0$ implies the scaling regime of the corresponding field theory i.e. power–law dependence of its Green functions.

3. Renormalization group and conformal theories

Such scaling regimes are typical for critical phenomena and indicate usually that the perturbation theory fails because strong coupling prevails. In the Fourier–conjugate to the rapidity space t one should consider the theory at large values of t. According to the renormalization group equations one gets [2]

$$\frac{d\ln d}{d\ln t} = -\mathscr{B}(d) \qquad (10)$$

where d indicates the factor connecting the total Green function D with a free one: $D = dD_f$ and function $\mathscr{B}(d)$ is sometimes determined from perturbation series expansion. The power–like behavior of d can be obtained if one assumes the constancy of $\mathscr{B}(d) \equiv \mathscr{B}$ at large $|t|$. Actually one needs only the asymptotic value of $\mathscr{B}(d)$ at large d to be constant since one deals with a strong coupling regime for large d at large t.

Then

$$D(t) \sim |t|^{-\alpha} \text{ for } |t| \to \infty,\ (\alpha = 2 - \mathscr{B}) \qquad (11)$$

The similar arguments provide the following formula for the vertex function

*) The formula (8) reminds the expressions for spontaneous magnetic moment and higher derivatives of free energy over magnetic field.

$$\Gamma_q(t_1,...,t_q) \sim |t_1|^{-\gamma}\Gamma_q\left[1, \frac{t_2}{|t_1|},...,\frac{t_q}{|t_1|}\right] \text{ for } |t| \longrightarrow \infty \tag{12}$$

wherefrom one easily gets the Green functions

$$<\varepsilon(t_1)...\varepsilon(t_q)> = \delta(\sum^q t_i)D(t_1)...D(t_q)\Gamma_q(t_1,...,t_q) \tag{13}$$

or in y–space

$$<\varepsilon(y_1)...\varepsilon(y_q)> \sim |y_q-y_1|^{\gamma+2\alpha-1}|y_2-y_1|^{-1+\alpha}...|y_q-y_{1q-1}|^{-1+\alpha}+\text{perm.} \tag{14}$$

According to the formula (8), the factorial moments behave like

$$F_q(\delta y \longrightarrow 0) \sim (\delta y)^{-\varphi(q)} \tag{15}$$

with

$$\varphi(q) = q(1-\alpha) - \gamma + 1 > 0. \tag{16}$$

The linear behavior of intermittency indices with the rank of the factorial moment q is predicted in this approximation. Another important feature is the appearance of singularities in the correlation functions at coinciding arguments.

Both consequences result somehow from our assumption that we have to deal with conformal theories when studying the limiting behavior of propagators and vertex functions. The above formulae correspond to the leading order terms in the operator expansion and some corrections to them will appear from non–leading terms. Therefore, the experimental consequences which are discussed below are valid up to this remark.

4. Perturbative approach

It is well known that the scaling principles for correlation functions as formulated in renormalization group analysis or in the graph technique of a strong coupling field theory have replaced the hypothesis of analiticity of free energy used in Ginzburg–Landau approach to phase transitions. Nevertheless, the latter has still some heuristic value demonstrating where and how the perturbative approach fails and indicating some possible phenomenological fits.

Having in mind this purely pedagogical purpose we proceed to describe what would happen if the fluctuations were weak enough for the free energy in (4) to be represented by first terms in its Taylor series expansion i.e.

$$F(\varepsilon) = F_0 + \int dy\left[\frac{a}{2}\left(\frac{d\varepsilon}{dy}\right)^2 + \frac{b}{2}\varepsilon^2 + c\varepsilon^3 + d\varepsilon^4 + ...\right] \tag{17}$$

It is reduced to Ginzburg–Landau Hamiltonian if c=0, d≠0 and all higher terms are equal to zero. The free fields case corresponds to c=d=...=0 and provides

$$Z_f(J) = \exp\left\{-\frac{1}{2}\int dy_1 \int dy_2\, J(y_1)D_f(y_1-y_2)J(y_2)\right\} \tag{18}$$

$$D_f(y_1-y_2) = <\varepsilon_1\varepsilon_2>_f = v\exp\{-|y_1-y_2|/\xi\} \tag{19}$$

$$v = \pi\xi/b; \qquad \xi = (b/a)^{1/2} \tag{20}$$

wherefrom

$$F_2(\delta y) = 1 + v\xi(1-\exp(-\delta y/\xi))/\delta y \sim (\delta y)^{-\varphi(2)} \tag{21}$$

if

$$a = a_0(\delta y)^{2\varphi(2)} \text{ at } \delta y \longrightarrow 0 \tag{22}$$

In the Fourier–conjugate space t one gets

$$D_f(t) \sim (t^2 + a/b)^{-1} \tag{23}$$

To describe an intermittency phenomenon for higher factorial moments analogously to F_2 one should assume that the coefficients c,d,... differ from zero and behave in a proper way when $\delta y \longrightarrow 0$. It looks even more artificial then the temperature dependence of coefficients in Ginzburg–Landau Hamiltonian. Still one can guess that the exponential behaviour of the Green function will survive at large $|y_1-y_2|$ because it corresponds to small distances in the configuration space. Its behaviour for small $|y_1-y_2|$ is changed according to above treatment what shows the failure of the perturbative approach which has been used here as an illustrative example only.

It is interesting to note that we have got the exponential damping of the Green function (19) in the phase space and the power–like behaviour in the configuration space i.e. the dependencies reversed to the usual treatment. That is why we have to deal with the ultraviolet stable point instead of the infrared one even though we are still interested in large distances in the configuration space and small separations in the phase space volume.

5. Conclusion

Dealing with a scaling approach in the framework of the renormalization group equation (10) we have been able to derive general power–law dependence of correlation functions (14) and, therefrom, the intermittent behaviour of factorial moments (15), (16). To calculate the precise values of critical indices (or intermittency exponents) one should apply the graph technique or ε–expansion. However, even at this initial stage one gets some qualitative features of experimentally observed characteristics. If they are confirmed by experiment it provides an additional stimulus for more accurate treatment.

Among the results which could be experimentally verified let us mention four conclusions.

1. The intermittency indices are equidistant in q what follows from their linear q–dependence. Experiment indicates that some corrections to this statement are needed. They could be derived as higher order terms in the operator expansion.

2. The two–particle correlation function can possess a singularity at small separation $|y_1-y_2|$. This conclusion can be checked with high statistics and high resolution experiments fitting two–particle correlations by a function

$$c(y_1,y_2) \sim |y_1-y_2|^{-\alpha}\exp[-|y_1-y_2|/\xi] \tag{24}$$

3. In three–dimensional case the similar singularity can arise for particles with close values of transverse momenta indicating strongly correlated pairs.

4. The excess of soft and/or low–p_T mesons should be larger in those reactions (in particular, in e^+e^-) where intermittency indices are larger. This result follows from the previous ones considered for reactions with varying initial particles.

This paper is based mainly on the results obtained in our previous paper[1] though some conclusions and ideas have been changed.

REFERENCES

1. I.M.Dremin, M.T.Nazirov. Mod. Phys. Lett. A (to be published).
2. N.N.Bogoliubov, D.V.Shirkov. "Introduction to the theory of quantized fields", Nauka, 1957, p.350.

SUMMARY OF THE RINGBERG WORKSHOP ON MULTIPARTICLE PRODUCTION FLUCTUATIONS AND FRACTAL STRUCTURE

A.Bialas
Institute of Physics, Jagellonian University, Cracow

September 30, 1991

A summary of the results presented at the workshop is given.

1 Subject and objectives.

Fluctuations in a multiparticle system are described by correlations between particles: if one disregards purely statistical fluctuations, the concepts of fluctuations and of correlations are equivalent. Positive correlations generate peaks in a multiparticle distribution. Negative correlations generate holes.

To study correlations one can consider either the multiparticle distributions and correlation functions, or the factorial moments and factorial cumulants of the (local) multiplicity distributions. They are directly related by the schematic formulae

$$\int \text{multiparticle distribiution} = \text{factorial moment} \tag{1.1}$$

$$\int \text{correlation function} = \text{factorial cumulant} \tag{1.2}$$

One can choose either method and/or any combination of them. They are mathematically equivalent. But, as ususal, some may be more useful than others when applied to a specific system and to a specific physical question [1-3]

The problem addressed by this workshop is "fractal structure". In terms of factorial moments and factorial cumulants fractal structure means the existence of a power-law singularity at very small intervals of phase-space, as expressed by the "anomalous scaling law" [4-6]

$$F(\delta) \to F(\Delta)(\Delta/\delta)^f \tag{1.3}$$

where F is the factorial moment (or cumulant) evaluated at a given interval, and f is the intermittency exponent related to the fractal dimension of the system [7]. In terms of multiparticle distributions and correlation functions (1.3) implies a power law singularity at very small differences between particle momenta (as can be seen from (1.1) and (1.2)) [3,6,8].

Consequently, the quest for fractal structure is the quest for power law singularities in multiparticle spectrum. This problem was thus the central issue of the present workshop. Translated into experimental data it amounts to study the dependence of particle spectra observed in a given phase-space region on the size of this region. At the simplest level, the issue is to establish the power law dependence. At the next level one is interested in the strength and the structure of this singularity, i.e. to determine the intermittency exponents for a full hierarchy of particle spectra and correlation functions.

At this instance one should perhaps address one point which becomes sometimes a point of misunderstanding. In the present context, the observation of a "singularity" in a spectrum does not mean the actual singularity in the mathematical sense. Indeed, proving mathematical singularity with a real physical system is clearly impossible. Instead, the two physical questions one tries to answer are: (a) is there a region of phase space where the power law (1.3) holds? and (b) how small is the smallest phase-space interval at which (1.3) is still satisfied? In the papers presented at the workshop the problems of fractal behaviour were addressed in three ways: (i) new and abundand experimental evidence was presented; (ii) new phenomenological methods and interpretations were proposed; (iii) some dynamical schemes were developped and investigated. My duty is to summarize this work.

2 Experimental evidence.

Most of the experimental investigations were devoted to factorial moments, in an attempt to verify the Eq. (1.3). The e^+e^- data were presented by Delphi [9], Aleph [10] and Opal [11] from LEP and by Cello [12] from PETRA. The muon-proton data were presented by EMC [13]. The hadron-hadron data were shown by UA1 [14] and NA22 [15]. Hadron-nucleus and nucleus-nucleus data were presented by EMU01 [16] and NA35 [17].

All experiments see a clear rise of the factorial moments in two- and three-dimensional spectra up to the smallest interval considered. The one-dimensional spectra show flattening at small intervals.

For an unsophisticated observer the emerging picture can thus be considered as a rather strong case for intermittent behaviour of the spectra, the flattening of the one-dimensional plots being understood as the effect of the projection from 3-dimensional phase-space [18,8]. In any case, the evidence seems to be much better than at the time of the Santa Fe workshop. This generally optimistic view does not preclude, of course, further discussions and questions about physical significance and interpretation of these experimental results. One thing is clear, however: the effect is there, so at least we have something real to quarrel about.

Although the general features of the data observed in all experiments seem fairly universal, there are nevertheless some substantial differences in interpretation. The e^+e^- data can be explained by the adequate versions of the JETSET/PS Monte Carlo code [9-12,19], although some discrepancies remain in certain cases [9]. In contrast, other data cannot be accounted for by the existing models [13-15]. It is not entirely clear if this is due to the fundamental difference in physics between e^+e^- annihilation and processes involving hadrons in the initial state. It may well be (and is even very likely, in my opinion) that, given the complexity of hadronic interactions, the realistic models of hadronic processes are more difficult to construct and that more work is needed to achieve this goal.

Explanation of the e^+e^- data by a Monte Carlo program, which implements standard ideas of perturbative QCD and describes hadronization by decay of colour strings, is certainly a very important step in investigations of intermittency. First found by DELPHI collaboration [20], it is now confirmed by all groups working on this problem, and with much higher statistics. There is no question that the consequences of this finding were not yet fully explored. One usually emphasizes its "negative" implication, i.e. that "no new physics" is needed to explain intermittency. This is of course important but, as I already pointed out elswhere [21], by no means exhausts the potential interest.

The chance to pin down the physical origin of intermittency by investigating the JETSET/PS model was not entirely missed, however. Several studies were presented at the meeting [9-12,19]. The main conclusions are: (i) the Dalitz pairs are an important source of contamination at small intervals, but do not account for the bulk of the effect; (ii) Resonance decays do not seem to influence the variation of factorial moments with decreasing phase-space interval; (iii) The 3-jet events are a dominant source of this variation, i.e., in the e e annihilation intermittency seems mostly a reflection of the transverse structure of the jet emitted at the largest angle with respect to the thrust axis .

Very high statistics of ALEPH data [10] allowed to study the variaton of factorial moments versus position in rapidity axis. The results turned out to be spectacular. One observed a strong dependence of factorial moments on rapidity. This observation shows that the fractal properties of the distribution are by no means uniform (along rapidity axis) and that there is much more structure than hitherto accepted. It is clear that these results shall have a great influence on further work in the field. The interpretation proposed by the authors relates the observed structure of factorial moments to the similar structure in rapidity distribution (although the positions of the maxima in these two distributions do not coincide) and thus to the 3-jet structure of the events.

To summarize this part, it seems that the origin of intermittency in e^+e^- annihilation is the multi-jet structure of the events. More work is needed, however, to substantiate quantitatively this conclusion: some problems still persist. For example, it is not clear if the model can explain the observed multiplicity dependence of the intermittency parameters [22]. Also the $p_\perp$ dependence may be a problem [23].

Much less is understood in other processes (involving hadrons in the initial state). The main problem is that no model can explain the observed intermittency [6] and it is not known if the existing models can be radically improved in this respect. In view of the results from e^+e^-annihilation, one naturally suspects that the multi-jet structure of the events is not correctly accounted for in the models now on the market. It would be interesting to verify this suggestion, particularly for lepton-hadron interactions which are certainly simplest of this class.

In absence of a good model it is difficult to interpret the results. Consequently most of the work was restricted to presentation of data and comparison between different samples. This is rather difficult to summarize and the best way seems to refer to the original papers published in this volume [13-17]. As I already mentioned, the general impression is that the quality of the data improved substantially but probably some time is needed before it can be fully digested.

Let me just add few remarks. Although the data do show a clear rise of the factorial moments with decreasing bin size, the observed dependence is often inconsistent with a simple power law. Therefore the authors have tendency to avoid the linear fits (in the log-log plot) but they give instead the "local slope". This is of course a "safer" procedure but it makes comparison of different data much more difficult. An improvement here is badly needed. Perhaps the solution is in shifting to study of factorial cumulants [1-3,24,25]

Another interesting question to be investigated is the relation between one- two- and three-dimensional spectra. The detailed comparison of these spectra gives a chance to determine (at least approximately) the singularity structure of the correlation functions. Needless to say, it would be a great improvement in phenomenology of multiparticle production.

New data on hadron-nucleus and nucleus collisions were shown by EMU01 [16] and NA35 [17] collaborations. They confirm the decrease of intermittency with increasing multiplicity in a given reaction. Comparison of different processes (i.e. the A-dependence of the effect) is less conclusive, but also seems to confirm the presence of some collective phenomena.

3 More of technology.

Although the studies of factorial moments dominate at the present time the investigations of intermittency and of correlations, other methods and improvements are constantly being developped. Several ideas were presented at the workshop. In fact these problems were dominating the "theoretical" contributions. They are listed below.

(i) Factorial cumulants.

They were advocated since some time by Carruthers and collaborators

[24,26]. Their theoretical significance is expressed by the Eq.(1.3). The advantage of factorial cumulants is that cumulants of different ranks are independent of each other. The disadvantage is that they are more complicated to measure than the factorial moments. It is likely that the increasing statistics of experiments will allow to study the factorial cumulants with a sufficient accuracy. This may turn out to be very interesting.

At the present meeting the analysis of correlations in terms of factorial cumulants was discussed by Carruthers [26] and by Sarcevic [27].

(ii) Distribution and correlation integrals.

This is another proposal of the Tucson group, presented by Lipa [28]. It can be viewed as a generalization of the factorial moments and factorial cumulants based on Eqs (1.2) and (1.3) with a modified integration region. The integration of two-particle distribution in this new region has a very simple and and appealing physical meaning: it measures the total number of pairs with the distance in phase-space smaller than a given value d. There are two immediate advantages as compared to standard analysis: (a) there is no restriction on the value of d and (b) the number of pairs counted at a given d is much larger (and thus statistical error significantly reduced).

It seems to me that this new observation may help in future investigations of intermittency and correlations. I also suspect that this is not the "last word" in this respect and that other selections of the integration region in the Eqs (1.2 and (1.3) may be of interest. A systematic investigation of this issue seems well worthwhile.

(iii) Other distributions.

The distribution of holes in the spectrum was discussed by De Wolf. This issue remains almost unexplored, although it is potentially very rewarding. It turns out, not surprizingly, that the distribution of holes contains essential information about the full distribution. How to use this information is of course a problem and I refer the reader to the De Wolf report [29] (where also other interesting relations to the general theory of fluctuations can be found) for details.

Simak presented an interpretation of the entropy distribution it in terms of the fractal structure of the spectrum [30].

Charlet [31] discussed the NA22 data on factorial correlators and studied their relation to factorial moments suggested recently by Peschanski and Seixas [32].

(iv) Determination of intermittency parameters.

Fialkowski [25] pointed out that the measured values of the intermittency parameters may be much smaller than the actual ones. He argues that the large background of uncorrelated particles and/or of lower order correlations has a tendency to decrease significantly the observed intermittency effects. If this interpretation is correct, it may also provide a natural explanation for a fast rise of the factorial moments for very small bins in some spectra. This is another point which is worth to be carefully studied experimentally.

(v) The Ochs-Wosiek relation between different moments [33].

This was discussed by Ochs [22]. The evidence for the linear dependence

$$\log F_q = a_q + b_q \log F_2 \tag{3.1}$$

looks still very good (better than that for Eq.(1.3)). It can thus be efficiently usef for studying the relation between intermittency exponents of different orders. The data for leptonic reactions and those for heavy ion collisions seem to follow the same pattern, indicating absence of any signal for phase transition in nuclear collisions.

(vi) A warning on the use of G-moments.

Florkowski presented a model example showing that the suggested universality obtained earlier in the analysis of the G-moments [34] may be simply a reflection of statistical fluctuations. This emphasizes again the well-known fact that the G-moments do not remove statistical fluctuations and thus a certain care must be taken when conclusions are drawn. The subject was also discussed in the papers presented by Chiu [35] and by Hwa [36] where more attempts to define universal properties of G-moments and of fractal spectrum were tried. It was suggested that universality can be achieved if appropriate cuts on multiplicity are made. Further work is certainly needed to clarify this point.

(vii).Other variables.

Traditionally the spectra are investigated in terms of rapidity (or pseudorapidity), azimuthal angle and transverse momentum. It is not obvious, however, if these are the best variables, i.e. the variables in which the effect is most cleanly seen. For the two-particle spectrum, e.g, a natural variable may be the invariant mass of the system (or, equivalently, the difference of four-momenta squared) [8], as is used in the description of the Bose-Einstein correlations. One may also consider generalizations to systems of more particles. This was suggested recently by Biyajima et al [37].Some interesting data on this problem were shown by Buschbeck [38]. I feel that this is a very interesting subject which deserves further attention. Indeed, it would be of great importance to identify the variables for which the intermittency effects are "as clean as possible" and not masked by projections in the multidimensional phase-space. The existing high statistics data should allow such an analysis which is undoubtely worth to try.

4 Search for Dynamics.

As was discussed by many authors [3], the phenomenon of intermittency occurs naturally if the physical system in question is scale-invariant. Among many scale-invariant physical systems the self-similar cascade seems to be a particularly natural candidate for the description of intermittency in multiparticle production. Indeed, as pointed out already in many papers [3], the cascade nature of particle production is supported by QCD, although we are still far from understanding the details of this process.

This point of view was accepted by most of the authors contributing to the present workshop. I already described the results from e e annihilation which strongly suggest that the QCD-inspired cascade is indeed the source of intermittency. Accepting this conclusion does not, however, close the story. On the contrary, new questions are generated. One would like to know, e.g., (i) why the jet decay does not seem to influence the correlation structure of the distributions; (ii) how the resonance decays conspire not to destroy the short-range correlations; (iii) what is the effective cut-off parameter which determines the range of the observed correlations. This list can be of course continued.

Investigations of cascade models were presented by Giovannini, Chiu, Hwa and Ochs.

Giovannini reported a study of multiplicity distribution in a Simplified Parton Shower model [39]. It was shown that not only quantitative but even qualitative properties of the distributions are sensitive to the detailed properties of the vertex in the cascade. In particular, the negative binomial fits are obtained only if the splitting vertex is cut strongly enough at small virtualities.

Chiu [35] presented a study of the JETSET MC code. He discussed the dependence of intermittency indices on energy, rapidity and multiplicity.

The conclusion is that the energy and rapidity dependence is a reflection of multiplicity dependence which seems therefore more fundamental.

Hwa presented a Geometrical Branching Model for soft hadronic collisions [36]. The model accounts for global features of multiparticle production as well as for "local" fluctuations. The calculation shows that in a realistic model it is by far non-trivial to obtain rising factorial moments, even in presence of the cascade (the authors had to assume singular splitting function to achieve the result). This is most likely a consequence of a substantial smearing of the original effect by superposition of many components required in "realistic" models. It looks like an important result which should be further investigated.

Ochs [22] discussed the local parton-hadron duality, an empirical property of the spectra. It is an intringuing property, not yet understood but necessary to explain the data. It is of course related to the question (i) above.

Another classic possibility of a scale-invariant physical system is the system at the critical point of the (second) order phase transition [40] (it was even recently suggested [41] that this mechanism of generating intermittency may be relevant at the phase transition between the unconfined and confined strongly interacting matter). Investigations into statistical field theory of density fluctuations were presented by Sarcevic [27] and by Dremin [42]. Sarcevic showed that in the nucleus-nucleus collisions only two-particle correlations are significantly different from zero. Higher order correlations vanish within errors. This can be interpreted in terms of the simple Landau-Gizburg version of the statistical field theory. Dremin presented an attempt to introduce the interacting terms into the Landau-Ginzburg action. This seems to be a promising direction but it is clearly only at its beginning.

Wosiek discussed fluctuations in the Ising system and showed that the theoretical intermittency indices can be explicitly derived from a critical exponent [43]. The relation to percolation transition [44] was also analysed.

Bopp presented an attempt to understand intermittency in the framework of the dual parton model [45]. It seems that, although the short-range correlations are qualitatively described in the model, it is not easy to reproduce quantitatively the observed effects at very small intervals. The most likely conclusion is that a "singularity" of some sort must be introduced ad hoc. This is of course an important result, indicating that the model should probably be supplemented by a kind of cascading mechanism.

Finally, Weiner gave a broad review of the application of the classical and quantum statistics as applied to the multi-pion systems which was developped over years by him and his collaborators. As I see no possibility to summarize this work in few sentences, I refer the reader directly to his contribution [46].

5 Outlook.

The workshop was largely dominated by the experimental contributions: new, good quality data were presented. They confirmed beyond any doubt the existence of the empirical phenomenon of "intermittency", i.e., of the rising factorial moments. No doubt this is a real progress, as compared even to the recent Symposium at Gut Holmecke [47], not to mention the Santa Fe workshop [48]. The main result is that we can now be fairly sure that we are talking about a "real thing". However, many questions are still unanswered and require further studies. The most immediate one is the question of "universality": can the observed effects be understood as reflection of one mechanism, common to all processes of particle production? I think this is still not excluded, although at the moment there is not much in the data to support this hypothesis (except perhaps the validity of Eq. (3.1)). The investigation of this problem seems to me of the highest priority for the immediate future, as I feel that already the existing data, when carefully analysed, shall provide significant clues.

Another important experimental problem, which should give a crucial hint for understanding the dynamics, is the determination of the minimum scale at which the "intermittency" ends. This can only be done by a careful analysis of the 3-dimensional spectra. Our hopes that this information shall soon be available.

I find much less impressive the progress on theoretical side. Although some interesting work was presented on methodology, we have still no real understanding of the dynamics. This is thus obviously the main problem and the main challenge for the future theoretical work.

Although it is definitely too early to speculate about the dynamical origins of intermittency, some specific messages are certainly provided by the data presented at the meeting. It is clear, e.g., that the jet structure of the events accounts for a large part of the observed effects in the e^+e^- data. This obviously means that the reaction mechanism is important in understanding the phenomenon (this conclusion is supported by the inability of the models to reproduce the data in hadronic reactions).

What is even more mysterious (and therefore interesting) is the role hadronizaton plays in the game. This is most likely related to the mechanism of confinement and therefore it is not surprizing that there are great difficulties in understanding this part of the process. Here a message which is acknowledged by most of us (but nobody knows how to use it effectively) is the observation of the parton-hadron duality [49]. It clearly shows that the transition from perturbative to non-perturbative region is a smooth one (perhaps more smooth than anybody expected) but a real meaning of this fact is still not understood.

Let me close this report by listing three natural possibilities of further developments

(i) It seems now that the phenomenon of intermittency is present up to the scales significantly below Λ_{QCD} . If confirmed, this almost certainly shall have an impact on our understanding of chiral dynamics. It would thus be interesting to investigate this problem further.

(ii) If the power-law behaviour of the moments extends over a significant region of scales, it may indicate that a mechanism of "self-organized criticality" [50] is at work. Again, it would be interesting to investigate this possibility.

(iii) Random decay of coloured strings provides another natural (but little known) mechanism for creating intermittency at small scales [51] and it would be interesting to study it in realistic models.

ACKNOWLEDGEMENTS.

Special thanks are due to N.Geddes who explained to me the meaning of the LEP results.

REFERENCES

1.A.H.Mueller, Phys.Rev.D4 (1971) 150.
2.For an early review, seee e.g. A.Bialas, Report at III Multiparticle Symposium (Pavia 1973).
3.For recent developments, see P.Carruthers, Int.J.Mod.Phys.A4 (1989) 5587.
4.A.Bialas and R.Peschanski, Nucl.Phys. B273 (1986) 703;
5.A.Bialas and R.Peschanski, Nucl.Phys. B308 (1988) 857.

6.For recent reviews, see A.Bialas, "Intermittency '90", Report at the Conference "Quark Matter '90", Menton (May 1990); Nucl.Phys.A525 (1991) 345c and R.Peschanski, CERN preprint TH 5891/90, Int.J.Mod.Phys., to be published.
7.P.Lipa and B.Buschbeck, Phys.Letters B223 (1989) 465.
8.A.Bialas and J.Seixas, Phys.Letters B250 (1990) 161.
9.A.De Angelis and N.Damaria, these proceedings.
10.ALEPH Coll., CERN preprint PPE/91-121 and N.M.Lieske, these proceedings.
11.N.Geddes, these proceedings.
12.J.Podobrin, these proceedings.
13.N.Schmitz, these proceedings.
14.W.Kittel, these proceedings.
15.B.Buschbeck, these proceedings; H.Dibon, these proceedings.
16.E.Stenlund, these proceedings.
17.I.Derado, these proceedings.
18.W.Ochs, Phys.Letters B247 (1990) 101.
19.V.Raab, these proceedings.
20.P.Abreu et al. (DELPHI Coll.), Phys.Letters B247 (1990) 137; A.De Angelis, Mod.Phys.Lett. A5 (1990) 2395.
21.A.Bialas, Proc.Of the XX Int. Symp. on Multiparticle Dynamics, Gut Holmecke near Dortmund (September 1990), R.Baier and D.Wegener Eds. (World Scientific, 1991), p.669.
22.W.Ochs, these proceedings.
23.EHS/NA22 Coll., N.M.Agababyan et al., Phys.Lett. B261 (1991) 165.
24.P.Carruthers, Proc.of the Perugia Workshop on Multiparticle Production (June 1988), R.Hwa,G.Pancheri and Y.Shrivastava Eds., (World Scientific, 1989), p.190.
25.K.Fialkowski, these proceedings.
26.P.Carrutthers, these proceedings.
27.I.Sarcevic, these proceeedings.
28.P.Lipa, these proceedings.
29.E.De Wolf, Acta Phys.Pol. B21 (1990) 611 and these proceedings.
30.V.Simak, these proceedings.
31.M.Charlet, these proceedings.
32.R.Peschanski and J.Seixas, CERN preprint TH-5903/90.
33.W.Ochs and J.Wosiek, Phys.Letters B214 (1988) 617.

34.W.Florkowski and R.Hwa, Phys.Rev.D43 (1991)
35.C.Chiu, these proceedings.
36.R.Hwa and J.C. Pan, Oregon preprint OITS-463 (April 1991).
37.M.Biyajima et al., Progr.Theor.Phys. 84 (1990) 930.
38.B.Buschbeck, these proceedings.
39.A.Giovannini and R.Ugoccioni, these proceedings.
40.J.Wosiek, Acta Phys.Pol. B19 (1988) 863; H.Satz, Nucl.Phys. B326 (1989) 613.
41.A.Bialas and R.Hwa, Phys.Letters B253 (1991) 436.
42.I.M.Dremin, M.T.Nazirov, these proceedings.
43.Z.Burda, J.Wosiek, K.Zalewski, Cracow Preprint TPJU 11/91.
44.R.Peschanski, Festschrift L.Van Hove A.Giovannini and W.Kittel Eds., World Scientific, p.519; S.Gupta, P.La Cock and H.Satz, Bielefeld preprint BI-TP 90/21 (November 1990).
45.W.Bopp et al., Z.Phys. C51 (1991) 99; P.Aurenche et al., Orsay preprint LPTHE 91/31.
46.R.Weiner, these proceedings.
47.Proceedings of the XX Int.Symposium on Multiparticle Dynamics, Gut Holmecke near Dortmund (September 1991), R.Baier and D.Wegener Eds., (World Scientific, 1991).
48.Proceedings of the Santa Feworkshop on intermittency in high energy collisions (March 1990), F.Cooper, R.Hwa and I.Sarcevic Eds., (World Scientific, 1991).
49.A.Azimov et al., Z.Phys. C27 (1985) 65; C31 (1986) 231. G.Cohen-Tannoudji and W.Ochs, Z.Phys. C39 (1988) 513.
50.P.Bak, C.Tang, K.Wiesenfeld, Phys.Rev.Letters 59 (1987) 381; Phys.Rev. A38 (1988) 364.
51.A.Bialas et al. Phys.Letters B229 (1989) 398.

LIST OF PARTICIPANTS

Andersson, Bo
Department of Theoretical Physics
University of Lund
Sölvegatan 14A
S-22362 Lund
THEPBA @ SELDC52 (bitnet)
BO @ thep.lu.se (internet)

Bialas, Andrzej
Jagellonian University
Institute of Physics
ul. Reymonta 4
PL-30059 Krakau 16
BIALAS%CHOPIN.DECNET
@ UXPLGW.CERN.CH

Bopp, Fritz W.
Universität G.H. Siegen
FB Physik
Adolf-Reichwein-Str. 2
5900 Siegen 21
GG162 @ DSIHRZ51

Buschbeck, Brigitte
Institut für Hochenergiephysik
Nikolsdorfergasse 18
A - 1050 Wien
V2035DP5 @ AWIIEZ11

Carruthers, Peter
Department of Physics
University of Arizona
Tucson, AZ 85721
USA
CARRUTHERS @ ARIZRVAX

Charlet, Martine
Fysisch Laboratorium
Universiteit Nijmegen
Toernooiveld
NL-6525 ED Nijmegen
U632007 @ HNYKUN11

Chiu, Charles B.
Center for Particle Theory
and Dept. of Physics
University of Texas
Austin, TX 78712 / USA
CHIU @ UTAPHY

De Angelis, Alessandro
Istituto di Fisica dell'
Universita di Udine
Via Larga 36
I-33100 Udine
DEANGELI @ CERNVM

Derado, Ivo
Max-Planck-Institut für Physik
Föhringer Ring 6
8000 München 40
IWD @ DMOMPI11

De Wolf, Eddi
Department of Physics
Univ. Instelling Antwerpen
Universiteitsplein 1
B-2610 Wilrijk
DEWOLF @ BANUIA51

Dibon, Heinz
Institut für Hochenergiephysik
Nikolsdorfergasse 18
A-1050 W i e n
V2033MU9 @ AWIIEZ11

Dremin, Igor
P.N. Lebedev Physical Institute
Academy of Sciences of USSR
Leninsky Prospect 53
M o s k a u 117924 / USSR
THEORDEP @ FIAN.MSK.SU

Fiałkowski, Krzysztof
Jagellonian University
Institute of Physics
ul. Reymonta 4
PL-30059 K r a k a u 16
UIWOSIEK @ PLKRCY11

Florkowski, Wojciech
Gesellschaft für Schwerionenforschung
Planckstr. 1
6100 Darmstadt 11
TP00 @ DDAGSI3

Geddes, Neil
C E R N
PPE Division
CH-1211 Geneva 23
GEDDES @ CERNVM

Giovannini, Alberto
Dipartimento di Fisica Teorica
Università di Torino
via Pietro Giuria, 1
I-10125 T o r i n o
GIOVANNINI @ VAXTO.INFN.IT

Hwa, Rudolph C.
Inst. of Theoretical Science
and Department of Physics
University of Oregon
E u g e n e , OR 97403 / USA
HWA @ OREGON

Jancso, Gabor
Max-Planck-Institut für Physik
Föhringer Ring 6
8000 München 40
GJA @ DM0MPI11

Kittel, Wolfram
Fysisch Laboratorium
Universiteit Nijmegen
Toernooiveld 1
NL-6525 ED N i j m e g e n
U632004 @ HNYKUN11

Lieske, Natalie
C E R N
PPE Division
CH-1211 G e n f 23
LIESKE @ CERNVM

Lipa, Peter
University of Arizona
Physics Department
Tucson, AZ 85721
USA
LIPA @ ARIZVM1

Ochs, Wolfgang
Max-Planck-Institut für Physik
Föhringer Ring 6
8000 München 40
WWO @ DM0MPI11

Podobrin, Oliver
D E S Y
Notkestr. 85
2000 Hamburg 52
F14POD @ DHHDESY3

Raab, Volker
Max-Planck-Institut für Physik
Föhringer Ring 6
8000 München 40
RAABV @ ALWS

Sarcevic, Ina
Department of Physics
University of Arizona
Tucson, AZ 85721
USA
INA @ ARIZRVAX

Schmitz, Norbert
Max-Planck-Institut für Physik
Föhringer Ring 6
8000 München 40
NOS @ DM0MPI11

Seyboth, Peter
Max-Planck-Institut für Physik
Föhringer Ring 6
8000 München 40
PXS @ DM0MPI11

Šimák, Vladislav
Chechoslovak Academy of Sciences
Institute of Physics
Na Slovance 2
CS-18040 Praha 8
SIMAK @ CSPGAS11

Stenlund, Evert
Cosmic and Subatomic Physics
University of Lund
Sölvegatan 14A
S-22362 Lund
KOSU_TREVE @ KOSFY.LU.SE

Weiner, Richard
Fachbereich Physik
Universität Marburg
Mainzer Gasse 33
3550 Marburg
WEINER @ DMRHRZ11

Wosiek, Jacek
Jagellonian University
Institute of Physics
ul. Reymonta 4
PL-30059 Krakau 16
UIWOSIEK @ PLKRCY11

www.ingramcontent.com/pod-product-compliance
Lightning Source LLC
LaVergne TN
LVHW082255241026
850861LV00008B/49
* 9 7 8 9 8 1 0 2 0 8 1 0 3 *